Warship 2017: Naval Submarines and UUVs

Bath, United Kingdom
14-15 June 2017

ISBN: 978-1-5108-8302-4

Printed from e-media with permission by:

Curran Associates, Inc.
57 Morehouse Lane
Red Hook, NY 12571

CURRAN ASSOCIATES INC.
proceedings
.com

The Royal Institution of Naval Architects
8-9 Northumberland Street
London, WC2N 5DA
United Kingdom

Phone: 020 7235 4622
Fax: 020 7259 5912

publications@rina.org.uk

Additional copies of this publication are available from:

Curran Associates, Inc.
57 Morehouse Lane
Red Hook, NY 12571 USA
Phone: 845-758-0400
Fax: 845-758-2633
Email: curran@proceedings.com
Web: www.proceedings.com

CONTENTS

SET-BASED REQUIREMENTS, TECHNOLOGY, AND DESIGN DEVELOPMENT FOR SSNX

M C Parker, Naval Sea Systems Command, USA
M D Garner, Naval Sea Systems Command, USA
J T Arcano, Naval Sea Systems Command, USA
N Doerry, Naval Sea Systems Command, USA

SUMMARY

Maintaining affordable undersea capability will require modern design methodologies to minimize costly design changes while continuing to address evolving threats. An aggressive timeline and broad uncertainty around design requirements has led the U.S. Navy towards a set-based strategy for requirements, technology and design development for future submarines [1]. Set-based methods enable informed and defendable decisions by systematically understanding tradeoffs prior to commitment [2], and have successfully been implemented on several ship programs [3] [4] [5]. This paper will provide an introductory background on set-based methods including: definition and clarification of terminology, a history of application within the Naval Sea Systems Command, and a qualitative comparison to traditional point-based methods. Next, the paper will discuss the motivation for adopting set-based methods for SSNX. Finally, a detailed overview of a set-based strategy for future submarine requirements, technology, and design development will be presented.

NOMENCLATURE

DSE	Design Space Exploration
PMS 450	VIRGINIA Class Program Office
RSDE	Rapid Ship Design Environment
SBD	Set-Based Design
SSBN	Ship, Submersible, Ballistic, Nuclear
SSCTF	Small Surface Combatant Task Force
SSN	Ship, Submersible, Nuclear
SSNX	Future Nuclear Attack Submarine
VCS	VIRGINIA Class Submarine

1. INTRODUCTION

The United States Navy operates three classes of attack submarines: LOS ANGELES, SEAWOLF, and VIRGINIA. The early-stage design work for these classes began in 1963, 1978, and 1988 respectively. The long-range naval shipbuilding plan calls for the continued block procurement of the VIRGNIA Class, with a plan to support construction of a follow-on class in 2034 [6]. These timelines indicate that future VIRGINIA blocks will still be in service post 2070, 80 years after design began, and a follow-on SSN, SSNX, could be in service post 2090. The timelines also suggest that requirements, technology, and design development are already behind the curve for the next generation submarine, Figure 1.

Requirements, technology, and design development for submarines facing a 2070-2090 world is a major challenge, even greater than that for recent classes and blocks. The Block I VIRGINIA Class submarine (VCS) development effort benefitted from clear requirements, recent SEAWOLF Class technology research and development, and a very experienced workforce. Blocks II-IV have focused on evolutionary improvements and cost reduction. Block V adds a significant payload module amidships, but is technologically evolutionary. The COLUMBIA Class SSBN in development incorporates many new technologies, but is a single mission platform with a clear requirement. In contrast, the requirements future VIRGINIA blocks and SSNX will be expected to meet and the technologies to do so are yet to be defined, and could significantly change throughout their lifecycle. Uncertainty in requirements and rapidly advancing technology defines a high risk development environment, further compounded by a workforce reduced in size and experience through attrition.

To meet this challenge, a set-based strategy for submarine requirements, technology, and design development is being formulated. Traditionally the Navy has used iterative methods for submarine design. Though flexible and accurate, iterative methods require starting points, i.e. fixed requirements and technologies, and end with single design solutions. Iterative methods are far from ideal when the starting points are uncertain or likely to change. Set-based methods require starting sets, i.e. ranges of requirements and technologies, and end with a range of design solutions. The variability of requirements and technologies is captured from the outset. The objective is to enable timely and robust acquisition decisions by systematically understanding tradeoffs prior to commitment.

This paper provides an introductory background on set-based methods, a history of application within the Naval Sea Systems Command, and a qualitative comparison to traditional point-based methods. Next, the paper discusses the motivation for adopting set-based methods for future submarines. Finally, an overview of a set-based strategy for requirements, technology, and design development is presented, including an example specific to SSNX.

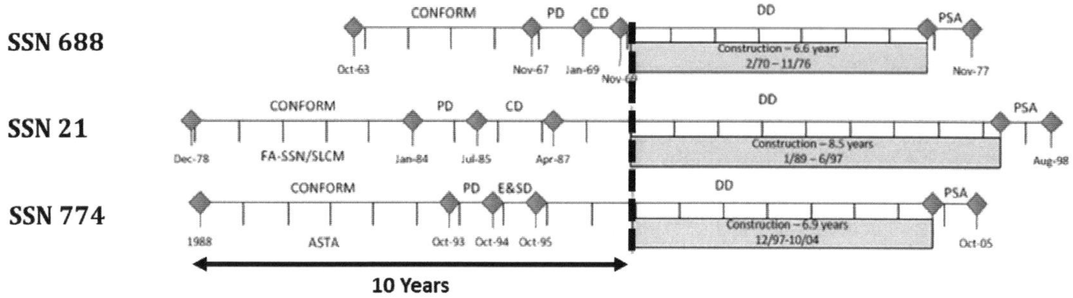

Figure 1: U.S. Navy SSN Design Development Timelines [1]

2. SET-BASED DESIGN

2.1 DEVELOPMENT OF SET-BASED DESIGN

In the late twentieth century Japanese manufacturing firms were extensively studied because of their higher efficiency and quality relative to firms in the United States. A special focus was placed on the Toyota Motor Company because it was also outperforming other Japanese automotive companies. By traditional measures Toyota's production system seemed inefficient, yet put high quality automobiles on the market faster than the competition. This was termed the "First Toyota Paradox" [7]. Part of the paradox was identified as strong parallel rather than sequential links between traditionally separate disciplines, e.g. manufacturing and product development, consistent with a Concurrent Engineering approach. Recognizing and studying the importance of Toyota's product development method to the company's production success, Ward, Liker, Cristiano, and Sobek identified what they called the "Second Toyota Paradox"; the overall design phase was shorter and more efficient, yet purposefully withheld decisions, communicated ambiguous specifications, and produced excessive numbers of prototypes [7]. The term set-based design (SBD) was used to define this product development method, and quantitative research later verified that Toyota's success was not in spite of the second paradox but in part because of it [8]. Research at the University of Michigan beginning in the late 1990's explored naval design applications for the method, and introduced it to the U.S. Navy [9]. Since then, the Navy has applied versions of the method on several surface ship and vehicle programs [3] [4] [5] [10]. Set-based methods have also been applied commercially outside the automotive industry, notably in aerospace [11]. Within academia submarine set-based design has also been explored, but never applied [12].

2.2 TERMINOLOGY

The terminology used within design research often varies. The following definitions based on the work of McKenney are used for clarity [13]:

- Design Method: The way in which design alternatives are understood, analyzed, and selected
- Design Process: A series of structured steps to implement the design method
- Design tool: In support of design methods, tools are used to provide information that enables decision making as part of a design process.

SBD is a design method. It is implemented in the context of a design process that may or may not employ design tools. In applying SBD, as is true with most design methods, a customized design process is usually developed to implement the method.

2.3 A SET-BASED DESIGN PROCESS

SBD arrives at a design solution by a process of elimination rather than iterating around a small number of potential design solutions. A single solution is determined by sequentially excluding regions of the design space that are either infeasible or highly dominated. A feasible solution is one that meets known hard design constraints, such as neutral buoyancy, stability, and validated minimum performance requirements. It is important to exclude "desirements" from feasibility criteria to avoid over constraining the design space. A highly dominated solution is a feasible solution with properties (typically cost) that are greatly inferior to another feasible solution; and therefore, a highly dominated solution is not likely ever to be a preferred solution.

An SBD process enables groups of domain experts to evaluate the design space semi-autonomously. Each domain provides its evaluation of the design space in terms of feasibility, dominance, and preference. Typically, the feasibility evaluation is communicated as regions in the design space where the solutions are feasible, not feasible or uncertain with respect to the domain analysis. As each group completes its evaluation, the results are integrated with the results developed by the other groups. Once combined, if the uncertain region is still relatively large (and the uncertain region is not evaluated as infeasible by any domain), higher fidelity analysis is typically performed to reduce uncertainty and refine the borders of the feasible and infeasible regions. An SBD process has the following characteristics:

- Communicating broad sets of design values
- Developing sets of design solutions
- Evaluating sets of design solutions by multiple domains of expertise
- Delaying design decisions to eliminate regions of the design space until adequate information is known
- Documenting the rationale for eliminating a region of the design space

Figure 2 depicts an example with 3 domains. In this example, Domain 2 performs higher fidelity analysis to refine its uncertain region. Domain 3 takes advantage of the knowledge that certain regions of the design space have already been shown to be infeasible to limit the design space it explores. Once all the domains have completed their analysis, any solution in the remaining feasible region in the integrated design space is acceptable.

2.4 DESIGN SELECTION STRATEGIES

Selecting a single design solution from the feasible region can be done with a number of different strategies. It is uncommon for the ultimate decision authority to rest on those with technical authority, thus it is necessary to provide clear and defendable technical information to the stakeholders responsible for the decision. When many options are under consideration as is common with SBD, this quickly leads to the challenge of data visualization. Ultimately decision makers are looking for a cost vs. capability curve, but a single plot does not readily convey the tradeoffs in a multi-dimensional design space.

2.4 (a) Qualitative Stakeholder Decision Making

In the end, the ideal is to provide insight on the differentiating characteristics within the feasible design space to stakeholders; allow them to negotiate among

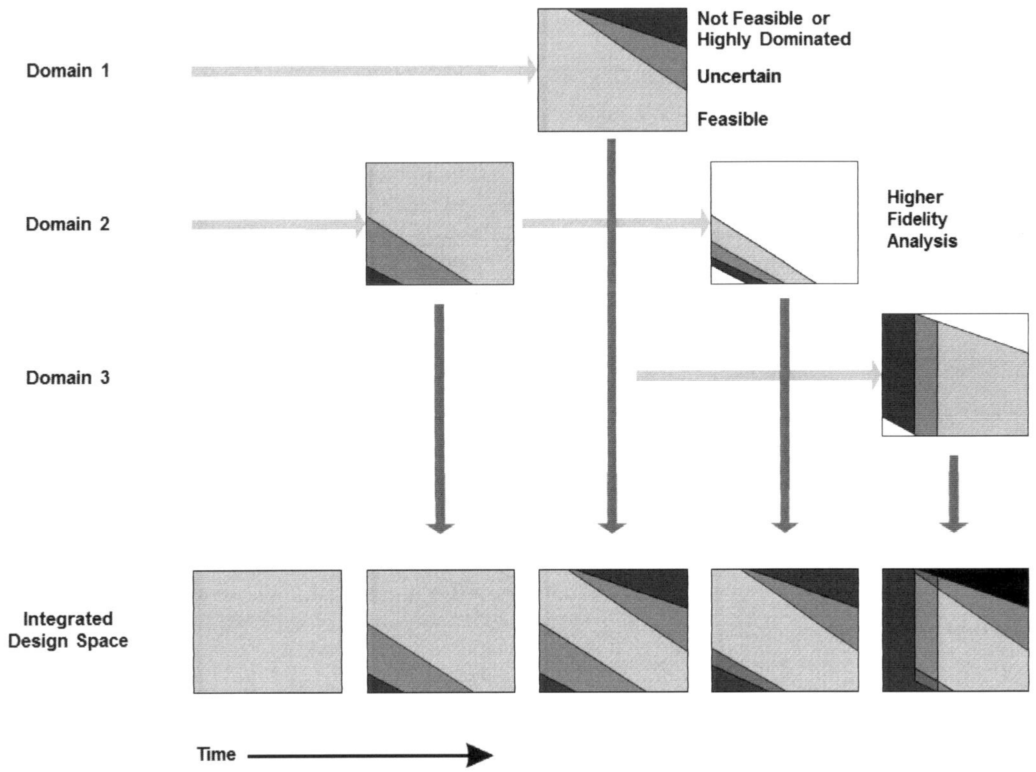

Figure 2: Set-Based Design Process Example

themselves to select a single design solution. When there are legal source selection procedures that must be followed the selection cannot be perceived as arbitrary. In this case the differentiating characteristics must be more formally documented, i.e. specifications, such that the decision is clearly defendable.

2.4 (b) Quantitative Stakeholder Decision Making

If decision makers are more interested in a quantitatively defendable single point solution, there are several familiar methods to narrow the feasible set.

Traditional optimization tools are one option, based on preferences or utility functions obtained from stakeholders. Caution is required to hold the use of optimization until selecting from an all feasible set, as optimizers seek a best solution as opposed to the SBD principle of selection by elimination.

Given the qualitative nature of preferences but a desire for quantitatively based decisions, fuzzy logic tools have been developed to aid set-based decision making [14]. Though fuzzy logic is not necessarily required, the general idea is for stakeholders to designate regions of the design space into one of three categories: highly preferred, preferred, or not preferred. Using techniques similar to those identified for Figure 2 in establishing feasibility of the integrated design space, an integrated design space depicting regions of preference is constructed. The design space is narrowed to the highly preferred region if it exists. If a combined highly preferred region does not exist, stakeholders may negotiate to select a solution.

Risk based methods can also be appropriate. The objective is to eliminate until a robust subset of the feasible design space is revealed. A robust subset is one that is impervious to remaining risk, but minimizes the cost of continuing multiple design alternatives.

Multiple designs from the subset are continued until remaining risks are resolved. A risk is resolved when it has been reduced to insignificance, no longer warranting tracking or mitigation, or realized. The robust subset contains solutions which are not impacted by the realized risks; these solutions can be continued and the design will progress in a risk-tolerant manner.

2.5 WHEN TO APPLY SET-BASED DESIGN

During the past decade, most Naval Sea Systems Command applications of SBD have focused on requirements development and early-stage design. This differs from later stages in that the main focus is on understanding requirements, the design problem, and potential solutions. This stage is analogous to "Requirements Elucidation" as defined by Andrews [15], and occurs prior to the creation of a formal acquisition program. The majority of a program's life-cycle costs are locked in during this time; early decisions have a greater influence on outcomes than later decisions [2] [11]. As a result, the penalty for mistakes is greatest during requirements elucidation. Set-based design has the most benefit early on in the design process when critical design decisions must be made well.

Set-based design is not always the right method, but is a good candidate when there is large trade space that is not fully understood:

- A large number of design variables
- Tight coupling among design variables
- Conflicting requirements
- Uncertainty or flexibility in requirements
- Uncertainties in technologies
- Design problems that are not well understood
- Learning is required for a solution

U.S. Navy warship designers will attest that most recent early-stage design projects have all of these attributes.

The purpose of using set-based design is to make the right decisions the first time by methodically understanding tradeoffs prior to commitment. The benefit of good decisions during the early-stages is better designs, at lower costs, on shorter schedules.

2.6 COMPARISION TO POINT BASED DESIGN

The traditional design method is point based design. The method proceeds as follows: requirements are defined as specifically as possible, several concepts are generated, and there is some interpolation between concepts. A single concept (point design) is selected early to lock in major characteristics, and the concept is further developed until it can meet all requirements. If the concept fails to meet requirements or the requirements change, the process repeats. For naval architecture, the process to implement point-based design is commonly illustrated as a design spiral [16], often credited to Evans in 1959 [17].

The distinguishing feature of the design spiral is that each discipline (e.g. combat system, propulsion, arrangements, weight, displacement) is analysed for the concept in a sequence, handing results "over the wall" to the next discipline. The sequence is repeated (spiralling) to increase design fidelity or to adjust if the initial result is infeasible. The challenge that Evans was addressing in developing the design spiral was the high cost and limited bandwidth to produce and communicate design information. The "over the wall" approach minimizes the amount of communication required between disciplines, and spiralling on a single concept reduces the calculation required at any point in time. These were very important considerations in 1959 when design calculations were largely done by hand and information was communicated on paper.

The sequential nature of a manual point-based design process reduces the number of design points that can be evaluated, translating to a limited understanding of tradeoffs [1]. For design projects that do not have a large trade space, e.g. little requirements uncertainty and an experience base to select a good starting point, the design spiral remains very effective. It is in these cases where set-based design can be fairly criticized as slower, less efficient, and potentially unnecessary. However, even in these situations set-based design will still produce a "more defensible design with greater resilience to requirement changes [13]." If the requirements don't change set-based design may appear wasteful, but the data set can be maintained for potential future evolutions.

It should be noted that many other design approaches, methods, and processes have been developed for early-stage design, but they cannot all be described here. McKenney provides a more comprehensive overview of these methods and how they related to SBD in a chapter on early-stage design [13].

2.7 NAVY APPLICATIONS OF SET-BASED DESIGN

The Navy has applied various degrees of set-based methods to several programs including the: Ship to Shore Connector, Amphibious Assault Ship, Amphibious Combat Vehicle, and a Small Surface Combatant Task Force (SSCTF) [3] [18] [4] [5].

2.7 (a) The Small Surface Combatant Task Force

The SSCTF example is the most relevant to future SSNs because requirements, technologies, design, and acquisition were considered in a holistic manner. The following brief description is derived from Garner et al. [5].

The task force was created in 2014 in response to a Secretary of Defense directive to develop alternatives for a "capable and lethal small surface combatant generally consistent with the capabilities of a frigate." The SSCTF was tasked to establish a requirements trade space, requirements, evolutionary and clean sheet designs, and a cost and acquisition schedule for each design. The task force was focusing on potential changes to the Littoral Combat Ship program which had already delivered ships to the fleet, meaning the time horizon was present day. There was a very short schedule in order to produce results in time to inform the FY16 budgeting cycle. Set-based methods were adopted for requirements development and clean sheet design. Given the limitations of current design tools, set-based methods were not applied as broadly to evolutionary designs.

For requirements, the SSCTF defined four primary mission areas with varying levels of possible capability. A specific combination of mission areas and the associated concept of operation was defined as a Capability Concept. The initial set included 192 capability concepts, a full factorial expansion of four primary mission areas and between three and four capability levels for each. The initial set was reduced down to eight through several rounds of set reduction.

In parallel, separate teams were also working to define and reduce the trade space for combat systems and platforms (ships) to carry those systems. More than 2000 discrete possible combat system architectures were mapped to the 192 capability concepts. Space, weight, power, and cooling requirements to support each combat system were generated as part of this exercise.

Rather than sequencing the platform design after capability concepts and associated combat systems were generated, the platform team generated more than 15000 ship configurations in parallel. Parallel development required the naval architects to decompose potential combat systems into variables for space, weight, power, and cooling that spanned the range under consideration. Each ship and combat system configuration were assessed for cost.

Within the overall trade space of capability concepts, combat systems, and ship configurations, extensive use of regression was used to make estimates for in between designs, increasing the numbers further. After several rounds of set reduction a cost vs. capability trade space was produced, displayed as scatter plots. Ultimately this was simplified into a two dimensional cost vs. capability curve on which existing, modified, and clean sheet designs were placed. A single solution was selected from this curve by senior Navy leadership. The set-based approach provided the information necessary to make defendable acquisition decisions through a full exploration of the trade space in a short period of time. In the event that those decisions are revisited or warfighting requirements change, the full data set is maintained.

2.7 (b) Navy Design Tools

One of the many misconceptions about set-based design is that it is a tool, tool suite, framework, framework of tools, or that it requires one of the above. Set-based design is merely the specific method by which a trade space is analyzed, understood, and a single design selected. There is no explicit requirement for tools.

However, recent U.S. Navy implementations of set-based ship design have started by generating a broad trade space through design space exploration (DSE), including the SSCTF. DSE was conducted using the Navy developed Rapid Ship Design Environment (RSDE) software. The purpose of the software is not to automate a design method, but to populate a trade space of many ship designs and evaluate them for basic naval architecture feasibility criteria. In short, software served to broaden the trade space that could be explored.

There is no submarine capability within the Rapid Ship Design Environment. Recognizing the broad trade space that SSNX presents, a submarine design tool is the focus of ongoing development to enable set-based submarine design [19]. Submarine DSE presents unique challenges, namely that submarines are not surface ships and have very different fundamental equations. A small change in one area of a submarine design has farther reaching and greater impacts on other elements of the design. This stems from a near zero margin for error, mandated by physics and safety. The design fidelity required to account for those interactions is often higher.

3. FUTURE SUBMARINE DEVELOPMENT

3.1 MOTIVATION FOR A SET- BASED STRATEGY

Maintaining affordable undersea capability will require timely acquisition decisions that minimize costly design changes while continuing to address evolving threats. This is a high risk development environment, further compounded by a workforce reduced in size and experience through attrition.

Both academic research and practical experience with set-based design have demonstrated that methodically understanding tradeoffs prior to commitment leads to better, faster, and cheaper programs. Additionally, the structured communication, negotiation, and information transfer between domains is well suited for a smaller and less experienced workforce [2]. As a result, the U.S. Navy is working toward a set-based strategy for requirements, technology and design development for future VIRGINIA Blocks and SSNX. The ultimate objective is to enable informed and defendable decisions by clearly presenting tradeoff information between cost, schedule, and capability.

3.2 SET-BASED STRATEGY

The trade space for future submarines includes both Block development for the VCS and SSNX, spanning near, mid, and far term time horizons. This is a broader scope than previous Navy implementations of set-based design, a reflection of the broad early-stage decisions that must be supported. These decisions include:

- What capability gaps do future SSNs need to fill in the near, mid, and far term?
- Which technologies should be invested in to provide capability, and when does that investment need to occur?
- Which capability gaps can evolutions of the VIRGINIA Class fill?
- Which capability gaps require SSNX, and when does design development need to begin?
- What is the cost of technology and design development to provide required capabilities when needed?

Answering these questions requires that a broad decision space be produced of capability concepts, feasible enabling technologies, feasible submarine designs, and feasible acquisition plans. As a result, the four primary domains for a holistic future submarine set-based strategy are:

- Capability Concepts
- Technology Concepts
- SSN Concept Design
- Program Assessment

3.3 SET DEFINITION

3.3 (a) Capability Concepts Set Definition

The Capability Concepts Set Definition will define the range of warfighting capabilities future submarines could be required to have. The initial inputs to form this set will include: historical data, fleet engagement, and other subject matter expertise. To generate the trade space, those inputs will be used to conduct alternative futures studies, future threat assessments, mission engineering, and war gaming. The primary output of these activities will be a fully documented set of capability concepts. It is important to specify that capability concepts should be independent of specific technology or platform assumptions.

3.3 (b) Technology Concepts Set Definition

The Technology Concepts Set Definition will define the range of technologies future submarines could employ and the development timelines needed to mature them. Inputs to this set include known science and technology frontiers, research and development frontiers, and current technology. To generate the trade space, those inputs will be used to conduct technology assessments and generate representative architectures and design integration information, forming a future SSN technology database.

3.3 (c) SSN Concept Design Set Definition

The SSN Concept Design Set Definition will define the range of whole ship architectures possible for future submarines. Pre-studies will be conducted to define the naval architectural characteristics and boundaries for these architectures forming a future SSN hull, mechanical, and electrical database.

3.3 (d) Program Assessment Set Definition

The Program Assessment Set Definition will define the range of acquisition strategies, cost, industrial base limits, and design and build timelines applicable to future submarines.

3.4 SET REDUCTION & COMBINATION

The overall trade space for future SSNs is construction start date independent, but each time horizon will use unique set reductions. The design timelines shown in Figure 1 indicate that any near term fleet impacts can only be realized with a VIRGINIA Class evolution.
For purposes of example a VCS Block development will be considered, imposing a specific set of technology, platform, and programmatic constraints.

3.4 (a) First Set Reduction

The first round of set reduction can be conducted individually by each domain, eliminating infeasible options with documented and defensible analysis. Within the Capability Concepts set this would be alternatives that lose when war gamed for current or near term scenarios. The capability concepts set is almost time independent, as a Capability Concept that is infeasible today is highly unlikely to be feasible in the future.

For Technology Concepts, only currently available technologies, at a minimum ready for prototyping and at-sea testing, can be considered in the near term. Within the SSN Concept Design set, alternatives outside of VIRGINIA Block evolutions would be infeasible, in addition to any that violate the laws of physics. For the Program Assessment set, infeasible acquisition plans would include those outside the authority of the VIRGINIA Class program office (PMS 450), costs in excess of Congressional authorization, or design and build plans outside of the industrial base's capability. At this point there are four distinct sets with individually assessed infeasible options removed, as shown in Figure 3. The outer boxes represent the bounds of the trade space. The shaded ovals represent the potentially feasible concepts, per that domain, within the trade space. Concepts outside the oval but within the trade space are infeasible per that domain.

3.4 (b) Combining the Sets for Feasibility

After the first set reduction the individual sets can be combined, as shown in the upper left of Figure 4. It is at this point that communication and subsequent analysis between and within domains really begins. Information flows in all directions, but the first pass is likely to be sequential. First, the Capability Concepts set is mapped to the Technology Concepts set. For example, a particular anti-submarine warfare capability concept requires a shipboard sonar array of particular sensitivity. If array(s) exist within the Technology Concepts set that meet the requirement, these two sets overlap. Second, the technology concepts set is mapped to the SSN concept design set. The shipboard sonar array(s) have specific weight, displacement, location, power, and processing requirements. If there are feasible VCS platform designs within the SSN Concept Design set to handle any one of the possible arrays, the three sets overlap. Finally, the SSN Concept Design set is mapped to the Program Assessment set. If concept design(s) are assessed to fall within cost, schedule, and industrial base capacity for the VCS, the four sets overlap and a feasible region exists, as shown in the upper middle of Figure 4. In this fashion a capability concept is now linked to enabling technologies, platform designs and acquisition plans with an assessed cost. This region is carried forward and further refined from subsequent reductions as described in Section 2.3 based on additional design, analysis, or preference.

3.4 (c) Combining the Sets for Infeasibility

Almost as important as a feasible region, infeasibility is also identified and can be assessed. The capability impacts of technology gaps, platform gaps, or unaffordable solutions can be demonstrated. This is particularly relevant to a strategy with multiple time horizons. Each horizon imposes a specific set of feasibility constraints and uncertainty bounds. Near term feasibility is tightly constrained but uncertainty is low, whereas far term feasibility is less constrained but uncertainty is high. In other words, the feasible region is a function of time. A primary purpose of using set-based methods in this context is to methodically and continuously identify infeasible and uncertain areas and invest in them to push the boundaries where possible. In order to do this smartly, investments should relate back to desired capabilities.

The upper right of Figure 4 represents an infeasible region where block development technology investment is required, i.e. a technology pull signal. This will occur if sufficient capability cannot be provided with available technology, but the existing platform and programmatics are sufficient. The lower right of Figure 4 represents an infeasible region where technology is available that does not directly map to a capability concept, i.e. technology push. This is not altogether uncommon, and the capability concept set may expand if a warfighting advantage is found for the new technology.

In the VCS block development example, perhaps the most important result outside of the existence of a feasible region is the identification of the SSNX pull signal. This is shown in the lower right of Figure 4. A new class is required if sufficient capability cannot be provided within existing platform or programmatic constraints; technology investment needs will also be apparent.

3.5 DECISION INFORMATION

Repeating the set-reduction and combination process for the range of time horizons will produce information on technology and design development options, schedules, and costs, to achieve capabilities. This information will enable informed decisions for the Navy's schedule driven budget cycle. The analytical link between capability,

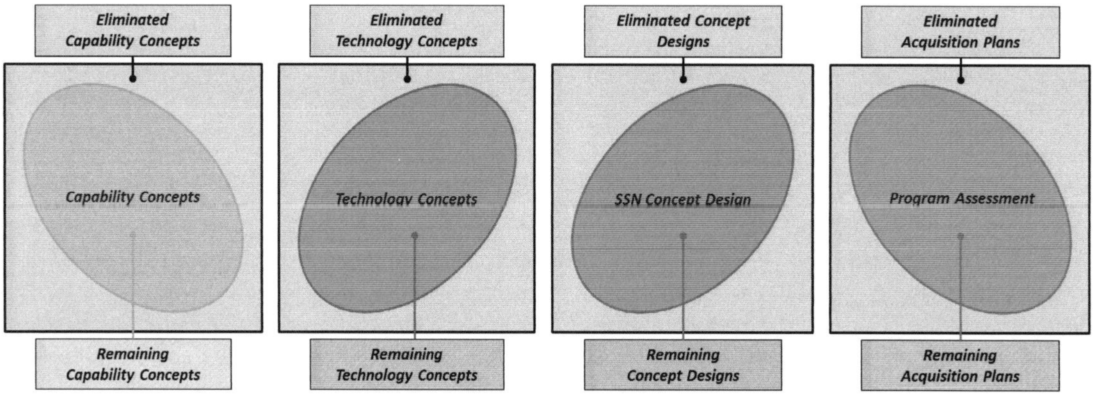

Figure 3: First Set Reduction

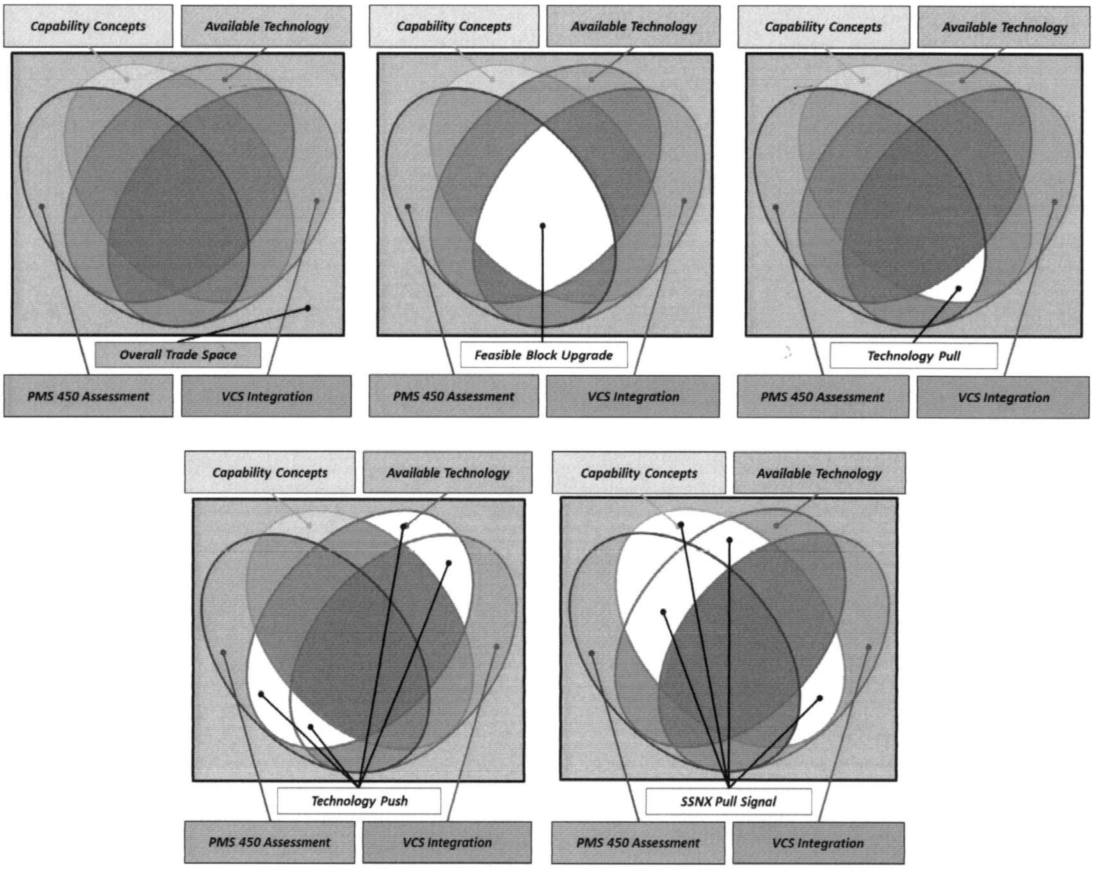

Figure 4: Set Combination "Kaleidoscope" for VCS Block Development

technical solutions, schedule, and cost for a broad trade space makes decisions defendable at the highest level, which is a valuable attribute in a fiscally constrained environment.

4. CONCLUSION

An aggressive timeline and broad uncertainty around design requirements has led the U.S. Navy towards a set-based strategy for requirements, technology and design development for future submarines [1]. The purpose of using a set-based strategy is to make the right decisions the first time by methodically understanding tradeoffs prior to commitment. The benefit of good decisions during the early-stages is better designs at lower costs on shorter schedules.

To achieve this, a set-based strategy has been outlined that considers capabilities, technologies, SSN concept designs, and program assessments to produce defendable decision information in terms of cost, schedule, and capability. In the near term this strategy will inform acquisition decisions on VIRGINIA Class Block upgrade developments, and technology and design development for SSNX.

5. REFERENCES

[1] J. T. Arcano, "Engineered Resilient Systems: Submarines in a New Era of Undersea Warfare [Brief]," *18th Annual NDIA Systems Engineering Conference*, Springfield, 2015.

[2] N. Doerry, D. J. Singer and M. E. Buckley, "What is Set-Based Design," *Naval Engineers Journal,* vol. 121, no. 4, pp. 31-43, 2009.

[3] C. Dowd, D. J. Singer, M. E. Buckley, W. L. Mebane and C. M. Carlson, "Set-Based Design and the Ship to Shore Connector," *Naval Engineers Journal,* vol. 123, no. 3, pp. 79-92, 2011.

[4] J. Burrow, N. Doerry, M. Earnesty, J. Was, M. Myers, J. Banko, J. McConnell, J. Pepper and C. T. Tafolla, "Concept Exploration of the Amphibious Combat Vehicle," *SNAME Maritime Convention*, Houston, 2014.

[5] M. Garner, N. Doerry, A. MacKenna, F. Pearce, C. Bassler, S. Hannapel and P. McCauley, "Concept Exploration Methods for the Small Surface Combatant," *World Maritime Technology Conference*, Providence, 2015.

[6] Deputy Chief of Naval Operations for Integation of Capbilities and Resources, *Report to Congress on the Annual Long-Range Plan for Construction of Naval Vessels for Fiscal Year 2017,* Office of the Chief of Naval Operations, 2016.

[7] A. Ward, J. K. Liker, J. J. Cristiano and D. K. Sobek II, "The Second Toyota Paradox: How Delaying Decisions Can Make Better Cars Faster," *Sloan Management Review,* vol. 36, no. 2, pp. 43-61, 1995.

[8] D. Ford and D. K. Sobek II, "Adapting Real Options to New Product Development by Modeling the Second Toyota Paradox," *IEEE Transactions on Engineering Management,* vol. 52, no. 2, pp. 175-185, 2005.

[9] D. J. Singer, *A Hybrid Agent Approach for Set-Based Conceptual Ship Design Through the Use of a Fuzzy Logic Agent to Facilitate Communications and Negotiation,* Ph.D. Thesis, University of Michigan, 2003.

[10] N. Doerry, M. Earnesty, C. Weaver, J. Banko, J. Myers, D. Browne, M. Hopkins and S. Balestrini, "Using Set-Based Design in Concept Exploration," *SNAME Chesapeake Section Technical Meeting*, Arlington, 2014.

[11] J. Bernstein, *Design Methods in the Aerospace Industry: Looking for Evidence of Set-Based Practices,* Thesis, Massachusetts Institute of Technology, 1998.

[12] M. C. Frye, *Applying Set Based Methodology in Submarine Concept Design,* Thesis, Massachusetts Institute of Technology, 2010.

[13] T. A. McKenney, *An Early-Stage Set-Based Design Reduction Decision Support Framework Utilizing Design Space Mapping and a Graph Theoretic Markov Decision Process,* Ph.D. Thesis, University of Michigan, 2013.

[14] A. W. Gray, *Enhancement of Set-Based Design Practices Via Introduction of Uncertainty Through the Use of Interval Type-2 Modeling and General Type-2 Fuzzy Logic Agent Based Methods,* Ph.D. Thesis, University of Michigan, 2011.

[15] D. J. Andrews, "Arts and science in the design of physically large and complex systems," *Proceedings of the Royal Society A: Mathematical, Physical and Engineering,* vol. 468, no. 2139, pp. 891-912, 2012.

[16] D. G. Watson, Practical Ship Design, London: Elsevier, 1998.

[17] J. H. Evans, "Basic Design Concepts," *Naval Engineers Journal,* vol. 21, no. November, 1959.

[18] A. MacKenna, "LX(R) Analysis of Alternatives [Brief]," *18th Annual NDIA Systems Engineering Conference*, Springfield, VA, 2015.

[19] M. C. Parker and A. W. Gray, "Advances in Early-Stage Resilient Submarine Design Capability [Brief]," *19th Annual NDIA Systems Engineering Conference*, Springfield, VA, 2016.

6. AUTHOR BIOGRAPHIES

Dr. Morgan Parker is a naval architect in the Advanced Submarine Concepts Division of the Naval Sea Systems Command. He is responsible for developing future submarine concepts and design methodologies for the U.S. Navy, and supports the COLUMBIA Class program.

Matthew Garner is the Director for Submarine/Submersible Design and Systems Engineering, Naval Sea Systems Command. In this position, Mr. Garner provides technical leadership and the delivery of operationally superior war-fighting capabilities for submarines and submersibles for the Current Navy, Next Navy, and Navy after Next. Mr. Garner was selected for the Senior Executive Service in August 2013, and earned a Bachelor of Science and Master of Science degrees in Naval Architecture and Marine Engineering from the University of New Orleans. Additionally he holds a Master of Science in Systems Engineering Management from the Naval Postgraduate School.

Dr. Joseph T. (Tim) Arcano, Jr, is the Technical Director for Naval Surface Warfare Center Carderock Division who leads more than 2,000 employees who provide the Navy a broad range of technical support specializing in hull, mechanical and electrical engineering. Dr. Arcano served for 30 years of active and Reserve commissioned service in the US Navy. He served as Technical Authority for Advanced Submarines and as Technical Director/Technical Authority for VIRGINIA Class Submarines at the Naval Sea Systems Command. He is a graduate of the U. S. Naval Academy, the Massachusetts Institute of Technology, the National Defense University Industrial College of the Armed Forces, and the University of Maryland.

Dr. Norbert Doerry is the Technical Director of the Technology Group in the Naval Sea Systems Command of the U.S. Navy. In addition to leading special projects, Dr. Doerry facilitates the transition of technology from industry and academia into naval warships. He retired from active duty in the United States Navy in 2009 as a Captain with 26 years of commissioned service, 23 years as an Engineering Duty Officer. In his final billet, he served for nearly six years as the Technical Director for Surface Ship Design. He led Set-Based Designs for the United States Marine Corps Amphibious Combat Vehicle and for the Small Surface Combatant Task Force. He has published over 50 technical papers and technical reports; and has participated in the development of over ten technical standards.

MANAGING EARLY-STAGE DESIGN UNCERTAINTY AND ESTABLISHING POLICY FOR A LOW RISK DESIGN

S F Daley, Naval Sea Systems Command, USA
R E Dvorak, Naval Sea Systems Command, USA
M C Parker, Naval Surface Warfare Center, Carderock Division, USA

SUMMARY

Establishing a submarine concept design that can endure a rigorous twenty year development process and forty year service life is not without risk. Over that period the class will be refreshed with modern technology and new systems. The early-stage concept design must be robust enough to accommodate these changes, while continuing to satisfy Archimedes and originally specified requirements.

Modern U.S. Navy submarine programs have a low tolerance for risk and put a high value on risk management due to the cost of submarines and critical performance requirements. Risk management is not only a formal process, but is engrained in the culture of U.S. submarine designers. The U.S. government's role in submarine design is to set requirements and manage the acquisition. It includes concept design naval architects who lead the early-stage design effort. A major focus of the naval architect is to establish policy to ensure the design configuration is managed with an acceptable level of risk.

This paper does not focus on the formalities of ship acquisition, but will describe the early-stage processes and policies developed and used by naval architects in the development of the COLUMBIA Class, including setting key ship characteristics, margins, and inter-related system trade-offs to develop a low risk design.

NOMENCLATURE

A-1	Light ship condition without lead
AMR	Auxiliary Machinery Room
AWE	Accepted Weight Estimate
CDD	Capability Development Document
CMC	Common Missile Compartment
FWD	Forward
GDEB	General Dynamics Electric Boat
GFE	Government Furnished Equipment
GFI	Government Furnished Information
IOC	Initial Operating Capability
KPP	Key Performance Parameter
LCG	Longitudinal Center of Gravity
MBT	Main Ballast Tank
MTM	Missile Tube Module
NAVSEA	Naval Sea Systems Command
NPH	Non-Pressure Hull
NSWC-CD	Naval Surface Warfare Center, Carderock Division
PMS397	COLUMBIA Class Program Office
SLAT	Ship Length Assessment Team
SSBN	Ship, Submersible, Ballistic, Nuclear
SWaP-C	Space, Weight, Power, and Cooling
SWS	Strategic Weapons System
VCG	Vertical Center of Gravity

1. INTRODUCTION

Submarines are some of the most complex structures ever engineered, and with rapid leaps in technology, the complexity continues to increase. U.S. nuclear powered submarine design and construction typically spans 15 to 25 years, from early-stage concept development to Initial Operating Capability (IOC). The most recent design, COLUMBIA Class, which will replace the OHIO Class SSBN will achieve IOC in 2030, nearly 25 years after concept exploration was initiated. Establishing a submarine concept design that can endure a rigorous twenty year development process and forty year service life is not without risk. Over that period the class will be refreshed with modern technology and new systems. The early-stage concept design must be robust enough to accommodate these changes, while continuing to satisfy Archimedes and originally specified requirements.

Modern U.S. Navy submarine programs have a low tolerance for risk and put a high value on risk management due to the cost of submarines and critical performance requirements. A formal risk management process is integral to all U.S. Navy submarine design programs and Key Performance Parameters (KPPs) for all major Navy acquisition programs are required for space, weight, power and cooling (SWaP-C) margins to ensure platforms remain relevant as threats and opportunities emerge over a platform's life. But, risk management is not only a formal process; it is engrained in the culture of U.S. submarine designers. The U.S. government's role in submarine design is to set requirements and manage the acquisition. It includes concept design naval architects who lead the early-stage design effort. A major focus of the naval architect is to establish policy to ensure the design configuration is managed with an acceptable level of risk.

This paper will describe the early-stage naval architecture processes and policies used in the development of the COLUMBIA Class to establish a low risk design, including setting key ship characteristics, margins, and inter-related system trade-offs. Balancing the design to minimize ship size and cost while meeting often competing operational and design requirements is a challenge. Figure 1 depicts an overview of the design cycle and drivers. Costly rework or ship modifications often result if the design is not managed from the beginning by naval architects who establish design policy, support key decisions, and characterize associated risks. These early policy and ship configuration decisions focus on basic ship characteristics such as: ship size, reserve buoyancy, navigational draft, static equilibrium, and operational flexibility. The Program Manager (PM) ultimately decides how much risk their program can tolerate.

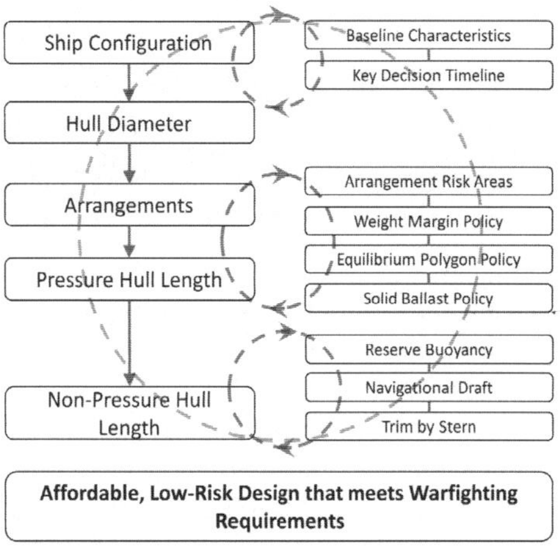

Figure 1: Design Cycle and Drivers

The efforts described in this paper are not ground breaking in naval architecture or design methodology, but are based on consciously quantifying all foreseeable outcomes and applying discipline to manage risk in the making of all design decisions.

2. HIGH-LEVEL REQUIREMENTS

Many of the overarching requirements were provided early-on from higher authorities, including an Operational Mode Summary / Mission Profile, developed by the Submarine Warfare Division of the Office of the Chief of Naval Operations, which provided key operational boundaries including sea state, sea water temperatures and ports (including maximum draft).

A Navy Capabilities Development Document (CDD) provided the performance attributes to meet the general capability needs identified in the Initial Capabilities Document. An early draft CDD methodically added detail in critical areas of performance and cost trade analyses, and was eventually approved by the Chief of Naval Operations. The CDD is the primary driver for the design. The resulting threshold and objective values for each performance metric were then approved by the Joint Requirements Oversight Council to support entry into the Engineering and Manufacturing Development acquisition phase.

KPPs critical to the design include: Materiel Availability, Operational Availability, and Reliability. Materiel Availability is the force level needed to be at sea and ready. Operational Availability is measured at the platform level and is the average percent probability that an individual boat will be operationally fully ready when expected. Achieving these KPPs and the necessary operating cycle leads to continuing OHIO Class SSBN operational and maintenance philosophies such as dual crewing, rotatable pool maintenance, and a 40 year service life. All of these high-level requirements guide the design and are primary considerations when assessing risk.

3. RISK MANAGEMENT PROCESS

Department of Defense and NAVSEA policies require PMs and other acquisition managers to continually assess and reduce risks. Risk management is an organized, systematic process that identifies and prioritizes risks, develops and documents contingency plans and mitigation strategies, and provides decision-makers with the necessary information at the appropriate time to make sound decisions.

For each identified risk area, an assessment quantifies the risk by considering both likelihood (probability of occurrence) and consequence (or severity) as illustrated in Figure 2. For example, a technical performance risk with minimal consequence and probability will be deemed Level 1, low risk. To be assigned a high risk designation a combined score of 15 or greater is necessary, representing a significant reduction in technical performance (level 5 consequence) combined with a probable occurrence (level 3 likelihood). Specific guidelines are provided to assess the level of impact and probability of occurrence.

These assessments are often subjective, but translate into a quantified score that can be used to prioritize efforts to mitigate risks, which may require investment.

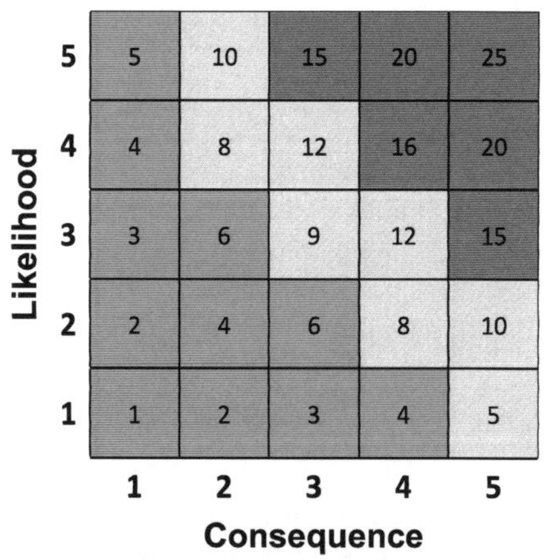

Figure 2: Risk Matrix Example

4. CONFIGURATION MANAGEMENT OF SHIP ATTRIBUTES AND BASELINE CHARACTERISTICS

Balancing the design to minimize ship size and cost while meeting often competing operational and design requirements is a naval architectural challenge that requires thoughtful planning, diligent monitoring, and constant assessment.

During the early-stage design of the COLUMBIA Class, a disciplined process was established to configuration manage key ship attributes and to methodically manage and reduce the associated trade space. Those ship attributes eventually became the "Baseline Characteristics" which were invaluable at keeping the hundreds of designers, engineers and managers from several diverse organizations, on the same page prior to completion of the ship specifications.

5. KEY DECISION TIMELINE

Early in the concept design phase the timing of key decisions was identified, prioritizing the work toward finalizing those decisions. The composition of the initial key decision timeline included inputs from several recognized submarine design experts. Eventually a path to incremental design decisions evolved, which started with a "full trade space," level 1, and progressed to a "final," or level 5 decision, as illustrated in Figure 3. To achieve a level 5 decision, the trade space was generally closed and the decision was formally documented as a maturing design area that supported development of the design products. Key decisions were typically focused on establishing a ship configuration and major systems that drove the ship's size. Dependencies between key decisions were mapped and documented, and a rigorous process was again used to manage the configuration.

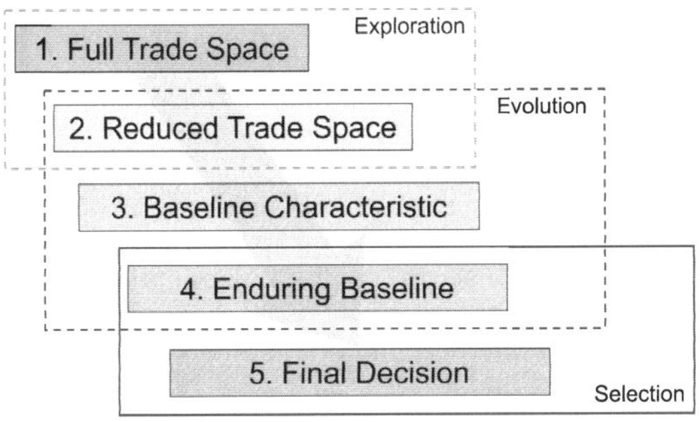

Figure 3: Key Decision Incremental Path

6. SPACE AND WEIGHT RESERVATIONS

For key attributes where the trade space remained open due to continued uncertainty, space and weight reservations were used to establish and mature a baseline configuration. The reservations were set so that the baseline design could accommodate a known range of alternatives in specific locations. While great effort and progress has been made in retiring several of the space and weight reservations, many will remain for incorporating potential alternative systems or equipment beyond the completion of the lead ship. Reservations are a risk mitigation strategy that set aside space and weight margin for specific items and locations in the future. But, in doing so, it becomes necessary to monitor and assess two design variant conditions known as the "Interim Ship" and "Ultimate Ship." The Ultimate Ship condition is the future ship condition that may result if all known potential changes occur (i.e. utilization of all space and weight reservations). The Interim Ship condition is the ship as delivered or as estimated at initial deployment.

7. HULL DIAMETER

Multiple concepts, large numbers of variants, and trade studies needed to be completed prior to selecting a hull diameter. Although the COLUMBIA is a single mission platform, several major configuration drivers remained in the design space. Selection of a hull diameter needed to accommodate the legacy payload, support modern submarine modular construction techniques, and provide the buoyancy necessary for an efficient design. Remember our ancient friend Archimedes...

The trade studies evaluated impacts on the ship characteristics including: missile tube integration, forward compartment arrangement, missile compartment arrangement, propulsion arrangement, and main ballast tank arrangement. Ship balance, ship length, displacement, solid ballast, and navigational draft were assessed by naval architects. Potential impacts to infrastructure and schedule were also considered. The final study evaluated a range of hull diameters from 42.0 feet (the OHIO Class hull diameter) to 43.0 feet.

The 43 foot hull diameter resulted in a ship length on the order of 10 feet shorter than a 42 foot diameter, with about a 1-2% greater submerged displacement. A-1 weight was near constant over the range of hull diameters from 42 to 43 feet.

A diameter of 43 feet resulted in a minimum length ship while accommodating 87-inch missile tubes, adequate clearbore for efficient modular deck arrangements, and Integrated Tube and Hull fabrication in the missile compartment. Selection of the larger hull diameter option was considered low risk due to the maturity of Baseline Characteristics and design flexibility that the larger hull diameter offered including arrangement flexibility and added weight margin.

Sustainment is also critical to the design of an SSBN which needs to meet maintainability goals by accommodating TRIDENT Planned Equipment Replacement flow paths with minimal interference. The

43 foot hull diameter recovered some of the deck area lost due to modular construction.

8. ARRANGEMENTS

Every platform in every compartment was arranged for the shortest possible length unless a substantial demonstrated cost savings dictated otherwise. Sustainment requirements also had to be met and arrangement risks well understood. In some cases, space reservations were included to accommodate quantified variations that were yet to be finalized.

The resulting FWD Compartment is longer than OHIO's primarily due to lost arrangement area that results from the use of modular construction. The Auxiliary Machinery Room (AMR) 1 is shorter than OHIO's due to changes in system architecture and equipment displaced to the missile compartment. The Torpedo room is longer than OHIO's for modern modular construction practices.

The Missile Tube Module (MTM) is longer than the OHIO equivalent (16-tube) MTM due to construction methods. The frame spacing was increased for producibility efficiencies, which increased the length.

AMR2 is longer than OHIO due to batteries and tankage displaced from the FWD compartment by modular construction.

The net result of the arrangement is that the COLUMBIA Class is nominally 560 feet overall. This is coincidentally close to the overall length of the OHIO Class.

9. PRESSURE HULL LENGTH

Each compartment and platform was analyzed to determine stack-up length drivers and arrangement deck area requirements for large equipment and spaces. Once all compartment lengths were determined, the weight margin was assessed to ensure that there was enough to comply with the margin policy. The weight margin at the time of setting ship length was ~10.2%.

9.1 WEIGHT MARGIN POLICY

Establishing weight margin policy early in the design process is necessary to ensure adequate margin through the stages of design, construction, and service life of each submarine to compensate for changes in weight and moment.

Since any change to a submarine's weight and moment must be compensated by an adjustment to solid ballast (normally lead), margins are reported and tracked in terms of solid ballast quantity and location. After allocating solid ballast to maintain minimum stability, list, and equilibrium requirements, the remaining solid ballast is available as margin.

Historically, margin policy is established by NAVSEA as declining minimum percentages of Condition A-1 weight for the different stages of design maturity. Prior to an Accepted Weight Estimate (AWE) milestone, the weight margin is divided among participating entities and is located vertically at the main axis. The AWE establishes the contractual values for weight margin and centers of gravity for each of the margin categories described below:

Future Growth margin is established by NAVSEA and is intended for post-delivery weight and moment changes over the life of each submarine in the class. The value is based on historic margin usage by similar submarine classes. Any excess margin not used prior to delivery is accrued to Future Growth margin after delivery. Future Growth margin may not be consumed for pre-delivery design changes.

Construction margin is established and used by the Construction Yard during ship construction for weight and moment changes due to raw material variances, construction process accuracy, errors in weight determination, variations in calculated versus actual hydrostatics, and tolerances for experimental variation in the inclining experiment and trim dive.

Design margin is established and used for changes and refinements in the design that result in changes in weight and moment.

NAVSEA development margin is established and used by the Program Manager for Navy mandated changes, or changes to requirements, during the design and construction of each ship and for changes to government furnished equipment prior to ship delivery. The allocated amount of this margin is based on historic usage and analysis of potential design changes.

A declining minimum margin was set for each phase of design and construction and for the service life of COLUMBIA Class submarines. COLUMBIA Class margin targets were informed by evaluating VIRGINIA Class design margin depletion. Margin consumption for the COLUMBIA is expected to be similar to the margin consumption for VIRGINIA due to the use of similar design processes and tools, with the exception of the Common Missile Compartment (CMC). The CMC design was more mature, in detailed design, than the rest of the ship when overall length was set. Therefore, margin consumption between setting ship length and AWE was anticipated to be less compared to the VIRGINIA Class. The allocation of design margin was contractually established at the AWE and negotiated by the participating organizations. This margin policy is based on each ship being delivered with the minimum Future Growth Margin, which was based on the analysis of actual lead and margin attrition for each ship of the OHIO Class, extrapolated to the 42 year service life.

The Program Office controls Future Growth margin, space and weight reservations, and any design margins not consumed during the design. The Program Office can consume this margin for design and requirements changes throughout the life of each ship.

In establishing the NAVSEA Development weight margin amount, PMS397 needed to control enough margin to manage the design risk for Navy responsible areas including GFI/GFE and potential requirements changes. At the same time, it is desirable to provide the maximum allocation of margin to the Design Yard for flexibility of development of the most affordable design with minimum risk, while achieving requirements. Several factors led to the conclusion that reducing NAVSEA development margin results in an acceptable level of risk. Considerable space and weight reservations are included in the design that will not be delivered on the first hull. In performing a bottoms up risk assessment, a small group of naval architects and ship integrators partially accounted for the weight reservations using judgment to assign a rough probability of occurrence to each potential change by applying knowledge of design maturity and the assessed level towards achieving a "final decision" for each key area. Since the probability of all of the changes occurring is low, a percentage of margin impact contributed by each area of uncertainty was assigned and the cumulative value was quantified as an upper limit of known potential margin consumption.

Through the qualitative to quantified risk assessment, the small team demonstrated that it was low risk to reduce the amount of NAVSEA development margin for the lead ship. For follow on ships, if significant space and weight reservations are eliminated by becoming a part of the baseline design, the risk to depleting NAVSEA margin could increase. But, by that time, margin depletion across all categories will be understood since the lead ship design will be well into detailed design. Depending on future design changes, PMS397 may need to revisit the weight margin policy and a revision to margin allocations, by conducting subsequent AWE milestone events.

Figure 4 shows the weight margin policy steadily decreasing to the Future Growth margin at delivery.

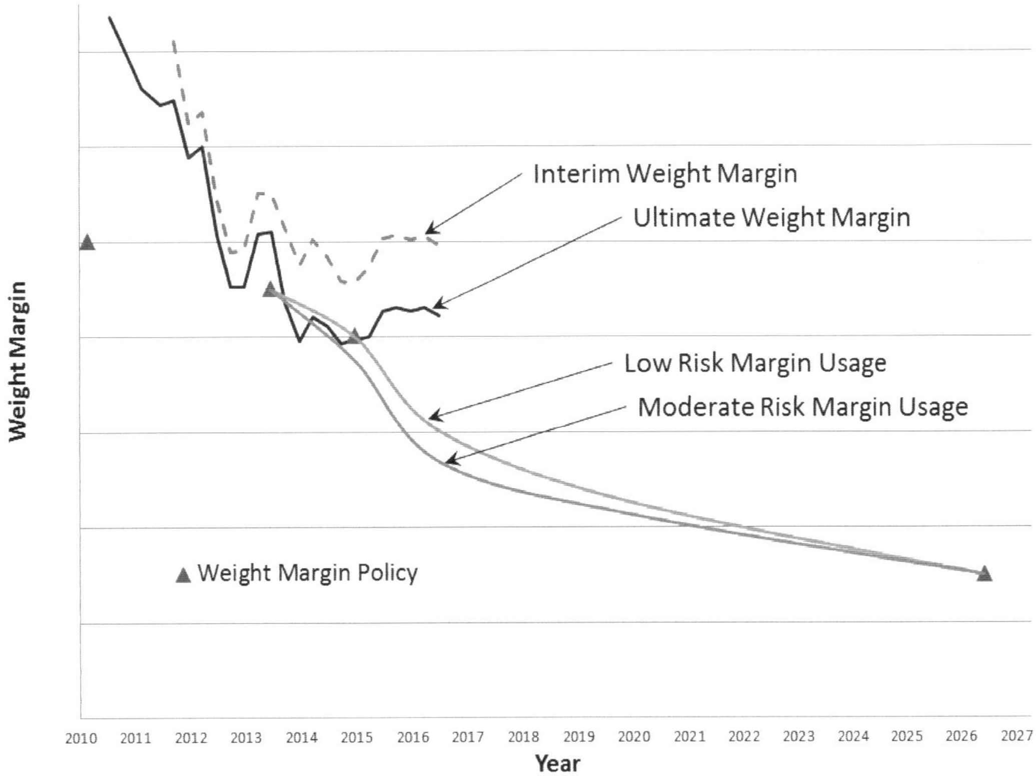

Figure 4: COLUMBIA's Weight Margin Policy and Trend

Weight margin policy dictates *minimum* margin to ensure a healthy ship and can influence ship size if the design is weight-limited. Weight-limited means any weight added requires an increase in pressure hull volume (e.g. a plug) to float it (i.e. weight increases correspond to pressure hull length increases). Increasing pressure hull length to ensure sufficient weight margin adds cost and has to be examined on a case by case basis. On the other hand, with a volume-limited ship, weight increases do not translate directly to increases in ship size because solid ballast can be consumed to offset the additional weight within the established weight margin policy. After ship length and diameter are fixed, margin compensates for weight changes. Margin levels below the policy minimum are cause to initiate weight reduction initiatives. Excessive margin represents an opportunity to design for other considerations (e.g. construction, maintenance, performance, etc.) and the service life margin allocation enables the fleet to incorporate changes throughout the ship's life.

A balanced design, with minimal ship size and cost, will be on the edge of weight and volume limited, including design margin commensurate with design maturity. The COLUMBIA design is close to this region, and remains relatively low risk. Figure 5 illustrates the COLUMBIA as a volume limited design, as margin is above the minimum required by policy.

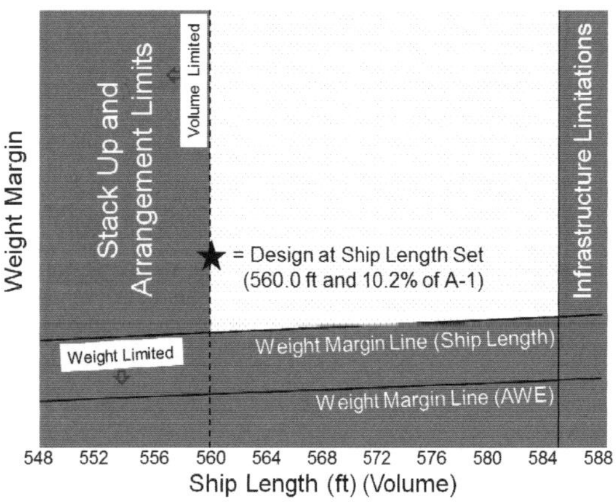

Figure 5: Weight Margin vs Ship Length

9.2 EQUILIBRIUM POLYGON POLICY

Requirements for the COLUMBIA polygon and equilibrium conditions are established to ensure that operational and mission capabilities are satisfied. These criteria ensure necessary in-service ballasting and operational flexibility.

General requirements for the standard equilibrium polygon conditions, defined in reference (1), are provided to the Shipbuilder through the Ship Specification. These points are modified for specific ship class mission requirements.

Variable ballast capabilities must be established to ensure that all design loading conditions and all operational requirements are satisfied. The guidance for establishing the loading conditions and polygon requirements for COLUMBIA ensures that the polygon properly encompasses the design loading conditions, as well as manages the variable ballast system throughout design and construction to ensure the delivered ship has the required variable ballast capacity.

To represent the actual variable ballast system capability, the equilibrium polygon reflects the COLUMBIA usable volumes of the variable ballast tanks, which is different than past practice which used net volumes, not accounting for entrapped air or residual water.

Past designs included a requirement to locate the heaviest equilibrium conditions above the lower boundary of the equilibrium polygon by a buffer. The buffer is intended to account for inaccuracies inherent in trim dive experiments, weight and moment changes between availabilities, and weight and moment variances caused by individual ship loading. There is also a buffer at the top of the polygon for similar reasons.

History shows that variable load weights tend to increase during a ship's life, both during design and in-service, which increases the spread of equilibrium points. Because of this understood trend, a margin of 5% of total usable tank capacity is included to ensure that the delivered COLUMBIA has the required variable ballast capability.

To manage reductions in variable ballast tank capacities during design (such as structural changes to tanks that reduce the variable ballast tank volume), a variable ballast Technical Performance Measure was created to ensure that the tank volumes are tracked throughout the COLUMBIA's design process.

As a result of the COLUMBIA polygon requirements, an inner, operational polygon is shown as a solid outline reflecting a buffer above the bottom of the full polygon and a buffer below the top of the full polygon. A variable ballast tank margin of 5%, of total usable tank capacity, was established at setting ship length and shown as an inner dotted line paralleling the top operational polygon.

All equilibrium polygon points will be within the operational polygon and margin line in the Final Weight Report. Figure 6 depicts a representative equilibrium polygon for the COLUMBIA Class design.

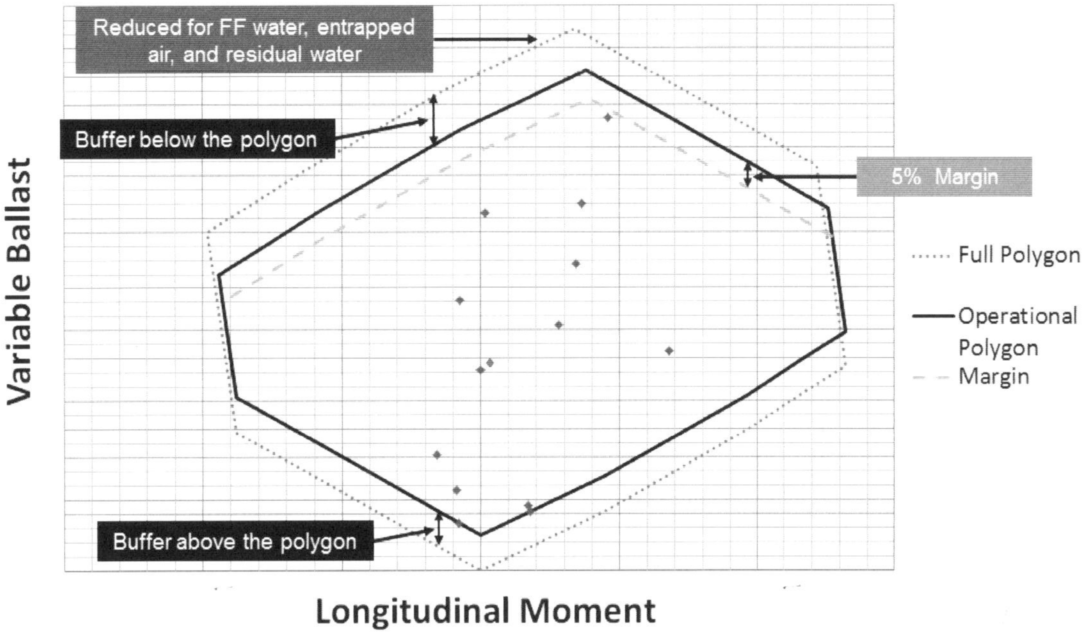

Figure 6: Representative COLUMBIA Equilibrium Polygon

9.3 SOLID BALLAST POLICY

The greater buoyancy of the COLUMBIA Class hull compared to OHIO Class requires more solid ballast for COLUMBIA than the OHIO. To ensure the design had adequate space available to stow all of the required solid ballast (for stability and weight margin policy requirements), a solid ballast stowage buffer was established early in the design. The solid ballast plot in Figure 7 is provided in each quarterly and depicts the stowage available and required for the whole ship and broken down between the pressure hull and non-pressure hull.

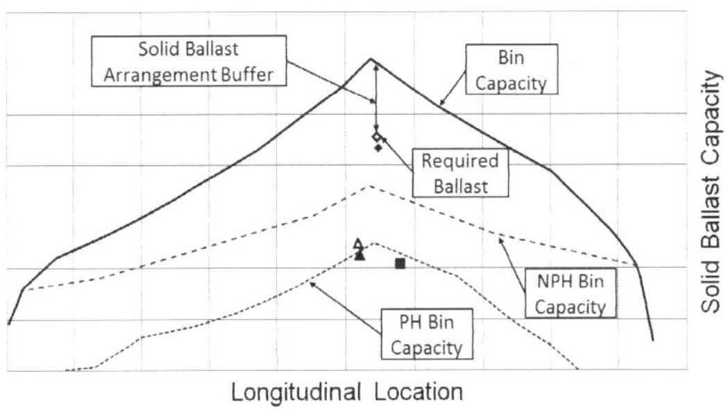

Figure 7: Representative COLUMBIA Solid Ballast Plot

The methodology for determining the distribution of the solid ballast in the pressure hull and non-pressure hull was agreed upon during the AWE. There are several competing objectives in a low risk solid ballast solution. They include:

- Solid ballast shall be stowed as low as practicable.
- Future Growth margin should be accessible during service life (in Main Ballast Tanks (MBTs)).
- Construction margin should be accessible in MBTs to correct potential weight underestimates after initial displacement check and trim dive.
- Adequate MBT empty stowage maintained to correct potential weight overestimates and longitudinal moment errors in weight estimate.

Based on past experience and weight margin minimums agreed to at the AWE, values were assigned to each of these competing objectives and the design was managed with sufficient solid ballast and empty stowage to ensure a successful delivery of COLUMBIA with minimal reballasting, especially changing the pressure hull solid ballast which comes at a large cost in both time and labor. The difference between the interim and ultimate conditions was also accounted for with solid ballast in the MBTs. If other design constraints mandate, such as trim and navigational draft, the distribution may be adjusted.

There were several changes to solid ballast policy in the COLUMBIA Class design. Based on past experience, the COLUMBIA Class added a new requirement for waterborne ballasting. Waterborne ballast are bins (non-MBT) that are accessible for pier-side ballast changes with minimum rip-out (no hot work). These bins have a defined capacity and allow for weight changes without incurring a longitudinal moment change and also allow for a moment change without incurring a weight change to provide the greatest flexibility. This should mitigate potential dockings for re-ballasting, saving time and money.

Since the COLUMBIA Class design is volume limited, there was the opportunity to increase structure to reduce solid ballast, thus reducing lifecycle cost and risk to achieving the materiel availability, and improving arrangement flexibility.

9.4 LIST REQUIREMENT

A new list requirement was established on COLUMBIA to address a lesson learned from USS VIRGINIA where lead was installed post-delivery to correct list. List shall not exceed one degree in all submerged equilibrium conditions and is reported and tracked in weight reports. The new list requirement reduces the risk that the COLUMBIA will be delivered with a list that would have to be corrected post-delivery (with additional lead bins at increased cost and schedule). Provisions are stored off center. So during deployment the transverse moment will vary. The list can be brought to 0 degrees by biasing the variable ballast water to one side.

10. NON-PRESSURE HULL LENGTH

10.1 RESERVE BUOYANCY

Trade studies were completed to identify options for reducing the length of the baseline design to understand the drivers of ship length and explore alternatives. Several opportunities to reduce the length of COLUMBIA were identified, including reducing the length of the MBTs, while maintaining the navigational draft design goal.

The navigational draft design requirement drove the size of the MBTs to a reserve buoyancy above the minimum criterion. A lower minimum reserve buoyancy would allow for reducing the non-pressure hull length as the design develops, if other constraints are relaxed or substantial changes are made to the ship configuration.

10.2 NAVIGATIONAL DRAFT

Based on the building of confidence with regards to risk, PMS397 increased the navigational draft design limit.

Generally, MBT volume is driven by the reserve buoyancy design criteria. However, for the COLUMBIA design, MBT volume was driven by the maximum navigational draft criterion. Increasing the navigational draft limit allowed the overall MBT volumes and lengths, balanced between the bow and stern, to be reduced.

Changes to the navigational draft design limit considered both the static ship condition and dynamic ship performance in the operating environment. It also considered the impact on docking at the construction yard and public ship yards.

NSWC-CD analyzed the days of availability for COLUMBIA to safely transit into and out of the most limiting port from the perspective of navigational draft. The Channel Analysis and Depth Evaluation Tool was used to determine the difference in days of accessibility between OHIO and COLUMBIA by using a database of the wave climatology at local ports, ship draft and trim, sinkage and trim due to forward speed, wave induced motions, and direction of transit. Overall, the study confirmed that the accessibility of COLUMBIA is equal or better than OHIO.

Review and analysis of public shipyard graving dock draft restrictions concluded that all routine SSBN maintenance and repair facilities have graving docks that can support the navigational draft limit.

10.3 TRIM BY THE STERN

There is no requirement for trim by the stern, but it is desirable for surfaced hydrodynamic performance and to keep the propulsor submerged.

Since navigational draft and trim are driven by anything that impacts MBT volume or displacement above the waterline, a thorough analysis and understanding of all potential changes was necessary.

10.4 NON-PRESSURE HULL

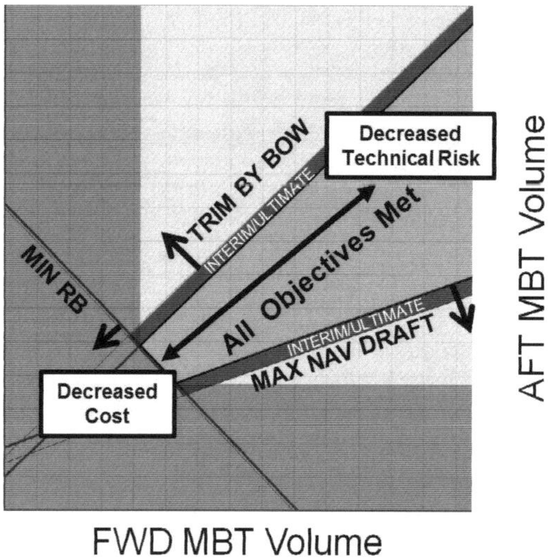

Figure 8: MBT Sizing Methodology

The MBT volume is sized to achieve reserve buoyancy and navigational draft requirements. Figure 8 depicts the NPH sizing methodology used to ensure all three requirements (draft, reserve buoyancy, and trim) were satisfied with the lowest possible MBT volume, which directly translated to NPH length savings. With the establishment of ship length, changes to displacement above the waterline or MBT volumes continue to impact navigational draft.

11. OVERALL SHIP LENGTH

Non-pressure hull and overall ship length recommendations are the result of team efforts between the Design Yard and the Navy. A detailed execution plan was developed to map out the path and provide the criteria to make the decisions that add up to overall ship length.

PMS397 analysis of the weights, displacements, arrangements, variable ballast, tankage, solid ballast, and hydrostatics, for the recommended ship configuration, were completed prior to approval.

Based on the confidence provided by the naval architecture processes, reviews and analysis, PMS397 concurred with the recommended NPH length and overall ship length and concluded that there is acceptable design maturity and manageable risk. The Program Office approved the COLUMBIA overall ship length of 560 feet in January 2014.

Design trade-offs to minimize ship size are part of any submarine design. Efforts to minimize ship length of COLUMBIA were ongoing for several years. Confidence was needed to allow the program to relax the initial navigational draft limit, minimizing MBT volume, and arranging each compartment for minimum stack-up length throughout the Concept Development phase. Reductions in ship length, along with all cost saving initiatives, were reflected in updated "Weight Based Cost Estimate" baselines, to reflect changes in associated weights that result from the ship length reduction. Savings were primarily attributed to the appropriate sizing of MBTs to achieve the maximum navigational draft for the finalized pressure hull length. All of the ship length decisions were the result of recommendations that quantified and evaluated risk.

Since the Navy only sets a submarine length about once every 20 years, an independent panel of submarine design experts was established to validate and assess the risks prior to a final decision and provide an assessment to PEO-SUB. That panel was termed the Ship Length Assessment Team (SLAT), and was established to assess the process and rationale of the COLUMBIA program and engineering team in its length recommendations. After weeks of briefings and independent analysis, the SLAT concluded that PMS397 followed sound processes and performed adequate due diligence. The SLAT supported the conclusion that reductions in the ship length below 560 feet are not supported by the current ship arrangements and would likely drive risk and cost into the program. They also stated that increases in length to mitigate emergent risks could be accommodated in the design although the benefits must be balanced with the additional construction costs.

12. CONCLUSIONS

While margins generally reduce risk, they are not free. Margins can sometimes overlap with a multiplying effect, and careful management is needed to avoid unnecessary overlaps so that risk reduction is affordable. Good margin policy and management is less costly than redesign at most stages of development.

The success of the COLUMBIA Class design is largely due to the efforts of early-stage designers and naval architects, commitment to detail, persistent rigor, and establishment of risk minded policies and processes that will endure the life of this design and influence the next. Some of the processes developed are already being applied to other submarine programs and the specifications will be a basis for the future. But, true success will only be apparent after the first boat is floated off in 2026 and achieves Initial Operating Capability in 2030, nearly 25 years after concepts were initiated. The inclining experiment will be the judging truth of the naval architecture commitment, discipline, and achievements applied over the years as real mass properties are compared to the "paper ship," and when the first boat begins its forty years of service.

13. ACKNOWLEDGEMENTS

The authors would like to thank the COLUMBIA Class Program Office, the members of the Naval Architecture Systems Integration Team over the duration of the early-stage development including those at GDEB, and all of NAVSEA 05U6 for ongoing support.

14. REFERENCES

1. S9086-C6-STM-010/CH-096R1, Naval Ship's Technical Manual, Chapter 096, Weights and Stability, 02 Aug 96

15. AUTHORS BIOGRAPHY

Mr. Scott Daley is a naval architect in the Advanced Submarine Concepts Division at NAVSEA. He is responsible for developing future submarine concepts for the U.S. Navy and was previously the Concept Design Manager for the early development of the COLUMBIA Class design and a Major Area Integration Team (MAIT) leader in the COLUMBIA program.

Mr. Robert Dvorak is a naval architect in the Advanced Submarine Concepts Division at NAVSEA. He is responsible for developing future submarine concepts for the U.S. Navy and was previously the Naval Architecture System Integration Team leader in the COLUMBIA Class program.

Dr. Morgan Parker is a naval architect in the Advanced Submarine Concepts Division at NAVSEA. He is responsible for developing future submarine concepts and design methodologies for the U.S. Navy, and supports the COLUMBIA Class program.

16. NOTE

The experiences and opinions of this paper are those of the authors and are from a naval architect's perspective. The technical work described was generally managed, performed, or supported by the authors.

SUBMARINE DESIGN IS NOT SHIP DESIGN

David Andrews, FREng, PhD, FRINA, RCNC (Vice President), University College London, London, UK

SUMMARY

The author's long involvement in submarine design and subsequently teaching submarine design and leading research into the design of these most complex of vessels provides the basis for this paper. The argument in the title is justified not just by the fact that the naval architecture of submarine is distinctly different to that for surface ships due to modern submarines operating largely in one medium rather than on the interface of two. But it is also the case that the design process followed in the earliest phases of designing a submarine and the over-riding design pre-occupation are also both quite different to that for a major surface warship.

A brief outline of the particular nature of warship design provides a benchmark to which to compare submarine design practice. Then the very specific nature of submarine naval architecture is outlined, given it tightly bounds the scope for novelty in submarine design options. This is followed by consideration of certain critical submarine design issues that effectively constrain the conduct of submarine concept design studies, in a manner quite different to the very open exploratory process that should be adopted for proper Requirements Elucidation of complex (surface) ship design. Finally some further wider policy issues that reinforce the thesis of the paper conclude the argument.

1. INTRODUCTION

As can be seen from the brief biography at the end of the paper, the author started his professional career in the RCNC in submarine design. This included in-service and repeat building conventional submarines, with practical stability and structural challenges. After a short spell on in-service SSBNs and future submarine options studies, his remaining basic experience was on the structural design of the building SWIFTSURE Class SSNs and *ab initio* structural design for the SSNOY design. The structural work included supporting the work led by Kendrick and Faulkner on the first elasto-plastic theory of overall pressure hull collapse due to cold bending of extruded ring frames [1]. Returning to new design submarines some nine year later he was for two years the Section Head on the VANGUARD Trident Submarine Programme responsible for the structural and hydrodynamic design aspects. His subsequent directing of the UCL Submarine Design Course clearly drew on this wealth of submarine design expertise and the comments in this paper reflect that.

UCL has had, since the RCNC sponsored courses went there in 1967/8 from RNC Greenwich, a unique emphasis on both the specialist aspects in submarine naval architecture (as part of the MSc in Naval Architecture) and undertaking studies in early stage design to give post graduate naval architects a feel for what makes submarine design different to the main ship design exercise they first undertake in their Masters Course. Both these aspects of submarine design are addressed in showing its distinct and demanding nature.

A final introductory point concerns the inevitable simplification in this paper of specific scientifically demanding engineering concepts. In addition topics are dealt with sequentially which can falsely under-emphasise that most submarine design issues are highly interrelated. More detail on submarine naval architecture is provided by the late eminent submarine designers, Roy Burcher and Louis Rydill, the originators of the post-graduate UCL Submarine Design Course in that course's textbook [2], as does the more technically detailed and specifically American paper [3]. Although the latter was produced almost six decades ago it still provides an in-depth technical overview of certain key aspects. Reference 2 also list some of the specialist papers dealing with specific topics in more depth, though for none attendees of the UCL MSc in Naval Architecture, some of these on the fundamentals of structural design and dynamic control may prove quite mathematically challenging. However the current paper is primarily focused on the design issues and process rather than the theories behind the naval architecture.

The next section briefly outlines the general nature of the design of naval vessels, flagging up where for submarines there are distinct differences. Following that the aspects of submarine naval architecture, which dominate submarine design to a greater extent than is the case for most surface vessels, given the latter's greater degree of design flexibility in contrast to the highly tuned nature of submarine design. After that those issues seen to be critical aspects in submarine design are addressed starting, as with all such complex systems, with choice of design style. The next section covers some further remarks, which characteristically of submarine design if not properly addressed in the earliest stages of a new design can prove extremely problematic. Finally the paper concludes with some wider design related issues, where once more submarine specific concerns can mean that even in the Concept Phase, if unclear can mean the eventual overall design will be un-balanced.

2. THE NATURE OF DESIGN OF COMPLEX NAVAL VESSELS

In order to see just how different submarine design is from general naval ship design it is necessary to briefly summarise in this section general warship design practice pointing out certain aspects that are common with submarine design. These are largely in the aspect of the design process,, where the generic process for military vessels differs so markedly to that for the generality of merchant ships.

2.1 GENERAL WARSHIP DESIGN RELATED ISSUES

Aside from some remarks below on combat systems related ship design issues, weapon and sensor design will not be addressed. A convenient taxonomy in considering how at least surface warship design issues differ from those for commercial shipping is to use Brown and Andrews "S^5" categories [4], see Figure 1, which notably did not contain any submarine examples.

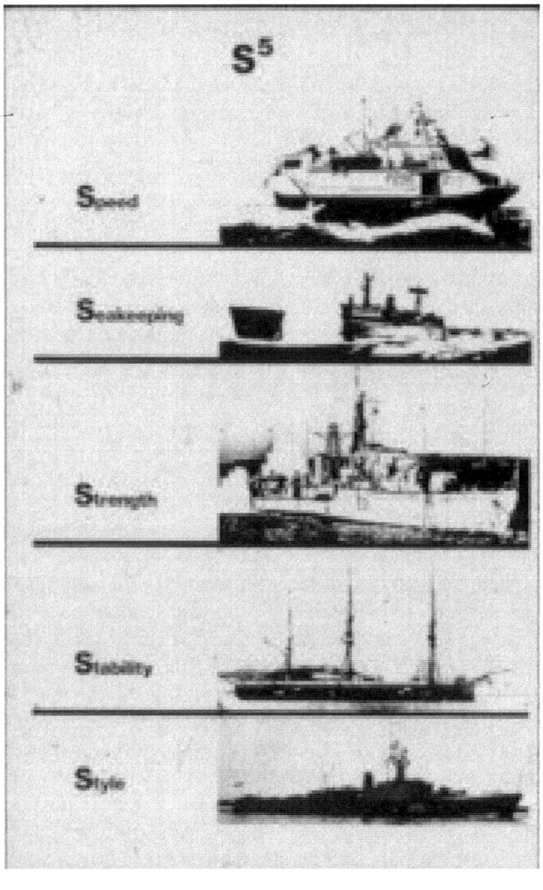

Figure 1: S^5 Warship Design Examples

2.1(a) Speed

The term Speed really encompasses resistance and propulsion, plus the need for endurance. Warships are characterised by needing to operate across a spectrum of speeds. Thus a high top speed (and very good manoeuvrability at top speed) is required for action to pursue enemy units or to take avoiding action. This then governs the choice of main machinery and the hull form, which for a surface combatant results in a long, slender (L/B > 8 or 9) and finely shaped underwater form. However such high speeds (typically around 30 knots) cannot be sustained for long distances, so a combatant's endurance is usually defined as (say) 6,000 nm at an endurance speed of 18 or 20 knots, so other engines than those providing the power for full speed are also usually required for this much less resistful speed, if the size of the fuel tanks is to be keep as small as possible. If there is also a need to reduce underwater noise signature (see below) this can lead to very expensive machinery arrangements, which further increase the cost of acquisition and through life support, already demanding due to the irregular operation of naval vessels compared to the very prescribed operations of most merchant vessels. The speed issue for submarines is quite different, as most thermodynamically efficient prime movers require continual access to oxygen.

2.1(b) Stability

The need both to resist the most extreme seas (weather avoidance may not be an operationally acceptable option) and to survive extensive hull damage means recent combatants have proportionately increased maximum waterline beam, with the consequence of reduced top speed. The need for large radar antennas at the top of high masts to increase range of missile detection coupled with light-weight machinery exacerbates this trend. Emerging new threats or new technological developments, which due to electronics, in particular, means the weapon fit is likely to have shelf life much shorter that that of the ship. The latter without an extensive life extension is typically 30 years (so nearer 50 years for a new class from initiation to last of the class going out of service). The need to survive extensive damage means that particular attention is paid watertight integrity, Thus each watertight section can only be accessed by through deck hatches, closed in higher action states with scuttles providing a secondary means of escape from compartments below. All this contributes to the complexity of design, construction and maintenance with significant cost implications, which is taken to the extreme in submarines.

2.1(c) Strength

The structural design of naval vessels is complex and costly, when compared to merchant ship practice but less so again than that for submarines. Surface warship design usually invests in light and structurally efficient scantlings with closely spaced longitudinal framing and extruded or fabricated "Tee bars", which are much more structurally efficient than asymmetric or flat bar sections provided in merchant ships. This naval practice has been adopted both to keep the structural weight fraction as low as possible (as the largest single component of ship and submarine displacement) and ensure the structure resists explosions, in particular those underwater. Also, for both

Table 1: Listing of style topics relevant to a naval combatant design

Stealth	**Protection**	**Human Factors**	**Sustainability**	**Margins**	**Design Issues**
Acoustic signature	Collision resistance	Accommodation standards	Mission duration	Space	Robustness
Radar cross section	Fire fighting	Access policy	Crew watch policy	Weight	Commercial standards
Infra-red signature	Above water weapon effect	Maintenance levels	Stores level	Vertical centre of gravity	Modularity
Magnetic signature	Underwater weapon effect /shock	Operation automation	Maintenance cycles	Hotel Power	Operational serviceability
Visual signature	Contaminants protection	Ergonomics	Refit philosophy	Ship Services	Producability
	Damage control		Upkeep by exchange	Design point (growth)	Adaptability
	Corrosion control		Replenishment at Sea (R.A.S)	Board Margin (upgrades)	Aesthetics

structural efficiency and shock resistance, sophisticated joints are adopted where transverse frames and deck beams meet and where orthogonal stiffeners cross good connections intersect. So achieving a robust navalised structural design costs money in design effort, fabrication and TL support. For this reason navies are increasingly using classification societies to see if some degree of commercial practice (as already has been adopted for some naval auxiliary vessels, such as fleet replenishment tankers) can be adopted in naval combatants but not for submarines.

2.1(d) Seakeeping

Good seakeeping is a virtue in all surface ship design, since warships are characterised as not being able to adopt weather routing if missions require immediate deployment regardless. Thus even before the adoption of computer simulation of ship motions in a (real) random seaway, surface warships were designed for good seakeeping. This could not be said to apply to modern submarines, since they can avoid rough weather by simply going deep.

2.1(e) Style

This term was adopted to cover a host of design aspects, most of which have been identified in Table 1. They have been grouped into six broad categories, some of which are quite coherent (Stealth) others less so (Sustainability and Design Issues). Despite this, most are either unique to naval vessels or a topic common to other ships but the issue or the design approach is quite different in the case of warship design. Some examples have been discussed in a recent paper [5]. Of interest here is that many are also applicable to submarine design but that the emphasis can be quite different, thus R.A.S. is relatively little considered. Whereas minimising a submarine's acoustic signature can be considered dominating, in not just for the Stealth category but also in the combat system design considerations. Another interesting difference in submarine design is in the Margin category where not just vertical c. of g. but also longitudinal centroid location is critical (see Section 3.1).

2.2 COMBAT SYSTEM RELATED WARSHIP DESIGN ISSUES.

Considering first the sensor side of the combat system suite, namely radars, sonars and communication systems, there one can safely predict on-going growth in capacity and ever greater levels of integration within a warship, together with a growing emphasis on the warship as an *information* node in "network centric warfare". There will be further implications of any new broader defence posture, to which future warships must be responsive, given the range of potential threats, including asymmetric warfare from terrorists and rogue states. This may present particular demands on submarine combat systems in moving beyond largely anti-submarine warfare (ASW) roles.

Sonar systems are particularly difficult to accommodate in ships and are major design and cost drivers. Firstly, sonars are likely to be required to be fitted in specific locations (well forward or towed from gantries over the stern, which can considerably constrain surface combatant and submarine design, particularly in way of propulsors. Secondly, to operate effectively, both the vessel's underwater form and its form of propulsion, as well as the self noise of the mass of the machinery onboard, has to be designed or mounted to ensure the performance of the vessel's own sonar systems are not degraded. The demands of the submarine environment and configuration can mean this is a dominant design feature.

Another feature, which has a large impact on surface combatants, is that of aviation facilities, yet does not come into consideration in submarine design. However, new vehicles and systems, such as Unmanned Underwater Vehicles (UUVs), may well be significant drivers in future submarine design [6].

2.3 PERSONNEL ISSUES AND SHIP TECHNOLOGIES.

Given that the running costs of all warships are dominated by personnel costs, current and future designs will focus heavily on means to further reduce personnel numbers despite mission creep. This is not just due to personnel related costs being the dominant component of warship through life cost, but also that the need to provide ever better accommodation demands most of the vital "real estate" in modern surface combatants. Again this matters in submarines but the density issue means the accommodation is considerably tighter and standards still very constrained, even on the largest military submarines.

Turning to consideration of the ship technologies, while Stealth in general matters for the surface vessels the acoustic element clearly dominates ship system design in submarines. Thus features for noise treatments may require mounting and special treatment of equipment, systems and structure to reduce noise transmissions to the sea as has already been emphasised, but unlike most surface vessels it is hard to put in margins for future growth.

To counter to the growth in surface combatant size, with the consequential reduction in fleet hull numbers, there has been a renewed interest in fast unconventional vessels (such as the fast mono-hull and trimaran versions of the US Navy Littoral Combatant Ship). With the variability possible, even in the mono-hull configuration, the surface vessel provides the concept designer with a very diverse choice of potential solution (see References 7 and 8) – quite unlike the submarine designer.

2.4 THE UNIQUE NATURE OF THE WARSHIP DESIGN PROCESS

Before reviewing the very specific nature of submarine design characteristics, it is considered worthwhile to point out the distinct "political" nature of all military vessel design and acquisition. Thus the naval ship environment does not have a direct commercial measure of performance that most merchant ships have with Required Freight rate as the commercial Measure of Merit. The mercantile approach starts with a very market oriented requirement exploration by a prospective ship owner (sometimes with an in-house team but more often using a design consultancy). This outline study and emergent, and very specific, requirement set is offered to the shipbuilding industry, who normally only have just six weeks to respond with a single solution, including a build cost and delivery time, to which the selected shipyard is then contractually committed. This means the shipyard has to produce quick and "commercially safe", and hence conservative, proposal based very much on their own previous built solutions

The naval vessel concept design approach couldn't be more different, particularly in the major navies such as the US Navy and the Royal Navy, where the process has been written about extensively (see References 9, 10 and 11)). There is firstly, the highly political nature of the process, well captured by Benford's [12] biting description:-

> "Multi disciplinary, multi million dollar Navy design extravaganza where every decision must be analysed, traded off, massaged and documented to the point that the basic design issues are often lost in the process"

and, secondly, the vast sums involved in the acknowledged complexity of concurrently developing and integrating a mobile system of systems with major weapon and sensor sub-systems (caught by Graham's claim [13]): "It is understandable that today's warships are the most complex, diverse and highly integrated of any engineering system"). All this means the warship early stage design process is particularly protracted, given the search for innovation to solve what can be seen as the squaring of the circle of (impossible) needs with the initial procurement cost squeeze resulting from tax payers' dislike of an exorbitant defence "insurance premium". This then results in a concept phase, which is distinctly different to even the downstream design process for most warships and quite unlike that for most merchant vessels [14]. In this regard the submarine concept process can be just as protracted that as that for major surface vessels, but not so much due to parallel weapon developments and many diverse solution choices but more so arising from the submarine's inherent density of complexity. The latter applies through out the whole of submarine design and is largely due to the submarine's unique naval architectural aspects, which will now be outlined.

3. THE NATURE OF SUBMARINE NAVAL ARCHITECTURE

3.1 STATIC CONTROL

Because a submarine is neutrally buoyant, when fully submerged, and has very little excess of potential buoyancy over its weight when on the surface, in comparison with most surface ships, its ability to remain upright, even in calm seas or deeply submerged, is far less than that for surface ship. On the surface the submarine, like a surface ship, must float at its design draught and attitude, or trim, and have sufficient transverse stability to avoid turning over, in extreme beam seas. However when submerged, in addition to achieving an exact balance of weight and buoyancy to be neutrally buoyant, the submarine must be both transversely stable and longitudinally stable. The latter is a condition easily achieved and maintained in a surface ship, due to its very large water-plane effect against plunging by either the bow or stern. Without a water-plane, when submerged, the submarine is highly sensitive to any change in longitudinal weight distribution and hence attitude, which is also an issue of utmost importance when manoeuvring at speed, as is discussed further under Dynamic Control.

The usual way to achieve the change from submerged to surface state is to have floodable Main Ballast Tanks (MBTs), within the external envelope, from which the water is blown by stored high-pressure air and then low pressure air blowers, to regain positive buoyancy in order to surface. Now this additional buoyancy could be thought of as added buoyancy to that provided by the pressure hull (plus a minor contribution by solid items in the free flood spaces), however it is normal to treat the full hydrodynamic envelope as the boundary of the displacement volume. Thus flooding the MBTs on the surface, by opening vent valves at their crowns, in order to submerge is seen as an addition of weight. So the submerged displacement of a submarine is greater than that on the surface by the weight of water now in the MBTs. This means that to get neutral buoyancy, when submerged, the submarine must start with the precise "diving trim" draughts fore and aft. The additional surfaced "reserve of buoyancy" is small compared to that of a surface naval vessel, i.e. typically 10% of displacement, compared to 200% or more for a surface combatant.

Submarine transverse stability on surfacing is a potentially dangerous manoeuvre, due to the circular hull form. So the transitional surfacing state, with MBTs partially blown, can be the critical condition, should the surfacing submarine then be subject to a large heeling moment (from, say, a beam sea). This has believed to be the cause of submarines being lost in instances when their metacentric height has been reduced due to high weight growth in late life modernisation. In modern submarines this transition condition on surfacing remains a concern, since the free flood space under the casing takes longer to drain out when surfacing, due to smaller free flood openings necessitated by concerns at the flow noise that these openings can induce, when the submarine is operating at ultra quiet states submerged. (An interesting instance of the interaction of different design aspects, typical of much of submarine design.)

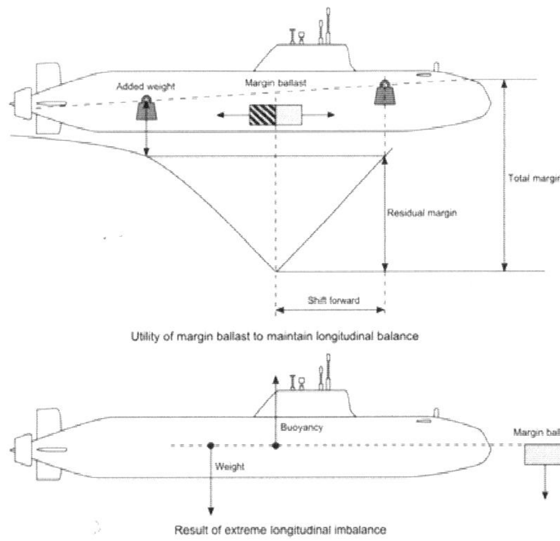

Figure 2: The basis of Longitudinal Margin Ballast (UCL / Rachel Pawling)

In considering submarine stability it is convenient to do so by looking at the role of ballast, which then has a significant bearing on how the life of a submarine might be amenable to additional equipment being fitted, as this is far more heavily restricted than in the case of a surface vessel. Thus with the submerged submarine neutrally buoyant, weight equals buoyancy but also the centre of gravity (G) must lie below the centre of buoyancy (B) (which is at the centre of the pressure hull cylinder) (i.e. the distance B to G must be positive (B above G)). Furthermore, the longitudinal positions of the centre of buoyancy and the centre of gravity must coincide. Essentially, ballast is made up of fixed ballast (usually pig iron or lead stowed very low down and mainly external to the pressure hull, in a box keel or the external structure) and variable (by pumping and flooding) seawater ballast in tanks, inside the pressure hull and known as Trim and Compensating (T & C) Tanks. Considering the fixed ballast firstly, this has two functions: Stability Ballast to keep BG satisfactorily positive (where satisfactory means there is a margin in hand to allow for variations, including the surfacing transition mentioned above), and Margin Ballast, which is intended to be removed during the submarine's operational life as extra equipment is added within the pressure hull. The Margin Ballast is disposed throughout the length of the submarine to maintain a coarse longitudinal balance of weight with buoyancy and then it can be removed as new equipment is added, ensuring the balance is maintained (see Figure 2).

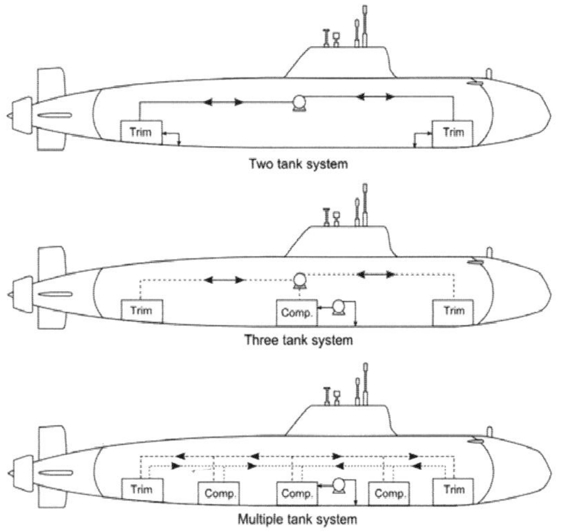

Figure 3: Submarine Trim and Comp Tanks Arrangements (UCL / Rachel Pawling)

In addition to the fixed ballast, there is a need to operationally control the weight/buoyancy balance, including the longitudinal balance, in a way not necessary on a surface vessel. For example the pressure hull contracts with increased depth, fresh water and stores/payload are consumed, diesel fuel is consumed, onboard personnel movement (the submarine is that sensitive at very low speeds) and sea water density changes, with depth changes and particularly near freshwater estuaries. All these likely changes can be put on a plot showing the changes to weight (vertically) and longitudinal moment changes (horizontally). This then gives the envelope of change that the variable ballast (Trim and Compensating (T & C) Tanks – see Figure 3) must accommodate. In essence (practical arrangements can be more complicated), these tanks consist of a centrally located tank, dealing with bodily changes, and two tanks, one very well forward in the pressure hull and one very well aft, to respond to the longitudinal moment changes. The diagram, shows the ability of these tanks to accommodate all the likely changes in weight and longitudinal moment and is termed the Trim Polygon, see Figure 4. At the designed seawater density and the boat's operational condition's outfit of fuel, stores, payload and personnel, the Trim Polygon gives the condition of the T & C Tanks from which using the T & C Tanks' capacities, the submarine can carry out its range of missions. This tank condition is known as the Standard Trim and Comp Condition and demonstrates quite vividly how sensitive a submarine design is to its disposition of weights fixed and variable. In this respect, despite a submarine's far greater size, it is more like an aircraft than a ship. This sensitivity can have severe implications, due to design errors or omissions, both for safety and the ability to adapt a design to new roles and technologies, if considerable forethought has not been applied to these eventualities in the original design.

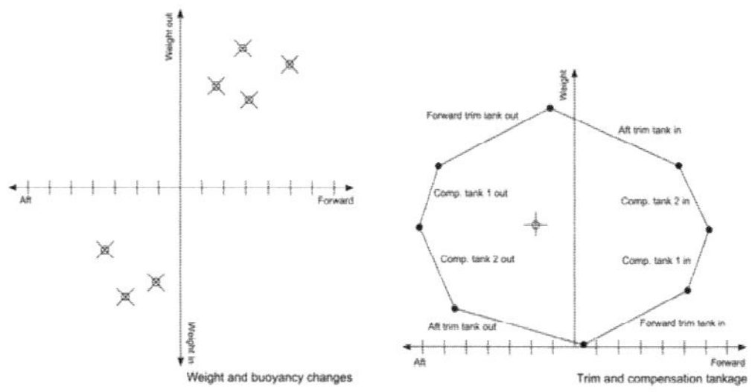

Figure 4: The basis of the Trim Polygon (UCL / Rachel Pawling)

3.2 SPEED AND PROPULSION

Given the desire to minimise resistance, and hence the installed propulsive power, as well as maximising endurance, propulsive considerations tend to dominate in the design of the naked hull envelope. However the completely separate operational consideration of minimising the underwater noise signature is also a significant hydrodynamic design driver. Once the submarine is well away from the water surface, the hull resistance is due to viscous effects of the water flow around the envelope. "Well away" is actually speed dependent and, for the moderate speeds likely when the vessel is close to the surface, twice hull diameter is probably a reasonable indicator of the limits of surface effects. Thus the hull resistance submerged is a combination of form drag and skin friction. The former depends on the rate of change of the sectional area of the body and is minimised by adopting a long thin form, while skin friction is dependent on the velocity distribution around the hull and the size of the surface area presented to the flow. The latter component of resistance is then minimised by the smallest surface area, i.e. a spherical form. So the combination of the form drag and skin friction has a minimum resistance to forward motion when the length to diameter ratio is about 6 to 1. However it is a shallow minimum and most modern submarines are nearer 8 to 1 or even more. They also usually adopt a long parallel mid body rather than a hydrodynamically ideal ellipsoidal form with a tapering to the stern (the teardrop shape). A parallel mid body has several advantages: not least, ease of construction; better internal space utilisation; and, avoiding the need for a tailor made docking cradle; while the resistance penalty of a longer simpler hull shape is very little.

In reality, the hull resistance is only part of the powering issue, as the normal form of submarine propulsor is a large single propeller, which is located at the stern on the axis of the pressure hull and envelope. This propulsive device then alters the flow regime ahead of it and the

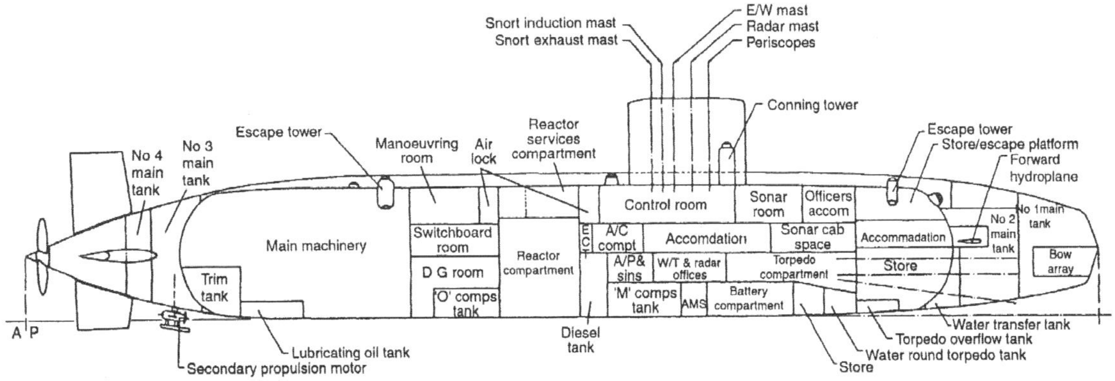

Figure 5: Profile of UK SSN [16]

after hull form and the propeller can be designed together to prevent or delay, at high speeds, the separation of the boundary layer from the hull. This not only makes the submarine more hydrodynamically efficient and, probably from a more operationally vital stance, quieter at higher speeds, as shedding eddies from the hull into the propeller can be delayed as speed is increased. The large slow running propeller, typically almost half the maximum hull diameter, can also then operate in the much more uniform flow regime resulting from the circular hull when compared to the case of a surface vessel (but the flow into the propeller will still not be perfect due to wake from the fin and the control surfaces). Modern nuclear submarines may be fitted with a "pump jet" rather than a propeller, primarily for delayed cavitation onset. The design of such large turbo-machines is quite complex as they both provide added propulsive lift and added resistance, from the pump jet casing.

The fin plus the rudders and hydroplanes are the major appendages to the submarine's hull form, with the fin typically contributing an extra 15% to the hull resistance, despite considerable analytical and model tank testing to make the fin shape and its attachment to the hull less resistful. Such shaping can hazard personnel access along the casing as this can be required when surfaced at sea and when coming alongside. There is also immense design effort devoted to reducing all other protuberances to a minimum and those still deemed necessary have to be carefully designed to align with local flow to minimise their effect. Shutters also have to be fitted to openings in the forward free flood structure, such as in way of the torpedo tubes' forward doors which then have to be opened before the weapons can be discharged from the torpedo tubes.

As stated in the section on submarine form, most non-nuclear powered submarines rely on diesel engines for primary power and they have to draw oxygen from the atmosphere. This is possible when the submarine is on the surface or, to substantially reduce detection by radar, at periscope depth via a "snort" mast, which incorporates a ball float to minimise the ingress of water into the diesel air intake system. The diesel generators both propel the submarine through the electric motor driving the propeller shaft and charge up the large batteries of lead acid cells, which provide air independent power for propulsion, through the electric motor, and auxiliary power to enable the air in the submarine to remain breathable and the operational equipment working. Despite the size of the batteries, typically two batteries of 220 cells, together giving 440 volts, their stored energy capacity is quite limited and this really inhibits the conventional submarine in speed and endurance, below periscope depth. This leads in conventionally powered submarine to the operational issue of "Indiscretion Ratio", which is the ratio of the time the submarine spends snorting to the time fully submerged (on battery or air independent propulsion (AIP) power). This is largely governed by the speed pattern the command adopts when fully submerged. Hence the attraction for the major navies of "limitless" nuclear power, despite the vast cost and nuclear safety implications which then tend to dominate a large portion of the design of such submarines, see Figure 5. There is the hope that the emerging technology of fuel cells will give the non-nuclear submarine greatly improved AIP endurance for both range and more sustained high speed at depth. It is likely that the introduction of fuel cell technology will drive up the size and cost of conventional submarines as such a new technology, at least initially, will still require diesel generators plus batteries in combination with any new fuel cell technology [15].

3.3 STRUCTURAL DESIGN ISSUES

This is a topic where submarine design is quite distinct from surface ship design. Apart from secondary structure, such as internal decks and the free flood spaces external to the pressure hull, little that is relevant to the structural design of surface ships applies to submarine structural design practice. Given that the sea loading on surface ships is essentially probabilistic and continually cyclical, it is difficult to predict. In comparison the main load on a submarine pressure hull, being that of external seawater pressure – governed by the depth of the hull below the surface, which increases by one atmosphere for every 10 metres depth – might seem to make the structural design task simpler. However, this is misleading because of the complexities arising from the compressive loading, which are best withstood by axi-symmetric structural configurations.

For military submarines, requiring large enclosed volumes for personnel and combat equipment and needing to be propelled at high speeds, Figure 6 shows the typical structural configurational features (taken from Reference 17, a comprehensive but quite technical summary of submarine pressure hull structural design). This figure shows possible variations from the main cylindrical shape, with transitions in diameter and conical sections usually terminated with hemi-spherical or torispherical domed ends, and with the cylinder single skinned and internally stiffened by "Tee" sectioned ring frames. The cylinder is then divided into compartments along its length by flat, stiffened internal bulkheads, each with a single watertight door. There are a few examples of more than one pressure hull but this is generally less efficient in both structural and overall arrangement aspects.

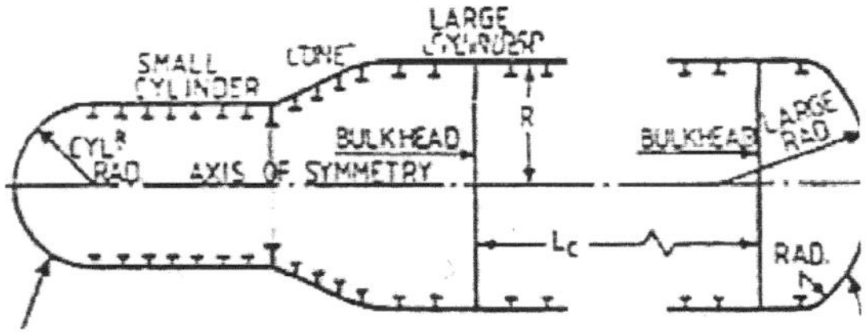

Figure 6: Typical Submarine Structural Configuration Features [17]

There are three principal modes of catastrophic failure of ring stiffened cylinders under external pressure:-
(a) Buckling of the shell between frames;
(b) Yielding of the shell midway between frames;
(c) Overall (cylindrical) collapse.

Burcher and Rydill [2] give more detailed descriptions of each mode but more specialist mathematical expositions are required for a full understanding of not just the collapse modes, but also the probabilistic nature of the justification of the choice of safety factors and the sensitivity to shape imperfections (such as 'out of circularity'). In addition there are further structural design issues:-
i. Design of transitions in diameter along the pressure hull's length and need to incorporate large penetrations in the pressure hull, such as escape towers and the conning tower down from the front of the fin, and even so-called "soft patches" for removal of major equipment in refit;
ii. Avoidance of sideway tripping of ring stiffeners under compression (a further collapse mode to be checked);
iii. Consideration of fatigue of a high stress low cycle nature, which is dealt with by careful detailed design, particularly in way of penetrations and pressure hull transitions;
iv. Transverse bulkhead design, best achieved by expensive ring extrusions and careful design of bulkhead stiffening;
v. Pressure hull cylinder to cone transitions in diameter and dome closure bulkheads, each with their own modes of collapse requiring avoidance of flatness beyond a tight tolerance, very akin to that for the main ring stiffened cylinder.

There remain other structural features beyond the pressure hull: decks which may need special connections to hull frames to allow for hull compression at depth and noise attenuation; external (soft) structure cantilevered off the end dome bulkheads; and many tanks. Some of the latter have to take full diving depth pressures internally, so they have to be designed as high pressure boilers located within the hull.

Thus a submarine's in-service design authority has to be well aware of the underlying assumptions and limitations in submarine structural behaviour, since specialist knowledge is necessary. Furthermore normal surface ship based structural analysis, computer based codes and finite element methods can be misleading, particularly if the design team lack an understanding of the full basis behind submarine structural design theory and the need to critically comprehend the implications of decisions impacting on the rest of the design based on the advice from the few deep experts.

3.4 DYNAMIC CONTROL ISSUES

Essentially these issues focus on why a naval submarine has rudders and hydroplanes and why they are situated where they are. This topic, like submarine structural design, is difficult to properly explain without recourse to some complex mathematics and specialist notation, only a flavour of which is given even in some general submarine references (e.g. Reference 2). The motion of a submerged submarine is quite distinct from that of a surface vessel, given its ability to manoeuvre in all three dimensions. So the design of the submarine has to take account of this distinction, which is both unique in marine vehicle design and complicated by the high density of the fluid medium and the relatively narrow vertical envelope of a few hundreds of metres, in which fast naval submarines operate, since any departure from that envelope can be fatal to the vessel and its occupants.

From the theory it is clear that the main vertical control of the submarine is by the larger aft hydroplanes with the fore 'planes best located far forward and on the main axis of the submarine for maximum effect. However most USN and Swedish submarine designs fit them on the fin or sail. The reason for this choice is primarily due to concerns at noise flow effects from the presence of the fore 'planes too close to the bow sonar arrays, given the latter are the submarine's primary sensor for navigation, detection and target acquisition. A further consequence of locating the fore 'planes on the fin means they are close to the surface when the submarine is operating at shallow depths and periscope depth, when the submarine is most difficult to control as a result of the water surface interface and wave effects. There are also choices on stern rudder and hydroplane configuration, see Figure 7.

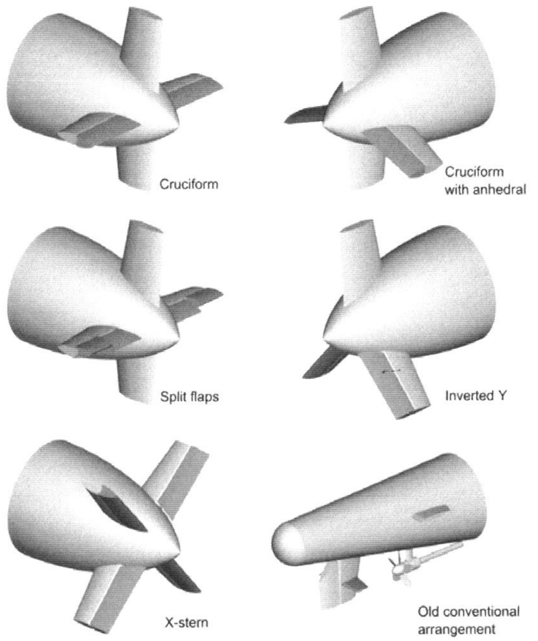

Figure 7: A range of Submarine Stern Configurations (UCL / Rachel Pawling)

These choices, on control surfaces design, heavily influence the manoeuvring envelope in which the submarine can safely operate. Close to the surface, as well as being hard to control, the boat is in danger from surface shipping and hazards like fishing nets – not to mention a greater likelihood of detection and attack when in threat environments. For most submarines the deepest depth to which they should operate is only a few multiples of the submarine's own length and excursions beyond that risk greatly increased probability of eroding tight fatigue limits and even resulting in actual pressure hull collapse. Thus the envelope in which the submarine is allowed to operate is restricted in terms of speed, depth and pitch attitude. Consideration of likely emergency situations due to floods (from burst "hard" pipework and valves) and hydroplane jams (due to power or mechanism failures) result in Manoeuvring Limitation Diagrams (see Figure 8), limiting the safety envelope in terms of depth and speed. Thus operating deep the submarine has to avoid low speeds to ensure that a defined "maximum critical flood size" can be recovered from, on the assumption of a specified response by the submarine command team (e.g. full speed and 'planes set to rise) can be undertaken in a defined and demonstrated reaction time. Furthermore, the danger of a 'plane jam at high speed, requiring recovery (by, say, operating the other set of 'planes, going astern, cycling the rudder and/or blowing main ballast tanks) leads to restrictions on top speed at the deepest operational depths. All this demonstrates the importance of minute attention to design, model testing, installation, full-scale trials and continual team training of submarine crews to avoid catastrophic failures in what remains a very dangerous vehicle operated at its limits and with very small margins for error.

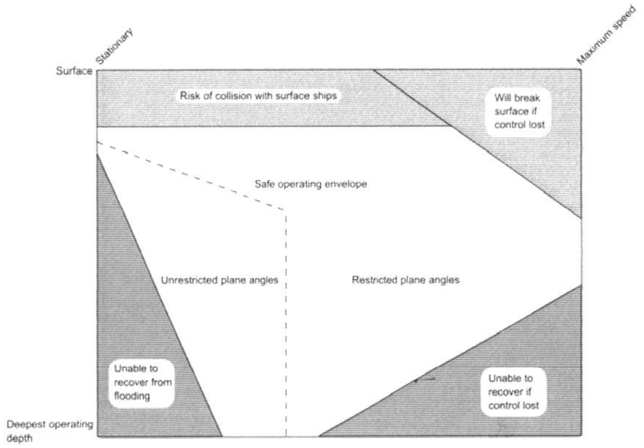

Figure 8: The basis of Submarine Manoeuvring Limitation Diagrams (UCL / Rachel Pawling)

4. CRITICAL SUBMARINE DESIGN ISSUES

4.1 FORM, ARRANGEMENT AND LAYOUT ISSUES

Clearly, submarines differ in overall appearance from surface vessels and modern submarines, whether conventional or nuclear powered, depart significantly from the ubiquitous Second World War submarine, which was actually more of a "submersible" than a true submarine. Radar and other modern surveillance techniques mean all modern naval submarines are designed to operate fully submerged and relatively deeply. So the modern submarine only comes close to the surface when absolutely necessary, such as to get a fire control solution on a surface target, communicate with other assets or, in the case of a conventional submarine, re-charge the large electric batteries by "snorting" at periscope depth. Thus the features of earlier submarines, with a high casing and a flared bow to operate efficiently on the surface in a seaway, are no longer attractive. This means the external hydrodynamically efficient form for submerged speed is the tear drop streamline shape dervied from the USS ALBACORE [18] and already discussed under Speed and Propulsion.

Unlike a surface vessel, where not only is the underwater hull shaped for minimal surface resistance at top speed and a generous freeboard provided above the waterline to the weather deck to ensure good seakeeping in rough weather, the modern submarine can appear very ungainly on the surface, as can be seen from the massive bow wave it generates at high speeds. It is also obvious that the submarine has only one distinct feature above the streamline hull, the tall and very slender fin or sail, unlike the surface vessel with large superstructures, masts and equipment on the upper or weather deck of the main hull. This mass of antennas is possible due to the modern surface warship having a relatively low overall density, which is impossible to achieve on the submarine, despite the fact that modern combat systems and indeed accommodation standards, which have largely led to this low density in the surface vessel, also affect modern submarine design.

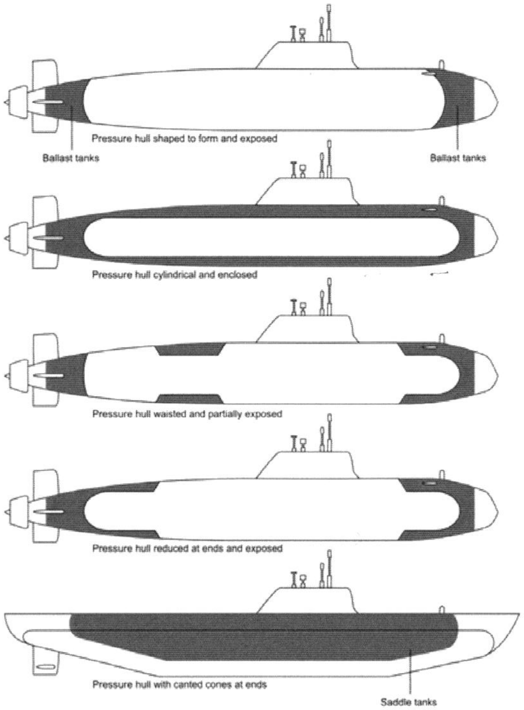

Figure 9: Choices for Pressure Hull and Main Ballast Tank Arrangements (UCL / Rachel Pawling)

So how is this dense, streamline form utilised to carry the capability needed to undertake underwater military missions? The major feature of the internal volume of the submarine's hull is the pressure hull, the over-ridding consideration of which is to resist high water pressure at depth, as discussed under strength considerations. Here the submarine designer is faced with a major design choice, which has significant impact on the subsequent design solution, namely, whether to totally enclose the pressure hull in an external, hydrodynamically efficient, envelope or have the pressure hull as part of that envelope, see Figure 9. The latter approach can reach an extreme where the only provision outside the pressure hull and fin is a minimal amount of top casing, for mooring equipment etc., and external structure forward and aft of the pressure hull. These spaces are not exposed to full diving depth pressures (i.e. they free flood at depth) but envelop various equipment external to the pressure hull: such as propulsion shafting; sonar transducers; torpedo tubes and access trunks projecting from the pressure hull; anchor equipment and the main ballast tanks. The choice of envelope arrangement is complex and it is noticeable that Soviet/Russian submarine design usually adopted a totally enclosed pressure hull, primarily for reasons of survivability of the pressure hull against torpedo attack. Whereas Western submarines, both nuclear and modern conventional vessels, have minimised the volume of the external structure, in order to keep the overall hydrodynamic volume as low as possible, hence reducing the hydrodynamic resistance and consequently the power required to propel the boat at the specified maximum submerged speed.

Maximising the pressure hull volume could be seen to be driving up the pressure hull weight, however it is consistent with the growing demands for space made by combat system equipment and personnel accommodation standards, already mentioned. From the above arguments, the pressure hull can be seen to meet two primary functions: to house crew and equipment at atmospheric pressure, regardless of submarine depth, and to provide the main buoyancy of the vessel to ensure the weight of the overall submarine is exactly matched, consistent with the need for the submerged submarine to be neutrally buoyant. Thus, for a typical conventional submarine, the structural weight (dominated by the pressure hull) is almost half the weight of the vessel, reflecting the compromise between the weight impact of increasing the maximum operational depth and, through buoyancy, the space available for people and equipment. It is worth noting that research vessels dive to extreme ocean depths, many, many times deeper than that to which any large naval submarine operates, with the pressure hull of such a small craft being a small and very thick sphere, as this is the most structurally efficient configuration. However, a sphere is inappropriate for the much larger and faster naval vessels, since wrapping a tear drop external envelope round it, would be very inefficient, as would be the utilisation of the internal volume of a necessarily large sphere, not to mention the problems of manufacturing of such a large structure. So the shape of a submarine pressure hull is essentially that of a circular cylinder with domed or, occasionally, flat end closures. This is a good compromise on grounds of structural weight, buoyancy provision and external envelope, although the resultant structural design is complex as is outlined in Section 3.3. In the long cylinder, space is more readily usable with length and diameter readily chosen to best fit with the hydrodynamic envelope. In fact shallow conical lengths of pressure hull can be used to improve this fit, especially to accommodate external tankage and projecting torpedo launching tubes, however any deviations from the cylinder incur penalties in weight and manufacturing costs.

As in any ship design, the internal layout is a complex web of compromises (see Figure 5). Simplistically, in a

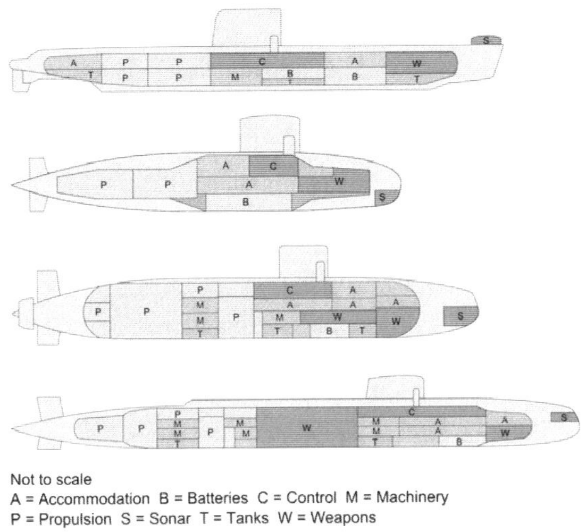

Figure 10: Typical Arrangements of Submarine Compartments (Two Conventional and two Nuclear Submarine examples based on Ref. 2)

submarine there are three main groups of spaces of roughly equal volume: propulsion; accommodation and support; and, control and operation. Most modern, conventional submarines are propelled by a single large propeller at the stern, so the aft end of the pressure hull is taken up with the propulsion machinery, normally consisting of an electric motor and diesel generators, plus auxiliary plant. The exception to this are the electric batteries, with many large capacity cells, each of which are heavy (about one tonne each) and fitted low down, to fit the curvature of the pressure hull below the main operational decks, centrally and forward. The operational decks are determined by human height, giving a close relationship between the number of decks and pressure hull diameter to maximise volume usage. While there is some mixing of the deck allocation between operational and accommodation spaces, traditionally the fore end of the pressure hull is taken up with the stowage of torpedoes (and nowadays sub-sea launched missile, as well as mines), given the forward most location of the tubes through which the weapons are discharged, see Figure 10. The other traditional location is the command or Control Room directly under the fin. This enables ready access through the conning tower hatch to the external bridge conning position, located at the top forward position of the fin and manned by the command when the submarine is on the surface. The location of the Control Room under the fin has also been traditionally driven by the ubiquitous periscopes, which are housed in the hull and the fin and raised above the fin when the boat is still submerged but close to the surface, i.e. at periscope depth. The latest submarines have non-penetrating masts, which electronically provide the command with the equivalent of the optical periscope capability. This opens up the ability to place the Control Room elsewhere; possibly a deck lower, at the widest part of the cylinder which would be far better for installing the control and combat system electronics than at the crown of the pressure hull. The latter location previously being necessary to maximise the hull's depth below the surface when the optical periscopes, penetrating the pressure hull, are fully raised.

4.2 NUCLEAR POWER DESIGN IMPLICATIONS

It would need a second paper to properly address the main aspects of nuclear powering a submarine. Such a significant feature has enabled almost complete independence of the sea air interface and hence the first true submarines. Furthermore in designing the submarine itself, this radically alters the design approach, from the conventional diesel electric machinery aft and large batteries central and forward. Locating the nuclear reactor plant forward of the propulsion secondary (steam turbine) plant introduces major design constraints (primarily the large pressure hull diameter – see Figure 5). The weight of the reactor plant and the associated dense radiation shielding, close to midships can cause a major longitudinal balance problem without the diesel electric powered submarines extensive and heavy batteries forward, if this is not addressed correctly from the very start of the design.

Design and location of shielding and structural containment design against a possible reactor incident leading to over-pressurisation of the reactor compartment are both aspects of a nuclear submarine design that the naval architect, as the overall submarine designer, has responsibility for. Having had that direct responsibility for the VANGUARD Trident Class design, the author is well aware what significance both of these aspects have on the viability of such large submarines (some 12,000 tonnes and 42 ft/12.8 m pressure hull maximum diameter, due to the missile tubes). That the nuclear plant requires back up with diesels and a small battery just adds to the direct price of the ultimate prime mover, and why only major navies make such an investment. In fact it could be said that style decision to be made in submarine design (see References 14 and 5 for the key importance of design style) is the choice to adopt nuclear power or not. This likely to be a major governmental decision rather than the submarine designer or even the navy and if the reactor plant is to be of a substantially new design itself, then it is likely to be the ultimate "long lead item", whose development and prototyping may drive the design and manufacturing programme. This does not mean that the overall submarine designer/team should not retain control of the design choices, beyond the nuclear/conventional principal choice, otherwise the danger is that the design could turn out to be unbalanced. The importance of ensuring the design is balanced, both in the sense of overall design coherence and economy but also very directly in the physical design balance – especially achieving longitudinal weight/moment balance and margin ballast allocation, as Section 3.1 has emphasised.

4.3 SUBMARINE SYSTEMS

A submarine could be said to be characterised by the extensiveness and complexity of the systems distributed throughout the vessel. This is in part due to the need to have systems throughout most of the boat to operate control surfaces, Trim and Compensation, Weapon discharge, Main Ballast Tank vents and external fairing shutters, periscopes and masts, battery ventilation, air purification, emergency and escape systems. All of these are in addition to the normal surface ship services of chilled water, air conditioning, fire mains, domestic systems and the electric power, together with the profusion of modern electronic data flow by wires and optical fibres, all to be installed within a very dense and highly integrated, enclosed space. The result is that the CAD based Integrated Product Model showing the composite systems arrangement for the actual physical installation on a modern naval submarine is truly mind boggling [19]. This has significant implications for the latter stages of the design, makes the building process very demanding and means the through life operation, support and up-dating is considerably more demanding and expensive than for almost any other type of vessel.

There is a growing use of electrical power to automate as much as possible, in the attempt to reduce the number of personnel required to operate a submarine. However, the inherent concerns for safety of the boat and the substantial forces involved to undertake many of the evolutions deeply submerged means that the submarine hydraulic system is still the prime means to operate control surfaces, due to its efficiency in power transfer. Because these demands are vital in terms of the submarine's ultimate safe operation, not only must the integrity of the pipework and its connections to the central power plant be of the highest standard, but also these vital systems usually have triple redundancy. (A more detailed explanation, with basic system diagrams, is provided in Chapter 9 of Burcher and Rydill [2]. Diagrams and guidance is given, not just for the hydraulic systems, but also for the other major submarine systems: high pressure air with a mass of air bottles; water distribution; ventilation, including fresh air production and emergency breathing systems; and electrical power and data distribution.)

The need for flooding in and pumping out of sea water at high (up to full diving depth) pressure, means there is an inherent concern with this as a potential source of critical flooding. For this reason the pipework for submarine compensation (taking on and losing weight) is kept to a minimum and a separate low-pressure system used to transfer water between the forward and aft trim tanks. The bilge pump needs also to be able to discharge against full diving depth pressure and so is usually able to be connected to or in common with the compensation system, acting as a backup to it. These are two obvious openings that require valves connected to the pressure hull and there are quite a few more necessary penetrations in the pressure hull, particularly associated with the cooling of the propulsion plant, and all need very careful attention as potential sources of threat to hull integrity. The fact that pipework associated with "hard" systems, tested beyond (maximum) Deep Diving Depth, will be affected by the compression of the pressure hull, when deep, leads to special flexible features which again need extremely careful design, installation, test and through life assurance, as do the penetrations and the associated hull valves, often with a second valve in the line as back up. Such specialist features need to be allowed for from the earliest design stages and any new significant features are likely to need their own research and development processes which then need to be integrated into the main design programme.

5. THE CONSEQUENCES OF CONDUCTING CONCEPT DESIGN INADEQUATELY

5.1 BUILD AND PROCUREMENT IMPLICATIONS

Turning to the specifics of submarine build, this is largely an extreme form of naval surface ship construction. The structural fabrication of the pressure hull and incorporation of penetrations into the hull is uniquely demanding and the quality standards for the distributed systems akin to those required in high performance aircraft, but on a physically larger and non production line scale. However, the bulk of the task of assembly of equipment and its installation is classic shipbuilding practice. The feature that makes that and the integration of all the systems extremely challenging is the submarine's inherent overall density (and hence complexity) being three times that of a comparable surface vessel, due to the balance of weight and submerged buoyancy. This has led in recent years to adopting modular construction placed on "floating decks" with fully assembled outfitting of equipment and systems. These decks are then slid into the end of the pressure hull before the major pressure hull units are finally welded together to close up the pressure hull. This form of construction requires a high degree of management and control of the myriad of sub-sub-systems, to ensure that the many suppliers involved deliver information and, subsequently, hardware to enable, firstly, a 3-D Integrated Product Model to be produced and, then, a sophisticated installation schedule to be maintained. (The UK example of the current ASTUTE Class SSN build problems, summarised in the technical press [20] well exemplifies this vulnerability; even if, in this specific instance, some blame could also be laid at wider procurement policy choices, taken for political reasons.)

Finally, on the direct submarine construction side, it is worth noting from the Australian COLLINS Class, that this non-nuclear submarine project had 426 sub-contracts (306 were Australian companies or Australian subsidiaries, 56 were UK companies, 21 US, 15 Swedish,

17 German and the remainder from 7 other countries [21]). So design and build is also a major project management task, not to be undertaken by the prime contractor or the customer's procurement project team lightly or under resourced. (The hands-off stance by the UK MoD was seen to be a crucial contribution to the ASTUTE programme over runs above, where almost 100% cost overruns have occurred with four years delay compared to the final 26 months delay for the delivery of COLLINS [20, 21].

5.2 CONTROLLING DESIGN DEVELOPMENT

Submarine operations are inherently dangerous, as the price paid for their unique military capability. Safety has always been a dominant feature in their design, acceptance, operation and maintenance through life but in last few decades there have been moves that have changed the inherent reliance, particularly in the three NATO navies with nuclear submarines, on a substantial highly capable and experienced in house technical capability. This has made them, in this specific regard, more akin to the smaller conventional submarine equipped navies, such as the Australian COLLINS Class experience. Greater responsibility placed with industry for design, build and maintenance has led to strengthening the formal safety regimes. Thus in the UK safety assurance for submarines has been incorporated in defined Hazard Case Assessment [22] and the responsibility for Design Authority issues has become more like that for aircraft design [23]. One major conventional submarine nation, Germany, has even followed naval surface ship practice with the national classification society (Germanischer Lloyd) producing Naval Ship Rules for Submarines [24]. Such split responsibilities, between government customer and prime contractor, can have significant implications for procurement practice, through life liabilities and apportionment of risk. It is noted that in the UK, even with the ASTUTE Class Design Authority being with the Prime Contractor (BAE Systems as the owner of the Barrow Shipyard), the design safety regime still requires the customer to have a competent Naval Authority. This owner organisation is charged with certifying the design for each of the key hazards (i.e. stability, structural safety, escape, fire and explosive safety), to employ Independent Safety Advisors and a named competent Duty Holder for the design (usually the Project Director), who is the formal signator of the Safety Certificate for each of the separately assessed Key Hazards [22].

5.3 NOT LIKE SHIP CONCEPT PHASE

Because it is dominated by interacting physical demands, producing a concept design study for a new submarine programme tends to go into far more detail than is usually considered necessary for a surface ship. However it is still characterised by the (implied) decision making process presented in the author's publication on Concept Phase for complex ship design [14]. Furthermore as Burcher & Rydill detail at their Figure (see Figure 11), the synthesis of a new submarine study can be undertaken in a similar manner to s naval surface combatant, however the latter is likely to be part of a much wider Concept Exploration investigation [14]. Although this may seem very like surface ship initial design, the above comment on the level of detail in a submarine final concept design, makes a considerable difference to the resources employed by the end of the Concept Phase when the decision to proceed has to be made.

Thus both in the level of detail and in the need to ensure that a sensible balance (not just in vertical stability and powering as for a surface ship concept) but importantly including longitudinal weight balance and adequate stability margins, could be said to really make submarine concept design distinct. The counter to this need for substantial detail is that the solution space is far less than that for even comparably expensive surface vessels. It is also possible, as Nordin [25] has shown, that quite detailed design decision making can be driven by extensive operational analysis, provide the concept of operations (CONOPS) is tightly defined. Whether this is applicable for a much wider general sea control and littoral warfare vessel, is seen to be questionable. Furthermore, Nordin still recognised that going from function to a specific (Swedish) form still requires the designer making a "style selection" [14].

6. SOME CONCLUDING ISSUES

6.1 SUBMARINE ACQUISITION ISSUES

While the introductory remarks suggest that the acquisition process for submarines does not differ in essential substance from that for other major naval combatants, there are some submarine specific issues, which are worth highlighting. Not least of these are the design and acquisition drivers and the issue of safety assurance, given the potentially catastrophic consequence of a major performance failure arising from almost any of the topics covered in this paper. In the specific case of the COLLINS Class, the wider procurement drivers were quite clear [21]:-

> "When choosing the consortium to design and build the submarine, issues of modern construction techniques, technology transfer and Australian industry involvement were as important as the design itself. "

In that respect it is worth appreciating the scope of the enterprise for these six conventional submarines: each had "almost 4 million parts, 75 km of cable, 200,000 connectors, 23.5 km of pipe, 14,000 pipe welds and 34.5 km of hull welds". It would seem, the public perception is that this procurement project was a vastly over budgeted disaster, all too common in defence acquisition worldwide. (See Yule & Woolner's valuable historical record [21], even if it is a little thin on the technical detail for a professional submarine design audience, when compared to, say, Wrobel's paper on the UK Type 2400 SSK [26].) However, it is relevant that the COLLINS Class acquisition outturn was within 3-4% of its 1987 price ($3,892M), if one ignores the subsequent McIntosh-Prescott "Improvements", which are considered further below. More significantly, the submarine elements in the design were made to work (even if there were instances, such as the Kockums' choice of diesel engines, which look highly questionable in hindsight) and the greatest problems in the project were those associated with the acquisition of the COLLINS Class combat system, rather than with the "boat' itself. The combat system problems would appear to have originated with the customer's ambition coinciding with a step change in computer technology, from bespoke centralised integration to the commercial explosion in PC technology, which could almost be put down to bad luck in timing.

6.2 SUBMARINE SUPPORT ISSUES

Another organisational aspect concerns that of through life ownership, with consequences for the safety regime. Ownership of indigenously, wholly or partially, designed and built submarines requires more than just naval support personnel and infrastructure, common to any first class navy, together with an investment in ongoing maintenance costs. (Yule & Woolner quote $322M maintenance costs for year 2007-8 for the six COLLINS Class as 38% greater than that for eight ANZAC Class frigates [21].) Thus there is a significant commitment required beyond initial procurement, which often has a substantial "improvement" component rarely recognised in the initial budget allocation. (Again Yule & Woolner quote $1.7B extra allocated to COLLINS Class following the McIntosh-Prescott Report of July 1999, yet just $143M of this was for "true fixes" to meet the contract, $300M for changes to the operational requirement (in the ten years from the 1987 contract) and $727M for ten years of technological obsolescence [21].) These substantial figures, reached after ten years, a period not untypical for contract to First of Class acceptance and shorter than most major procurement durations from start of concept work, show, in part, the challenge if considering the acquisition of a modified version of an existing submarine design, as is discussed further in the next section. The effort required to get an ill-thought through submarine support programme back on track is well spelt out in the Coles team's exercise for the COLLINS Clas [27].

6.3 NO SUCH THING AS A REPEAT SUBMARINE DESIGN

While submarine design might be considered a very evolutionary practice, given the highly constrained nature of the design due to the many aspects being interrelated and the small margins for error, this conclusion is not that obvious. If one is to take a given successful design and assume it can be easily up dated, then there are a series of issues, several of which have been highlighted in this paper, that mitigate against this. Thus only those systems and components that could be readily accommodated, on a one for one basis in space, weight, vertical and longitudinal location (for reasons explained in the section on Static Control) could be substituted for existing equipment in an extant design.

It is worth briefly considering the case of UK nuclear submarine development to understand the issue of the degree to which a new submarine design can be a direct variant of a previous design and be regarded as an "evolutionary" design. A new class of operational submarines rarely emerges directly by a radical change of design style. Thus the US Navy's SKIPJACK Class, which could be considered the first modern operational, as opposed to prototype, nuclear submarine, was a fusion of the nuclear prototype of NAUTILUS and of the teardrop configuration of the conventionally powered research submarine ALBACORE [18]. The UK's SSNs developed from SKIPJACK through the UK prototype DREADNOUGHT to its "variant", the VALIANT Class. Then the UK produced a significant step change in design with the SWIFTSURE Class, arguably a more radical design step from SKIPJACK than occurred in their USN equivalents. The SWIFTSUREs were followed by the TRAFALGAR Class, which were true variants with the same pressure hull design as SWIFTSURE plus the addition of three extra ring frame bays, accommodating respectively: improved noise reduction features in the main machinery space; an extra diesel generator (justified on nuclear safety grounds); and, an increased combat system installation in the accommodation block forward. Thus TRAFALGAR was a genuine variant or "evolutionary" design, with little change in most features and equipment and, it could be argued, was only achievable because the SWIFTSURE parent design had very generous margins built into that design. The issues outlined on the Static Control in Section 3.1, with regard to Margin Ballast, proved crucial in that case to enabling a successful variant to be produced. Providing such generous margins in a design is rarely done and would need to be clearly provided in any potential parent's design philosophy and design specifics, right from the cconcept design. Even then other aspects, such as a seamless maintenance of the procurement and build programme from parent class to variant is probably required to minimise differences due to equipment obsolescence, changes in legislation, operational concepts, technology advances, etc. If more likely the timescale between the two designs is some 20 years, then the obsolescence of the majority of the equipment procured from the large number of sub-contractors, already noted in the previous section, would make a repeat of the TRAFALGAR pattern highly questionable, just on this aspect alone.

Thus in considering the issue of an evolutionary design option, it should be appreciated that the choice is not analogous to surface ship design practice and, coupled with a likely generational gap in regard to obsolescence of equipment, the choice of a truly "evolved design" looks highly questionable. There may be some scope for, perhaps, some commonality in style choices to take from a successful "parent", but even this should not be assumed. Keane & Tibbitts even caution the risks with "repeat" designs for surface naval combatants [28]. Such issues would still need to be technically justified as being VFM against alternatives. What a "derivative" design will definitely not give a government is a cheaper, lower risk and less government resource intensive solution. The belief that it would be otherwise may well be the motivation by wider government authorities in seeing an "evolutionary design" or "variant" as an attractive option. If anything such a constraint on the design can be more awkward, more compromised, harder to prove for assurance, and end up being an even more risky and costly design than a bespoke, albeit largely "derivative", design. It is the author's view from being close to the evolution of the TRAFALGAR Class, that that design had nothing like the extent of "changes from its parent", that would seem inevitable in any successor. In fact TRAFALGAR still required a completely new design input and resource demands comparable to its "parent (SWIFTSURE) design". So an evolutionary design is not likely to be the route to achieving programme cost savings and is likely to add risk, arising from the mistaken belief that reduced effort could be employed – it could be a recipe for dangerously cutting corners. The whole point of this paper has been to emphasise that submarine design is NOT like surface ship design. It is highly interrelated and unforgiving and must be resourced properly, which means it has to be led by experienced submarine design professionals.

6.4 FINAL REMARKS

This paper has compared submarine design with that of complex naval combatants and while there are aspects, particularly in the overall design environment that make the acquisition of any sophisticated military vessel quite distinct from merchant ship practice, submarine design exhibits a further level of distinctiveness. This is to a large degree due to the physics of operating primarily well below the sea-air interface that dominates the operating environment of other marine vehicles.

However, it is also the fact (often used to say "ships are different") that the degree of interaction of the various design features, necessary to achieve the eventual requirement, is much greater in the case of submarine design than it is for most surface naval vessels. It is also the case that a submarine design is highly "tuned" or sensitive, in terms of requiring a precise balance – in weight vs. buoyancy and transverse AND longitudinal balance – that is not the case with most surface ships. This places a very great demand on the whole submarine designer. That the consequences of failure in a host of sensitive aspects in a submarine design is as catastrophic as in aircraft design is the final distinction submarine designers, and their political and bureaucratic masters, forget at their and the submariners peril.

7. ACKNOWLEDGEMENTS

The contributions made by the author's colleagues to on-going research into naval vessel design, especially Dr Pawling not just for her production of many of the diagrams, is gratefully acknowledged, while the argument presented remains the views of the author alone on behalf of UCL.

8. REFERENCES

1. Kendrick, S: "The Influence of Shape Imperfections and Residual Stresses on the Collapse of Stiffened Cylinders", Paper C10/79, IMechE, 1979.

2. Burcher, R & Rydill, L: "Concepts in Submarine Design", Cambridge UP, 1994.

3. Arentzen, E. S. & Mandel, P: "Naval Architectural Aspects of Submarine Design", Trans SNAME, 1960.

4. Brown, D K and Andrews, D J: - "The Design of Cheap Warships", Proc. of International Naval Technology Expo 80, Rotterdam, June 1980. (Reprinted in Journal of Naval Science April 1981)

5. Andrews, D: "The Key Ship Design Decision– Choosing the Style of a New Design", COMPIT 2017 , Cardiff, May 2017

6. Purton, I: "Concept Exploration of a Novel Submarine Using Innovative Computer-Based Research Approaches and Tools", PhD Thesis, UCL, 2016.

7. Andrews, D: "Does one size fit all? Or do different warship designs require different ship design methods?", Warship 2016: Advanced technologies in Naval Design, RINA, Bath, June 2016.

8. Andrews, D: "Unconventional Ships and Unconventional Design", HIPER Conference, Cortona, Italy, 17th Oct 2016.

9. .Gale, P: "Ship Design Process", Chapter 5 of "Ship Design and Construction" Lamb T (Ed): SNAME, Jersey City, NJ, Vol. 1, 2003.

10. Andrews D J: - "The Management of Warship Design", Trans RINA 1993 (reprinted in Journal of Naval Engineering 1993)

11. Andrews, D J: 'Marine Requirements Elucidation and the Nature of Preliminary Ship Design', IJME Vol. 153 Part A1 2011. (DOI No: 10.3940/rina.ijme.2011.a1.202)

12. Benford (1979)

13. Graham, C.: quote in Gates, P J & Rusling, S C (1982) 'The Impact of Weapons Electronics on Surface Warship Design', *Trans RINA, Vol. 124.*

14. Andrews, D: "The True Nature of Ship Concept Design – And what it means for the Future Development of CASD", COMPIT 2013, Cortona, Italy, 2013.

15. Donaldson. A J: "Submarine Propulsion and Operational Effectiveness", RINA Warship '99, Naval Submarines 6, London, 1999.

16. Betts, C. V: "Nuclear Submarines in the Royal Navy – The Design Challenge", Nuclear Energy, Vol. 38, No. 2, Apr 1999.

17. Faulkner, D: "The Collapse Strength and Design of Submarines", RINA Warship '83, Naval Submarines, London, 1983.

18. Jackson, H: "USS ALBACORE", RINA Warship '87, Naval Submarines, London, June 1987.

19. Willcox, N & Tasker, P.H.: The Royal Navy's ASTUTE Class SSGN", RNE, Vol, 53, No. 2, 1999.

20. Janes' Navy International: "Contrasting policy approaches divide French and UK submarine enterprises", JNI p. 26, Dec 2008.

21. Yule, P & Woolner, D: "The Collins Class Submarine Story", Cambridge UP, 2008.

22. Andrews D J:- 'Recent Developments in the Safety Regime for Naval Ship Design', Quality and Reliability Engineering International 2006 Vol 21.

23. Andrews, D: Discussion to "Design Authority of the DARING Class Destroyers" By: Gates, PJ, TRINA/IJME Vol 147, 2006.

24. Petersen, L: "Classification of Submarines", RINA Warship 2005, Naval Submarines 8, London, 2005.

25. Nordin, N. "A Novel Submarine Design Method", PhD Thesis, Chalmers University of Technology, Gothenburg, 2014.

26. Wrobel, P G: "Design of the Type 2400 Patrol Class Submarine", Trans RINA, 1985.

27. Coles, J. D: "Study into the Business of Sustaining Australia's Strategic COLLINS Class Submarine Capability", DoD, Commonwealth of Australia, Canberra, Nov 2012.

28. Keane, R. G. & Tibbitts, B. F: "The Fallacy of using a Parent Design – "The Design is Mature"", SNAME Expo & Ship Production Symposium, Bellevue WA, Nov 2013.

9. AUTHOR BIOGRAPHY

David Andrews was given a new Chair in Engineering Design at UCL in 2000, following his early retirement as a Director the UK MoD. His early career in the RCNC included postings in in-service and new construction submarine design sections. He was subsequently Head of Preliminary Design responsible for the initial studies, including on the Future Attack Submarine. In 1993-98 he was appointed MoD Professor of Naval Architecture at UCL where he directed the MSc in Naval Architecture and the post-MSc Submarine Design Course. In 2001 he set up a new Design Research Centre in the Department of Mechanical Engineering, which is focused on CASD and the design of complex systems. He is a Fellow of RINA and a Vice President, in 2000 was elected a Fellow of the Royal Academy of Engineering and is currently International Chair of the International Marine Design Conferences.

HUMAN FACTORS ISSUES FOR UNMANNED UNDERWATER VEHICLES

Dr Caren Soper QinetiQ, UK

SUMMARY

The use of autonomous vehicles is a significant growth area, widening in scope of application to deliver greater military effect for those dull, dirty and dangerous tasks. This brings a range of associated Human Factors (HF) issues, which can have significant safety, security, and legal / ethical implications. The focus of this paper is on the HF issues associated with more truly autonomous systems, extending to those that incorporate artificial intelligence. It draws on data collected during the mine countermeasures work within Unmanned Warrior 2016, the largest demonstration of its type ever undertaken in the world. The HF assessments centred on the control of unmanned underwater vehicles in live demonstrations and on the usability of the human computer interface features and recorded observed tasks, equipment used and levels of automation.

The paper goes on to posit that development of trust and building the operator's confidence will be more relevant as automated vehicles become orientated towards the higher end of the autonomy spectrum; designed to perform mission critical tasks in the underwater domain.

An approach is proposed to identify those design features that should be incorporated for highly autonomous vehicles to enhance trust, using recent advances in the literature associated with anthropomorphism (attribution of human related characteristics, such as the ability for rational thought and conscious feeling, rather than a humanlike face / body) and a three-layered model of trust. It is the ability for the autonomous vehicle to portray these features that leads to the development of a greater trust.

1. INTRODUCTION

Unmanned and autonomous vehicles are increasing in complexity and application to a widening scope of operational contexts and greater range of tasks. This usage causes a range of Human Factors (HF), issues which can have implications for the effective and safe delivery of the intended effect within legal and ethical boundaries. It is important that these HF issues are addressed. HF issues become particularly pertinent in situations such as failure and reversionary modes, mission termination and veto opportunities, all of which relate to the ability to accept risk.

The HF issues are extremely broad covering:

- Identification of the human / machine functional task distribution;
- Determination of the automation boundary, including the level of autonomy that is realistic and desirable in the UK mission systems context;
- Development of the relevant Human Computer Interfaces (HCIs) for the Command and Control (C2) of autonomous systems;
- The resultant impact on human tasks and roles, team structures, workload, skills and competencies required, manning levels, organisational structures, and career pyramids;
- The resultant impact on issues such as trust, and building the operator's confidence in autonomous vehicles and the data that they provide.

The focus of this paper is on the HF issues associated with more truly autonomous systems, extending to those that incorporate artificial intelligence. At the high end of the autonomy spectrum, Unmanned Underwater Vehicles (UUVs) are capable of working autonomously today, particularly with respect to vehicle control functionality and following search patterns to detect a particular type of object. However, there is still an element of human involvement required in the launch and recovery of the vehicle, termination of a mission and provision of a veto / override command especially where loss of life may be involved.

The Maritime Autonomous Platform Exploitation (MAPLE) programme is a significant programme of work, involving a large consortium that has been underway over a number of phases. Phase 3 delivered capability for Unmanned Warrior 2016 (UW16)[1], to demonstrate the use of unmanned vehicles in all three domains in a multi-national exercise, the largest demonstration of its type ever undertaken in the world.

[1]www.royalnavy.mod.uk/news-and-latest-activity/operations/uk.../unmanned-warrior

Part of the HF work undertaken within this programme developed a functional de-composition of those tasks associated with the C2 of autonomous vehicles, together with an assessment of realistic autonomy (feasible and desirable) function allocation to human, machine or a blend of each. This task derivation and function allocation was validated in workshop settings with military and unmanned vehicle Subject Matter Experts (SMEs).

2. UNMANNED WARRIOR 16 DATA CAPTURE

This paper draws upon data collected during the Mine Counter Measures (MCM) element of the work for UW16. The HF assessments centred on the C2 of UUVs in live demonstrations, the usability of the UUV related HCI features and recorded observed tasks, equipment used and perceived levels of automation. The tasks assessed for levels of automation were those from the functional de-composition; data captured whether they were performed by human, operator or a mix of each.
This was a real world first to collect HF data in relation to the C2 of autonomous vehicles operating with other nations in live sea trials. The data collected will be used to inform follow on phases of MAPLE. In particular, the task autonomy levels observed will be used as a baseline for task to role allocations and level of autonomy options, including adaptable autonomy, for testing in synthetic environments / live exercises. Figures 1 and 2 illustrate the MCM cabin.

Figure 1: At the Mine Counter Measure (MCM) cabin in Kyle for its key role in Unmanned Warrior 2016

Figure 2: Naval operators inside the MCM cabin

3. HCI FEATURES ASSESSMENT

With respect to the usability of the HCI UW16 observers recorded both their own assessment of the UUV displays and where possible the views of UUV operators.

Overall operators / observers rated the displayed information positively; it was considered useful, comprehensive and easy to understand. In general, the views were that the information provided with respect to planning the mission, operational conduct and data obtained enabled the operators to develop a good level of Situation Awareness (SA). SA is considered an essential element enabling the operator to start to develop trust.

The majority of useful features were contained in either the UUV management tote or the sensor tote. Through the UUV management tote operators were able to manipulate and modify the displayed UUV C2 information as well as call up data feeds associated with specific UUVs, as required. Of particular utility was functionality to enable UUV track position, show / hide range and show / hide plans.

The range indicator provided operators with an indication of range to mothership. However, observers /operators recommended that more management functions would be necessary for real world applications. For example, it would be crucial to be able to interface directly with specific UUVs and thereby direct current tasking and coverage once the UUV was deployed. This expressed view indicates a lack of trust in UUVs to assess their own progress on current tasking and re-direct themselves, which will need to be addressed as more highly autonomous UUVs come into service.

The sensor management tote also had many strong features, and provided an overview of available data outputs. Operators were able to select data output from

any currently active UUV using the drop down menu, either for multiple UUVs with information tiled in a single display or for an individual UUV and enlarged to fill the entire sensor tote. It was recommended that data sources not available should be clearly indicated.
To enhance trust it is important that the data output for each UUV should contain accurate and reliable metadata, including position reporting, that can easily be interpreted. Similarly, an accurate and well defined indication of endurance and mission state should also be immediately clear. This information should be available at-a-glance such that users can access key information quickly in an intuitive and usable form.

4. AUTONOMOUS RELATED TASKS

With respect to assessing the level of autonomy in relation to the functional decomposition of activities observers recorded observed tasks / subtasks, enablers (equipment used to enable task) and perceived level of automation associated with UUV operation.

It was found that not all tasks from the functional de-composition were observed during UW16. As the functional de-composition was orientated towards future UUV C2 operation (circa 2030), this was not unexpected. As the use of UUVs matures more UUV specific related tasks are likely to be required. In addition, for the same reasons a higher level of manual control was recorded. Indeed, very few completely autonomous tasks were observed during UW16. However, it is expected that as UUV technology develops and progresses higher levels of autonomy and more frequent blended control of UUVs will be realised.

Details of the level of autonomy were assessed in relation to the task decomposition taxonomy.

Vehicle level planning – goal based planning, selection of vehicles, accessing plans, understanding mission constraints / risks and accepting / amending tasks were all seen as a blend of operator and system tasks. Adaptive system mission control enabled a vehicle to take over from another when that vehicle became disabled e.g. out of battery.

Offboard vehicle C2 – all offboard C2 tasks were performed by UUVs including piloting of the vehicle and route and task direction. The UUVs themselves acted autonomously once in their allocated waterspace carrying out search activities for mines, as shown in Figure 3.

Figure 3: UUV on route to designated area to detect mines

Launch and recovery C2 – all launch and recovery C2 tasks, including approval to launch / recover, were performed by humans as illustrated in Figure 4. This involves manual manipulation of the UUV.

Figure 4: Recovery of a UUV following mission completion

Effects planning – all effects planning tasks, including understanding of the battlespace and UUV contribution to effect were performed by operators.

Payload planning and control – those payload planning and control tasks observed were performed by a blend of operator and system tasks using the planning aid.

Communications planning and spectrum management – observed communications planning tasks were either performed by operators, largely via briefs or by the system in planning communications between vehicles and the surface for single squads.

Space planning and management – all space planning and management tasks were performed by operators, including tactical space planning and space control.

Mission monitoring – all mission monitoring tasks were performed by operators.

Space de-confliction – all space de-confliction tasks were performed by operators, prior to the mission, as each UUV was allocated its own area within which to operate. Offboard sensor exploitation – those observed offboard sensor exploitation tasks were performed by operators. Thus it can be seen that today that there are relatively few tasks that are fully automated, although there is considerable scope to create greater UUV autonomy in relation to all aspects of UUV operations in the not too distant future.

5. UNMANNED WARRIOR 16 CONCLUSIONS

So given the data collected during UW16 it is clear that in relation to the various activities associated with UUVs that a large proportion of the tasks still fall to the human operator to perform. Nevertheless autonomy is leading in the underwater domain, largely due to communications bandwidth and regulatory influences.

The system can provide enhanced support information by providing a set of Apps / interconnected tools particularly with respect to the various planning related activities associated with the vehicles themselves, effects required, payload carried and utilisation of space as well as with those tasks associated with monitoring / influencing mission execution.

The MCM cabin provided a functional representation of the required elements of a Royal Naval platform's operations room. This representation included an adaptation of the in-service Combat Management System, enhanced with prototype applications for the planning and tasking of UUVs.

The Apps were used successfully in the planning, monitoring and post mission analysis of squads of unmanned assets throughout the trials. In particular, goal based plans for squads of unmanned vehicles comprising of underwater, surface, and aerial vehicles from three different nations (UK, US and Canada) were successfully generated. It was demonstrated that the operator could create a high-level mission plan and then automatically push this out to the vehicles via agreed interfaces.

In addition, a communications relay to the UUVs was successfully demonstrated via an unmanned air vehicle.

6. CRITICALITY OF TRUST IN THE FUTURE

Given the limited degree of autonomy implementation today as illustrated in UW16, in the near and mid-term, much can be done to ease the operator's challenge of managing multiple heterogeneous vehicles undertaking complex missions. In this context full autonomy will tend to be limited to very specific aspects of task execution and a toolset of integrated applications, designed to HF practice, will be required if operators are not to be overloaded.

Looking to the longer term where the scope and scale of autonomy can be expected to rise dramatically, a different approach is required as the autonomous vehicle will then become more of a collaborative team member. For high performing collaborative human teams, in mission critical situations, the development and maintenance of trust is crucial to deliver effective performance.

Therefore ensuring the development of trust, in autonomous vehicles is a key issue to be addressed as such systems are designed and developed. The development of trust and building the operator's confidence will be more relevant as automated vehicles become orientated towards the higher end of the autonomy spectrum; designed to perform mission critical tasks in the underwater domain.

The issue of trust has been examined extensively over a number of years, however, there are recent literature advances that seem to offer value, as they are less about analysis and more about developing and implementing design features to enhance trust. A combination of two recent approaches is proposed to identify those design features that should be developed.

7. APPROACH TO IDENTIFY DESIGN FEATURES FOR INCORPORATION IN AUTONOMOUS VEHICLES

An approach is proposed to identify those design features that should be incorporated for highly autonomous vehicles to enhance trust using recent advances in the literature associated with anthropomorphism (attribution of human related characteristics such as the ability for rational thought and conscious feeling rather than a humanlike face / body) and a three-layered model of trust. It is the ability for the autonomous vehicle to portray these features that leads to the development of a greater trust.

The three-layered model covers dispositional trust, situational trust and learned trust. This provides a new lens for conceptualising the human related aspects in developing trust and can be used as a basis in the design of autonomous vehicles. Dispositional trust relates to an individual's pre-disposed tendency to trust based on culture, age, gender and personality traits. Situational trust relates to the context including setting, difficulty of task and workload of the operator. It is the third element of learned trust, which is the area of interest as this relates to the automated system and design features that may affect perceptions of performance and level of trust.

It is proposed that incorporation of design features in five areas in conjunction with higher levels of anthropomorphism in some of these areas can significantly enhance trust development.

- Transparency. The explicit portrayal of the inner workings and logic of the autonomous system including traceability of the reliability of the autonomous system and explanations for any errors that might be made. This aspect is key as being able to interrogate an automated function to understand why it did something is crucial. Both the means of interrogation and response provided need to be intuitive and user friendly.
- Appearance. The provision of a well designed interface that is aesthetically pleasing with anthropomorphic features including name, gender and appropriate essential characteristics e.g. an autonomous system intended for lethal action should appear menacing and sinister.
- Ease of use. The provision of enhanced system usability and visual clarity of data with ongoing salient feedback on such aspects as progress with respect to task execution, system state and potential hazards. Layers of tote pages or pull down windows should be avoided as this will detract from the task in hand and likely result in a reduction in Situation Awareness for the operator.
- Communication style. The use of verbal communication, instead of text, with human voice rather than synthetic speech. This impact of the use of human voice can be enhanced by further using accents tailored to those of the individual operator. Verbal interrogation and response can minimise loss of focus in the task and is the natural means by which humans already relate and exchange information with other intelligent assets i.e. humans. In addition the demonstration of good etiquette and politeness by behaviours such as not interrupting and patience
- Level of control the operator has over the automated functions. For highly autonomous systems, the operator could be taken out of the loop, but trust may be maintained if ongoing information is provided to the operator. Conversely, part of building confidence would be keeping the operator in the loop in some way, for example knowing that should an operator glean information that the UUV does not have then the operator can still influence the UUV to achieve the tasks, auto-destruct or auto-home. An alternative may be the use of adaptive automation, which introduces a trade-off to level of control whereby the level of control / automation varies to meet the needs of the specific situation.

Incorporation of these design features should engender a higher degree of trust as anthropomorphic features, transparency, politeness, feedback and ease of use are enhanced.

8. CONCLUSIONS & RECOMMENDATIONS

It is clear that whilst we are not there yet there is an aspiration in the future for fully autonomous vehicles – indeed engineers wish to remove the human from the loop completely which brings with that decision significant ethical and legal implications especially where weapons / other lethal effects may be involved leading to loss of life. Given these ethical considerations for future naval operations, there will continue to be a need for personnel to be involved at some point in the decision making cycle with the ability and resourcefulness to respond effectively to surprise / unplanned mission responses that may be required.

Therefore, it will be important to design and implement features and functionality that enable a symbiotic collaborative relationship between the human element and the more autonomous UUV that could be capable of intelligent thought. The approach proposed should be used to develop design features that are then implemented within a real live UUV autonomous vehicle and tested in trials to determine the relative trust that is engendered compared with a baseline system that does not have these design features.

Given wider considerations adoption of highly autonomous trusted systems will have implications for a range of wider HF issues including manning levels, skills and competencies, organisational structures, and career pyramids together with recruitment and training required. Cognisance of these wider issues needs to be taken account of as such systems are bought into service.

9. REFERENCES

1 Hancock P.A., Billings D.R., and Schaefer K.E. (2011) A meta-analysis of factors affecting trust in human-robot interaction. Human Factors

2 Hoff K.A., and Bashir M. (2015) Trust in Automation: Integrating Empirical Evidence on Factors That Influence Trust. Human Factors: The Journal of the Human Factors and Ergonomics Society

3 Schaefer K.E., Chen J.Y.C., and Szaima J.L, (2016) A meta-analysis of factors influencing the development of trust in automation – implications for understanding autonomy in future systems. Human factors, May 2016

4 Waytz A., Heafner J., and Epley N. (2014) The mind in the machine: Anthropomorphism increases trust in an autonomous vehicle. Journal of experimental social psychology

10. AUTHORS BIOGRAPHY

Caren Soper BSc PhD C. Psychologist, QinetiQ**:** Caren is a highly experienced Human Factors practitioner. She is an Associate Fellow of the British Psychological Society. She has extensive experience in leading programmes and providing technical input, guidance and consultancy in a wide range of Human Factors (HF) areas. She has more than thirty years' experience in military human factors; over twenty where the maritime environment has been the main focus. She has worked on a wide range of collaborative and multi-disciplinary programmes including studies, experiments, trials and demonstrations. Recently she has been the HF lead on MAPLE cross-company programme, which included the derivation of new roles and tasks, determination of the automation boundary, identification of wider HF issues concerning UXVs and HF data collection and analysis during UW16.

DEVELOPMENT OF AN AUV LAUNCH AND RECOVERY SYSTEM FOR THE PORTUGUESE MANNED SUBMARINES VIA TORPEDO TUBES

Henrique Vieira da Silva, Naval Academy, Department of Sciences and Technology, Portugal.
Miguel Cavique, UNIDEMI_CINAV & Naval Academy, Department of Sciences and Technology, Portugal.
João Borges de Sousa, University of Porto & LSTS, Electrical and Computer Engineering Department, Portugal.

SUMMARY

The use of Autonomous Underwater Vehicles (AUVs) in the military Navy has increased in the last few years to perform missions including Mine Countermeasures (MCM), Search and Rescue (SAR), seabed mapping, and remote environmental assessment. The Portuguese Navy has an active dual-use Autonomous Underwater Vehicle (AUV) development program in which the capability of launch and recovery (L/R) of AUVs from manned submarines is under development. The development of this capability is receiving significant attention worldwide because it enables discreet operations in areas of interest. This paper discusses L/R systems for manned submarines and presents the design methodology of a L/R system for submarines. The design takes into consideration user and design constraints defined by the submarine flotilla especially in what concerns to the use of the torpedoes tubes to launch the AUVs. The proposed solution has to modular and minimally intrusive to enable launch and recovery with no major modifications to the submarine. This is achieved with an add-on L/R station that can be easily removed at sea to preserve all submarine capabilities.

NOMENCLATURE

HV	Vickers Hardness (kgf mm^{-2})
ρ	Density of water (kg m^{-3})
σ_y	Yield strength (MPa)
K_{Ic}	Fracture Thoughness (MPa m$^{-0.5}$)
m	mass (kg)
t	Thickness (m)
q	load (N)

1. INTRODUCTION.

The importance of AUVs for military operations increased in the last century because AUVs can access areas that manned submarines cannot. The US Navy UUV MasterPlan[1] describes AUVs potential missions according to the following priorities: Intelligence, Surveillance and Reconnaissance (ISR), Communication/ Navigation Network Node, Mine Countermeasures (MCM), Payload delivery, Anti-Submarine Warfare (ASW), Oceanography, Information Operations, Inspection/ Identification or Search and Rescue (SAR) and Time Critical Strike. Interoperability is a key characteristic in any military operation making AUVs to be supported in their missions, either from autonomous or manned vehicles. The launch and recovery systems (L/Rs) of AUVs from surface and submarine ships allows the missions to take place further away from the coastline [2,3]. Submarines are hardly detected when compared to the surface vessels which is an advantage in a hostile environment.

The Portuguese navy has currently dual-use AUVs program that preform missions of Mine Countermeasures, Search and Rescue, seabed mapping and remote environmental assessment [4]. The Navy accounts for a detailed experience on missions with submarines regarding the launching and recovering of AUVs. However, L/R relies on divers from the Portuguese navy that need to stay outside the submarine during the operation. This concept increases the danger of to the mission, and adds many tasks of concern to the personnel. The launch and recovery system from a submarine should enable fully automated discreet missions.

The aim of this paper is to describe the design of a launch and recovery system of an AUV, using the weapons tube of a submarine. Moreover, the design allows to remove the L/R system from the weapons tubes, making possible to use back the armoury in the tubes.

Section two of this paper discusses and compares existing systems of launch and recovery of AUVs from submarines, designed by enterprises and universities researchers. Taking into account the constraints of the current design, the comparison of the aforementioned systems allowed to define a design concept suitable for the Portuguese submarines. This section also mentioned the main functional requirements for the launch and recovery taking into accounts the Independence Axiom (Axiom1) of the Axiomatic Design method [5]. Section three presents a study of the materials that are suitable for the L/R system, taking into consideration the properties of the materials, using the CES Edupack software. Finally, section four explains the design and presents the calculation of the main piece of the L/R system. SolidWorks 2010 simulations presented in this section use the Finite Element Method (FEM).

2. CONSTRAINTS AND FUNCTIONAL REQUIREMENTS TO THE LAUNCH AND RECOVERY SYSTEM.

2.1. CONSTRAINTS.

Missions with AUVs have five different stages: mission preparation, AUV launch, AUV operation, AUV recovery to the inside of the submarine and the maintenance of AUV and of the L/R system.

Submarines have two different spaces for launching AUVs: the weapons tube and a hatch in the main tower. The weapons tube is the preferable choice to the flotilla because it already has systems to perform the AUV mission, needing no structural changes. Otherwise, using the main tower as a means to L/R the AUVs obliges to install an extra device that makes use of vital space needed to perform other tasks.

The recovery stage undergoes through three steps: rendez-vous, when the AUV is navigating to meet the submarine; approach, when the AUV is guided to the docking station; and, the docking, when the AUV is immobilized in safety and connected to the docking station.

The design has to take into consideration constraints for all the above mentioned stages that allows excluding some of the existing systems.

An important constraint from the flotilla is to have a removable L/R system, allowing the use of all weapons tubes in missions that do not require the employment of AUVs. Therefore, the U.S. Navy L/R systems *Boeing Torpedo Mounted Retractable Arm* [3] or the *Universal Launch and Recovery Module* [3] are not suitable as concept to the current design. The former system used at the *USS Hartford* and the latter in the *USS Ohio* has dedicated tubes to the AUV launch, therefore unavailable to other armouries.

Moreover, it is not possible to do any permanent physical changes in the submarine structure or in any compartment. Finally, the AUVs need maintenance actions between missions preformed in the inside of the submarine. Considering the three constraints, systems like the *Launch and recovery via dry/wet casing mounted hangar* [2] or the one of the *University of New Brunswick Design* [6] are not suitable too, as they make use of the submarine deck.

The funnel or cone design solution is able to recover the AUV using acoustic transponders installed in the cone, allowing to communicate with the vehicle during the approaching phase. The *MBARI Docking Station* [6], the *REMUS Docking Station* [7] and the EURODOCKER are examples of this type of conceptual design. However, the acoustic communication between the AUV and the host submarine it a matter of concern during hostile mission scenarios, as it may identify the position of the submarine.

It is more convenient to use lights [8] or electromagnetic homing [9] as a way to interconnecting the systems of communication of the AUV and the submarine. The optical approaching system is the best way regarding the stealth of the process, because of the limited area of identification.

The L/R system should be able to operate between periscope depth and 100 m depth, choosing to work out of the influence of surface weather conditions. Therefore, the design of the L/R system of the AUVs might not take into consideration the water turbulence, caused by wind and waves occurring at sea surface. Regarding the underwater current, the vehicle should do the underwater approach at currents lower than 3 knots [10]. Moreover, the system has to operate successfully in covert (stealth) and overt mission scenarios.

The system under design does both launch and recovery of the AUV using the weapons tubes, what is in accordance to the aforementioned constrains. The diameter of the AUVS is therefore limited to the diameter of the weapons tubes of the submarine [11]. The system under design uses a launching tube compatible to the weapons transport device, to take the vehicle until the muzzle door. Inside the launching tube a mesh of small tubes of polyethylene allows to shelter the AUV during launch and recovery, forming a funnel when opened outside the tube.

2.2. FUNCTIONAL REQUIREMENTS.

The functional requirements are the minimum set of independent requirements that completely characterize the functional needs of the functional domain [5].

To launch AUVs from submarines it is necessary to prepare the weapons tube, move of the launching system along the weapons tube and to separate the AUV from the launching system.

To recover AUVs to submarines it is necessary to guide the approach of the AUV, immobilize it and to move the recovery system along the weapons tube.

The first axiom of the Axiomatic Design theory is the independence axiom: "Maintain the independence of the functional requirements". According do this theory, the independence of the functional requirements is required during the design process in order to develop a good project. For the purpose, the physical pieces, the so-called design parameters, have to guaranty the independence of the functional requirements.

3. MATERIALS STUDY FOR THE L/R SYSTEM.

The study of the suitable materials for the launching tube should take into consideration the materials of the submarine tube, in order to avoid any damage to the submarine tube. The software CES EduPack 2010 allows doing the selection of materials, based on a database of 2954 materials that includes polymers, metals and ceramics. The program works in two phases described below: first it chooses the material and then defines the surface treatment to apply to the chosen material.

3.1. MATERIAL CHOICE.

The program uses the following parameters described below in close relation to the functional and operation requirements of the system.

- Water corrosion: the system operates in seawater environment, making the parts of the system to resist without losing the function they perform. To evaluate this parameter the software database was limited to the "acceptable" and "excellent" classification what reduced the available material to 2297.
- Friction: the launch and recovery of the AUV makes the launch tube to displace throughout the weapons tube, causing friction between the materials of both systems. Therefore, the Vickers Hardness of the material of the L/R system should be lower than the submarine tube in order to avoid abrasive wear on the weapons tube that is harder to intervene. Vickers Hardness was limited between 10VH and 55VH that is the hardness of the weapons tube material. Considering the two requirements, the materials left in the database reduced to 165 materials.
- Yield Stress: the system suffers a load when moved along the weapons tube. The material should have an elastic deformation, corresponding to a stress much lower than the yield stress. The limits established for the Yield Stress were between 100 MPa and 1000 MPa reducing the available materials to 129.
- Fracture toughness: is the resistance of the material to the propagation of a fracture. It has to be considered the water pressure at the operation depth. Fracture toughness between 20 MPa $m^{-0.5}$ and 100 MPa $m^{-0.5}$ which is at least 20 times superior to the pressure at the 100m depth. Database availability reduced to 49 materials.
- Density: it is important that the chosen material has negative floatability because if necessary the system might release and sink the L/R system, not denouncing the position of the submarine. At the same time, the density of the material has to give raise to a weight of the system compatible to be man moved inside of the submarine. Regarding the density, the acceptable range for the material is 1000 kg/m^3 to 3000 kg/m^3.
- Price/cost: material cost is taken into consideration in production cost of the system and in maintenance costs. The chosen range was from 1 to 3 €/kg.

By using the CES EduPack 2010, the six parameters presented above were considered in six different stages. Definition of operation limits reduced the data base only to the materials that satisfy the chosen limits.

3.2. SURFACE TREATMENT.

For this study, CES Edupack 2010 has 46 surface treatment processes in database and the considered parameters were:

- Wear resistance: evaluates if the surface treatment process has wear resistance properties. The wear resistance evaluation is qualitative in "true" or "false" resulting in 26 "true" processes.
- Corrosion protection (aqueous environment): Corrosion protection is important to retard surface damages. Seawater environment is particularly dangerous because of the high conductivity, enabling fast electro-chemical attack. This parameter is classified in "true" or "false" resulting in 11 true processes after intersecting to the previous requirement.
- Surface hardness: as the hardness acceptable values were defined in the materials choice study, it is important to confirm that the surface treatment does not increases the hardness surface. Accordingly, the defined limits were 10HV and 55HV like and the available processes reduced to 4 processes.
- Surface roughness: it is measured of the root-mean-square (RMS) amplitude of the irregularities on the surface. The lower the RMS of the roughness value, the smoother the surface. The evaluation of this parameter has three classifications from A to C where A is "very smooth". This study considerate only the "A".

3.3. RESULTS.

Out of the 2954 materials available in database 13 satisfied the selection criteria (all aluminium) as showed in Figure1.

Name

- Aluminum, 3103, wrought, H4
- Aluminum, 3103, wrought, H8
- Aluminum, 3105, wrought, H4
- Aluminum, 3105, wrought, H8
- Aluminum, 5005, wrought, H4
- Aluminum, 5005, wrought, H6
- Aluminum, 5052, wrought, H32
- Aluminum, 7075, wrought, T0
- Aluminum, A413.2: LM6-M, cast
- Aluminum, commercial purity, 1050A, wrought, H9
- Aluminum, commercial purity, 1200, wrought, H4
- Aluminum, commercial purity, 1200, wrought, H9
- Aluminum, S360.0 (a): LM9-M, cast

Figure1: list of the obtained materials.

Thus, making the study of the surface treatment two solutions are selected: Anodizing and Electro-plating. According to the previous studies, the chosen material is the aluminium 7075, Wrought, T0, because it has the lower hardness of all the 13 materials shown in Figure1. Both surface treatments are recommended, with the preferably choice being the anodizing as it avoids deposit other materials that may promote galvanic corrosion. The material to use in the launch tube design matches with the material of the torpedoes most common material. Therefore, the new application will not change the relation of electronegativity of materials that may cause any type of corrosion

4. PROPOSED DESIGN.

The proposed solution is composed of three main parts that are connected: The main piece, the gripper, and the funnel.

4.1. MAIN PIECE.

Weapons tubes have a weapons transport device installed inside in order to allow loading torpedoes or mines into the submarine. In what concerns to the launching system, torpedoes are self-propelled making the launching system not to require an additional pushing device. Regarding the mines, the aforementioned transport device of the weapons is able to push the mines through the submarine tube and make it fall by gravity outside.
In the case of small or light-AUVs, it is safer to activate their propulsion system near the tube door to avoid contact between the vehicle and the tube that may damage the AUV. Figure 2 depicts the proposed system showing the main piece that has the same diameter of the weapons tube and it is compatible with the mine coupling of the weapons transport device. This way, the weapons transport device push and pull the main piece along the weapons tube without needing any structural change in the submarine.

The mine coupling has two hooks to do the connection to the main piece (3 in Figure 2) in order to push and pull it. Also, if there is any emergency on board of the submarine, this coupling allows the drop of the main piece just as if it was a mine and the system will sink to the bottom. This piece has an open hole (2 in Figure 2) to reduce water resistance to the movement when the tube is flooded and pressurized. Moreover, it allows passing any electrical connection, as the recharge of the AUV battery is an example of. Power and electrical banks installation is possible from inside the submarine and connected to the inside of the tube by two openings, one in each side of the tube.

Figure 2: Main piece of the L/R system (front view and right view). 1- equally spaced holes to fix the polyethylene tubes. 2- open hole. 3- mine coupling joint. 4- security stop system.

The security stop system (4 in Figure 2) has a counterpart already installed in the weapons tube. This system prevents unwanted movements of the L/R system while the submarine navigates.

The minimum thickness of the plate (front view of Figure 2) was studied by considering a distributed load applied to the circular surface. This study considers the outer edge fixed and the inner edge free [12].

The weapons transport device allows transporting two mines each with a weight of 8000 kg. Consequently, as long as the force acting the main piece is lower than the friction of both mines, the weapons' transport device is able to move it. Figure 3 shows minimum thickness (t) as a function of load (q) varying the load in the range of 1000 N to 1600 N.

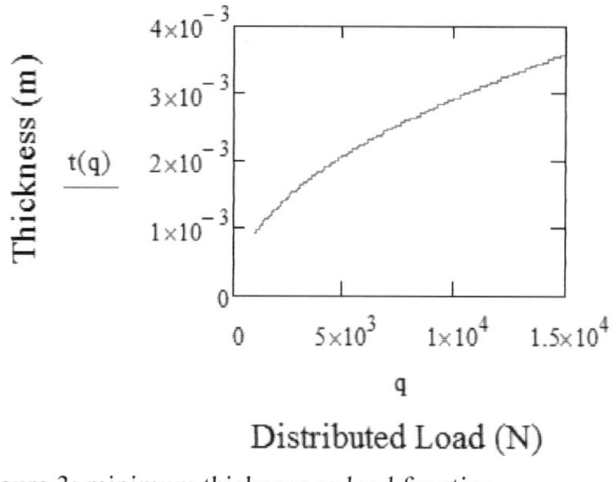

Figure 3: minimum thickness as load function

To calculate the thickness of the main piece two approaches were used: a simplified analytic solution that does not take into account with the holes; and an evaluation with a finite element program. In the ends, a thickness of 10 mm thickness was selected to the main piece using an open hole of 0.23m diameter. The calculation criterion included a safety factor of 5 taking into consideration the standard aluminium circular profiles available in the market.

The analytic approach calculates a maximum deflection of 9.28×10^{-6} m when the piece is under a load of 1600 N, using a 0.2 friction coefficient between aluminium and fibre glass. Using the SolidWorks finite elements analysis (Figure 4) the deformation is 6.67×10^{-5} m. The higher result depends on the effect of the holes to attach the polyethylene tubes, which cause stress concentration.

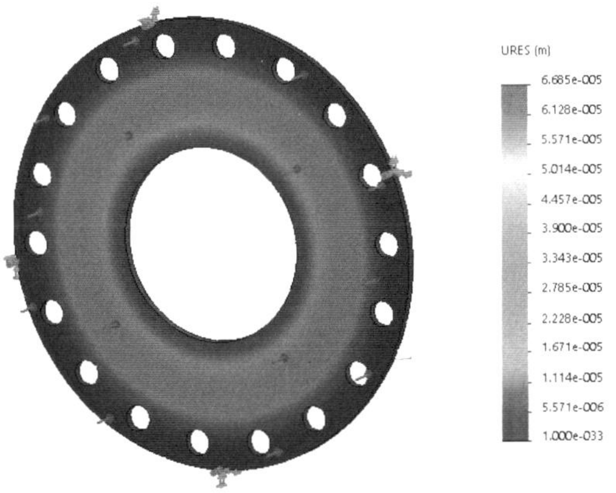

Figure 4: SolidWorks deflection simulation

Thus, the entire main piece (Figure 5) underwent a simulation using the SolidWorks program, using a cylindrical profile of 2 m length and 10 mm thickness. It achieved a maximum deflection of 3×10^{-5} m, lower than the plate calculation because of the reinforcement of the cylindrical profile.

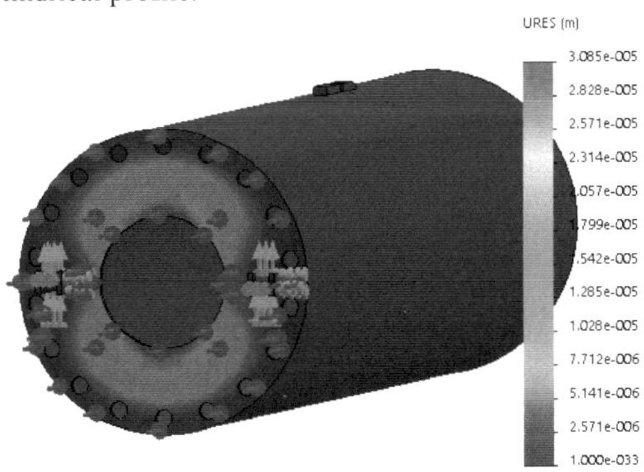

Figure 5: SolidWorks deflection simulation to the main piece.

4.2. GRIPPER.

The proposed system goal is to launch and recover large dimension or small dimension torpedo shaped AUVs. A gripper is not needed to launch and recover AUVs with large diameter between 0.45m and 0.35m. In this case the vehicle is placed in the main piece directly in contact with the polyethylene tubes. Regarding AUVs with smaller diameter, they have to be fixed to the main piece by a gripper (Figure 6) to be kept immobilized. The longitudinal position of the gripper in the main piece depends on the available section to grab the vehicle without damaging any of its sensors. An electric motor installed in the grippers support performs the grabbing and release operation. This electric motor is controlled either from the inside of the submarine by a human operator or from a distance sensor installed in the plate of the main piece that actuates the gripper when the AUV reach a pre-established distance to it.

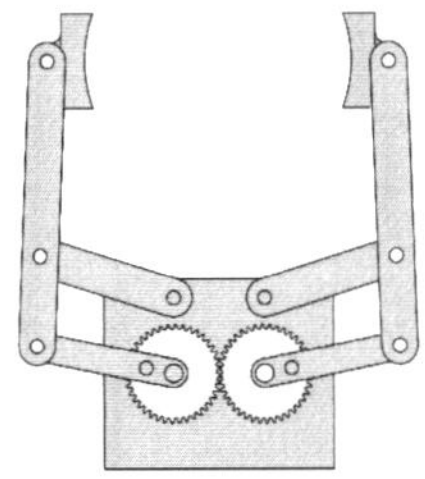

Figure6: Example of a gripper.

When the gripper is opened, AUV propulsion must be turned on in order to start the navigation and leave the main piece. There are two different operation modes to activate the propulsion of the Portuguese AUVs: One is an acoustic command activated by an operator inside the submarine and the other one is by a magnet that is in contact with the vehicle.

4.3. THE FUNNEL.

The main piece and the gripper are enough to launch the AUV from the weapons tube of a manned submarine. However, to recover the vehicle this project requires a method to prevent the collision between the AUV and the submarine during the approach.

Funnel main function is to increase the AUV approach area to the submarine protecting the AUV to collide to the platforms. It also protects the AUV from small collisions with the main piece walls that may damage the AUV.

The purposed material is high density polyethylene with tubes of 32 mm diameter connected to the main piece by electroplating welding joints. The diameter of polyethylene decreases the range the diameter of the docking station. So that, 32 mm tubes allow the recovery of AUVs until 0.45m diameter. Polyethylene tubes diameter also affect the diameter dimension of equally spaced holes in the main piece.

Eighteen tubes are equally spaced and create the small diameter ring of the funnel. Tubes go along the main piece length and at the end they are fixed by electroplating welding joints to a soft steel tube. The soft steel tube is in the elastic domain and is the larger diameter ring of the funnel (approximately twice the smaller diameter).

In this project, the most challenging step is to define how to open and close the funnel large diameter ring. It has to be open when outside the weapons tube and closed when inside it. For this reason, the purposed design uses the movement of the weapons transport device to open and close the large diameter ring. When inside the tube, the larger diameter ring is compressed and deformed. As soon as the funnel leaves the tube because the main piece has already got out and is on the launching position, the larger diameter tube opens and adopts a circular shape.
Likewise, during the AUV recovery, as soon as the vehicle is immobilized by the gripper, the weapons transport device starts to carry the main piece in the submarine's interior direction. The funnel is open and as it gets in contact with the tube, starts to deform and being in compression again. The opened funnel position is showed in Figure 7.The funnel when closed, might have more than two inflection points that provokes the large diameter deformation. That depends on the connections between the polyethylene tubes and the large diameter surface.

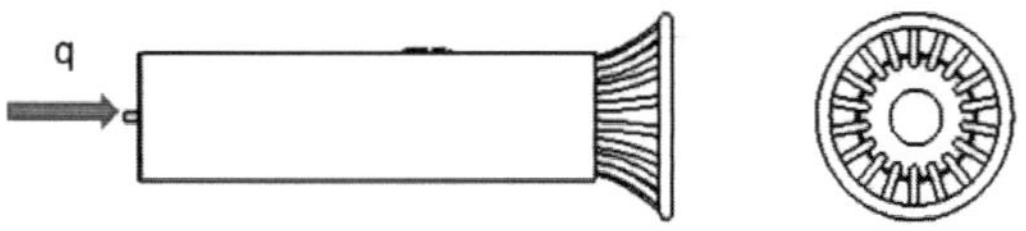

Figure 7: open funnel

4.4. DISCUSSION

The proposed system satisfies the functional requirements of the project without violating the independence axiom. The preparation of the systems tube (flood and pressure) is controlled by a system in submarine. Weapons transport device, already installed in the submarine, moves the L/R along the tube. The assurance of physical contact or physical separation between the AUV and the L/R system is either by the gripper or the polyethylene tubes depending on the dimensions of the vehicle. The approaching system that guides the AUV into the submarine can be installed on the funnel (optical or acoustic) or inside the main piece (acoustic).

According to the calculations with SolidWorks 2010 the proposed system weights 15 kg (considering 2 m length of the main piece). This means that it can easily be removed from the tube used for L&R if AUVs are not required for a specific mission of the submarine. Actually, by not requiring structural changes to the submarines, the proposed system does not affect operational and weaponry capacities.
In what concerns to the maintenance operations of the system, polyethylene tube substitution is easy in case of any major deformation.

5. CONCLUSIONS

The proposed system to launch and recover AUVs has two configurations according AUV size: with gripper and without gripper.

The dimensions of the system can be adjustable to the size of the AUV intended to launch and recover from the submarine.

The minimum thickness required withstanding the loads of the weapons transport device loads and the weight of the AUVs weight is less than 10 mm. Many standardized circular profiles in aluminium are available in dimensions that fit the weapons tubes dimensions. Moreover, 13 different samples of aluminium can be chosen for the main piece.

The approach of AUVs to the L/R system can be guided either by acoustic transponders or by an optical signal sent by lights installed in the funnel as long as the vehicle has appropriate sensors installed.

As future work, a prototype is being developed to test different configurations of the funnel larger diameter surface as well as different materials used in the same surface. A torque wrench will also be used to evaluate the force done by the deformed funnel on the tube walls.
Future work includes adapting this system to be used by surface vessels thus expanding the potential number of missions performed by AUVs employment in the Portuguese navy.

6. REFERENCES

1. U.S. Navy, "The Navy Unmanned Undersea Vehicles (UUV) Master Plan", Retrieved July 28, 2008.

2. HARDY, Tim, BARLOW, Gavin, "Unmanned Underwater Vehicle (UUV) deployment and retrieval considerations for submarines", INEC, Hamburg, Germany, 2008.

3. FEDOR, R,. "Simulation of a Launch and Recovery of an UUV to a Submarine", Master thesis in naval arquitecture from Royal Institute of Technology, Sweden, 2009

4. BRAGA, J., SOUSA, J., HEALEY, A. J., "Navigation Scheme for the LSTS SEACON Vehicles: Theory and Application", IFAC Proceedings Volumes, vol. 45 pp. 69 a 73, 2012.

5. NAM-PYO, Suh, "The Principles of Design", New York, Oxford University Press. 1990.

6. CURIE, Jason et al, "Dynamics of two active autonomous dock mechanisms for AUV recover", Canadian Society for Mechanical Engineering, Vol. 38, Nº 2, pp 213-226. 2014.

7. WIRTZ, Marius, HILDEBRANDT, GAUDING, Christopher, "Design and Test of a

Robust Docking System for Hovering AUVs", presented in Oceans Conference, 14-19 October 2012, Hampton Roads, VA, USA.

8. PARK, Jin-Yeong, et al, "Experiments on vision guided docking of an autonomous underwater vehicle using one camera", IEEE Journal of Ocean Engineering, vol. 36 pp. 48 – 61, 2009.

9. M. D. Feezor, F. Y. Sorrell, P. R. Blankinship, and J. G. Bellingham,"Autonomous underwater vehicle homing/docking via electromagnetic guidance," IEEE J. Ocean. Eng., vol. 26, no. 4, pp. 515–521, Oct. 2001.

10. AZEVEDO, Diana, "Conceito de operação dos veiculos submarinos autónomos SeaCon a partir dos submarinos da classe Tridente", Master thesis in military-naval Sciences from Portuguese Naval School, Portugal, 2013.

11. WATT, George D. et al, "A Concept for Docking a UUV With a Slowly Moving Submarine Under Waves", Ocean Engineering, IEEE Journal of Ocean Engineering, vol. 41 pp. 471 – 498, 2016.

12. YOUNG, Warren C., BUDYNAS, Richard G., "Roark's Formulas for Stress and Strain 7th edition", McGrawhill, New York, USA, 2002.

7. AUTHOR'S BIOGRAPHIES

Henrique Vieira da Silva is a student of Naval Engineering- Mechanical field and officer cadet at the Portuguese Naval Academy since 2012.

Miguel Cavique is Auxiliar Professor at the Department of Sciences and Technology of the Portuguese Naval Academy. He holds a PhD on Mechanical Engineering on Machine Design awarded by the New University of Lisbon. He graduate as diplomat Engineer on Mechanical Engineering and as a MSc on Mechanical Engineering in the Instituto Superior Técnico, of the Lisbon Technical University. Works on Design theories since 2005 and has before about fifteen years of experience on the design of mechanical equipment and fluid networks. During the eighties, he works for five years on the Lisbon gas network where he was head of the network project department. He has authored many publications on design theory and their applications.

João Tasso de Figueiredo Borges de Sousa is an Assistant Professor at the Electrical and Computer Engineering Department from Porto University in Portugal and the head of the Underwater Systems and Technologies Laboratory. He holds a PhD and MSc in Electrical Engineering both awarded by the University of Porto. Since 1997 he has been leading the design, implementation and deployment of advanced unmanned vehicle systems in projects funded by the Portuguese Foundation for Science and Technology, Portuguese Innovation Agency, Luso-American Foundation, NATO, Office of Naval Research and DARPA. In 2002 he was awarded the Luso-American Foundation Fellowship by the Portuguese Studies Program from the University of California at Berkeley. In 2003 the Underwater Systems and Technology Laboratory was awarded the Arca second Prize for the best technological realizations Respectful to Environment. In 2006 he received the national BES Innovation National Award for the design of the Light Autonomous Underwater Vehicle. In 2008 he received an outstanding teaching award from Porto University. He is a member of the Advisory Board of the Swedish Center in Marine Robotics. He has authored over 300 publications, including 30 journal papers.

METHOD FOR THE SIMULATION OF THE SWIM OUT OF A WEAPON FROM A LAUNCHING TUBE

F Pécot DCNS Research / Sirehna, FRANCE and **T Taillefet,** DCNS Ruelle, FRANCE

SUMMARY

Among the various missions expected from a submarine, weapons launching from a tube, either by pulse or swim out system, is of major importance. To secure these launchings and to determine the safe operating envelope, the use of a numerical approach constitutes an interesting alternative to expensive model or full-scale trials.

This paper focuses on the development of a numerical methodology based on a CFD code allowing to simulate swim out torpedoes launching from a tube. The methodology relies on the overset grid approach. This tool solves the strongly coupled URANS and 1-DOF weapon dynamics equations, with given propellers rotating velocities.

To validate this numerical methodology, full-scale trials of swim out launchings from a mono-diameter tube immersed inside sea water at rest were carried out with a torpedo-like drone. A satisfactory correlation was obtained between numerical and experimental results in terms of both weapon velocity and pressures inside the tube.

1. INTRODUCTION

Different kinds of missions are able to be assigned to a submarine: naval task force protection, commando squad transport for amphibious operations, mines laying, attacks against terrestrial, naval and even aerial targets by means of torpedoes and ballistic or tactical missiles... All the tactical weapons are launched from a tube either by pulse thanks to a mechanical device (directly as for a Pneumatic Ram, or through a fluid as for a Turbo Pump or a Water Ram), or swim out in the exclusive case of torpedoes.

At each weapon is associated a safe operating envelope. The success of a weapon launching and of course the safety of the submarine and its crew during this operation is guaranteed provided that the triplet [immersion depth, submarine velocity, sea state] belongs to the firing domain. To determine this domain, the use of a numerical approach able to predict the weapon hydrodynamic behaviour and identify the risks of launching failure (external collision between the mobile and the submarine, too low velocity at the bow shutter due to cavitation effects...) constitutes an interesting alternative to costly model or full scale trials.

In this context, numerical methodologies for the simulation of weapons launching and in particular for the swim out of torpedoes have been developed by DCNS, on the basis of the CFD code STAR-CCM+ [1]. Hydrodynamic calculations of the swim out firing of a torpedo-like drone from a tube were performed and the obtained results were compared with those of full-scale sea trials achieved with the objective of qualifying the developed numerical tool.

2. FULL SCALE SEA TRIALS

2.1 EXPERIMENTAL SET-UP AND MEASUREMENTS

Full-scale trials of swim out launching of a torpedo-like drone from a mono-diameter tube (closed at its bottom) were carried out in sea water at rest, to provide experimental data to validate the proposed numerical CFD methodology. The experiments were performed at an immersion of about 11 m on the tube axis, in order to avoid, or, at least, to limit cavitation inception.

The tube length was 6.6 m with an internal standard diameter of 730 mm (figure 1), but this diameter could be reduced to 580 mm by insertion of foam annular cylinder plates (figure 2). Thanks to four rails placed on a 537 mm diameter cylinder, the drone was guided in translation. In the upper rail, a groove had been machined to permit the drone vertical pin to slide and thus avoid any roll motion. At the tube exit, a "water droplet" shape improved sea water inlet. Indeed, due to volume conservation, water has to fill freed space behind the drone.

The drone length was 5.8 m (figure 3), with a maximum diameter of 533.4 mm, a mass in air of 1125 kg, and in sea water of -14 kg. The slight buoyancy of the body allowed to get it back easier after shot. Initially, the drone aft was located at 380 mm from the tube bottom. The mobile was propelled by a pair of counter-rotating propellers (figure 4), whose RPM were, for the upstream one, linked to the velocity command (Vmin = 11 knots or Vmax = 20 knots), and for the downstream one, determined in order to annul the total torque.

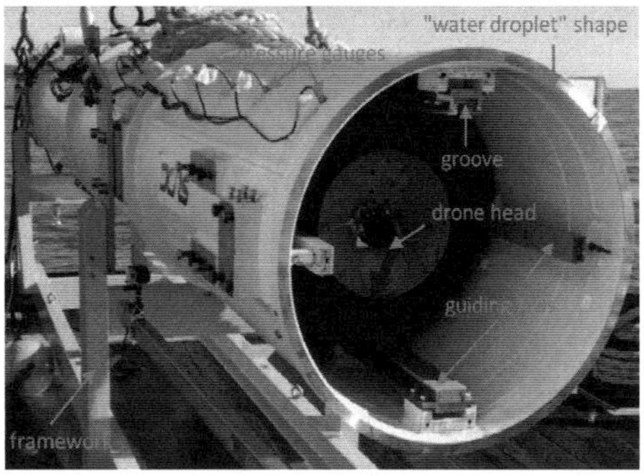

Figure 1: Torpedo-like drone within the 730 mm diameter launching tube

Figure 2: Launching tube equipped with foam annular cylinder plates (580 mm diameter)

Figure 3: Torpedo-like drone

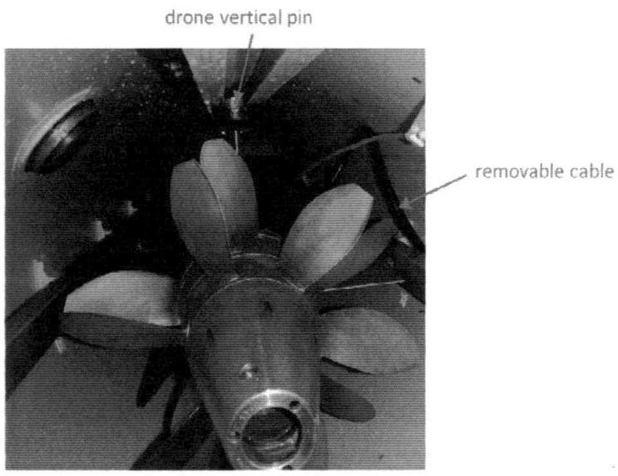

Figure 4: Rear part of the drone with its 2 counter-rotating propellers

Using a removable cable, an electrical impulse was sent to the mobile to trigger the propellers engine start. This cable was cut just before the drone departure by means of a guillotine-like device.

During the trial, a buffer within the drone recorded both the rotation speeds and the acceleration, thus leading to the mobile velocity and displacement. Pressure measurements along and at the bottom of the tube were also achieved, with a 10 kHz sample frequency.
At last, two fixed high resolution video cameras with 24 frames/s were placed at the bottom and outside the tube to shoot the propellers rotation and the drone motion.

2.2 TRIALS REPRODUCIBILITY AND STUDIED CONFIGURATIONS

In order to ensure results reproducibility, two trials were performed for all the three studied configurations:

- 730 mm tube diameter for both Vmin and Vmax velocity command,
- 580 mm tube diameter for Vmax velocity command only.

Figure 5 and 6 compare the evolutions of, respectively, the propellers RPM versus time, and the drone velocity versus the drone aft position, in both similar experiments, for the 730 mm diameter tube and Vmax velocity command. This shows that this reproducibility is quite satisfactory. On figure 6, the vertical line indicates the exit of the drone aft out of the tube.

Figure 7 allows to compare the typical evolutions of the RPM propellers versus the drone aft for all configurations. The vertical lines indicate the exits of the drone junction (limit between the conical rear part and the cylinder one) and aft out of the tube. In each case, the RPM upstream propeller increases quickly and reaches its final value after a drone displacement of about 0.5 m. Of course, this value rises as the velocity command increases. But it also depends a few on the tube diameter. The downstream propeller RPM evolution looks like the upstream one at the beginning of the drone motion. But, for the largest tube, this downstream propeller RPM continues to increase slightly (in absolute value), whereas, for the smallest one, it oscillates and seems to be affected by the strong tube confinement. In that configuration, RPM rises significantly, while the drone conical rear part leaves this tube.

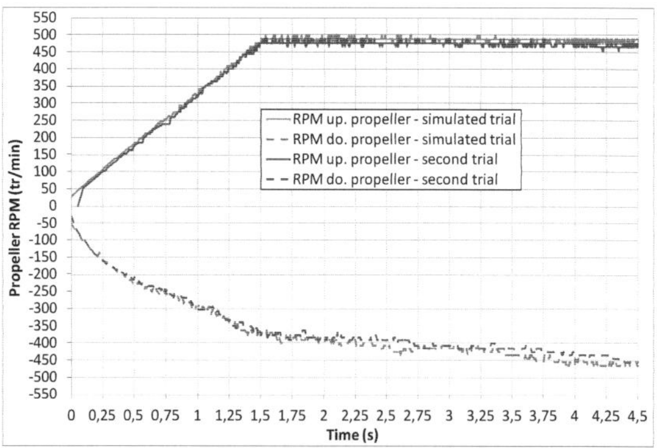

Figure 5: Trials reproducibility - both propellers RPM - 730 diameter tube - Vmax velocity command

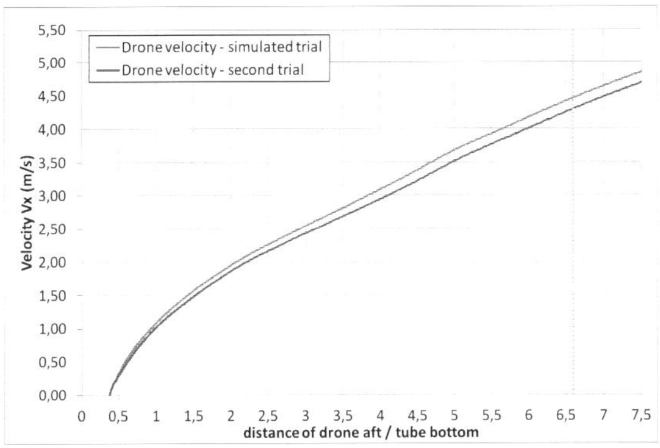

Figure 6: Trials reproducibility - drone velocity - 730 diameter tube - Vmax velocity command

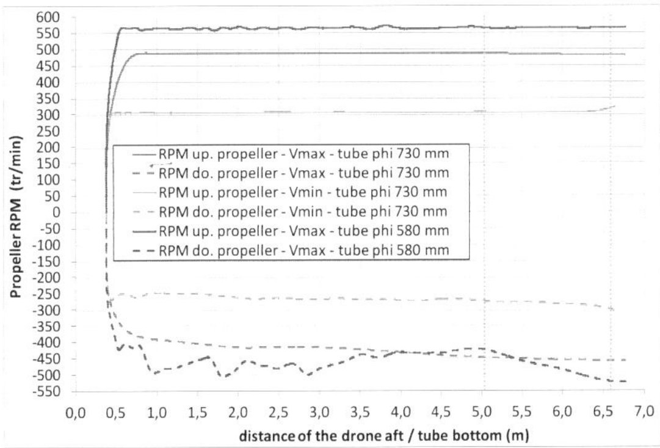

Figure 7: Both propellers RPM for the three configurations

3. NUMERICAL CFD APPROACH

With the ongoing computational capabilities and the CFD progress, the swim out launching of a weapon from a tube can now be simulated with a full CFD numerical approach. The selected code must be able both to deal with the time evolution of the calculation domain due to the weapon displacement, and to solve the strongly coupled URANS and 1-DOF weapon dynamics equations, with given propellers rotating velocities. In the current study, the ability to solve this problem and accuracy of the CFD code STAR-CCM+ is evaluated.

3.1 GEOMETRIC AND DYNAMIC MESH STRATEGY

Due the presence of the guiding rails, on one hand a very small gap (1.8 mm) is located between the drone and the tube, and on the other hand the tube cross section area is reduced. To avoid meshing this small gap and also to take into account the cross section area reduction, the real tube was modelled by an equivalent one which diameter is calculated as following: $D_{eq} = \sqrt{D_{real}^{\ 2} - \frac{4}{\pi} \cdot A_{rails}}$

where D_{real} and A_{rails} correspond to respectively the real tube diameter and the cross section area occupied by the rails. It can be noticed that, for the smallest tube diameter configuration, there was almost no difference between real and equivalent diameters. Moreover, the "water droplet" shape at the tube exit was modelled.

Unlike previous studies on weapons launchings [2, 3, 4], in order to avoid the remeshing of the evolving fluid calculation domain due to the drone displacement, the "overset" grid method available in the STAR-CCM+ was used. The accuracy of this method was demonstrated for many other applications, like simulations of bombs dropping from planes...

The "overset" method consists, for this current application, in superimposing two non deforming meshes:

- an overset mesh around the moving drone and its propellers, dealt with as a rigid body,
- a fixed background mesh (the fluid domain) inside, around, and outside the tube,

and exchange information data between them to solve both the flow and the drone dynamics.

The cells of these two meshes are of three types:

- the active cells (where the discretized flow equations are solved),
- the inactive cells (where no equations are solved). These cells can however become active, when the overset region moves,
- the "acceptor" cells of the background mesh and at the border of the overset region. They separate the active and inactive cells and constitute the link between the two meshes.

Data are transferred from the active cells of one mesh to these of the other mesh by interpolation (figure 8).

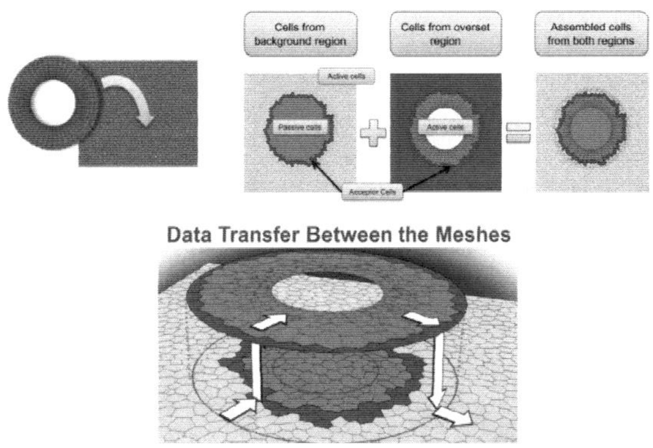

Figure 8: Data transfer between meshes [1].

3.3 MESHING

The numerical model included four different regions:

- the background region (fluid domain) made of two main parts: the inner launching tube and an outer volume cylinder, which were both meshed with trimmed hexahedral cells. The inner launching tube and its exit were homogeneously and finely meshed. But, outside, the mesh was then quickly released (figure 9).
- the overset cylindrical region around the drone and its propellers, which was meshed with polyhedral cells (in brown on figure 10). This region was finely meshed close to the drone ($y^+ \sim 70$) and inside a cylinder whose diameter is larger than the tube one. Further, the mesh was released.
- two other regions around each one of both propellers (in blue and grey on figure 10), included in the previous overset region.

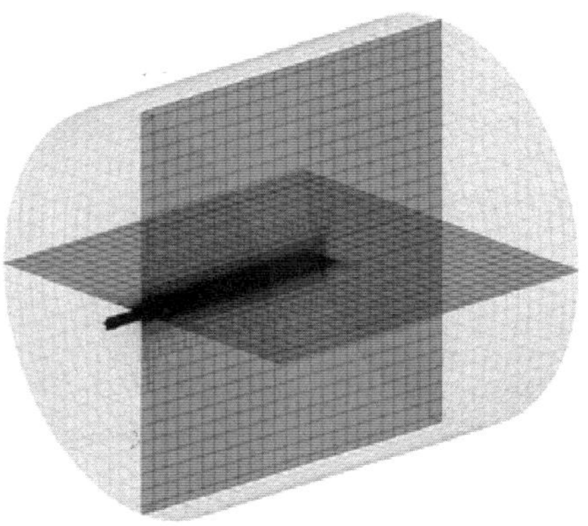

Overall view of the region

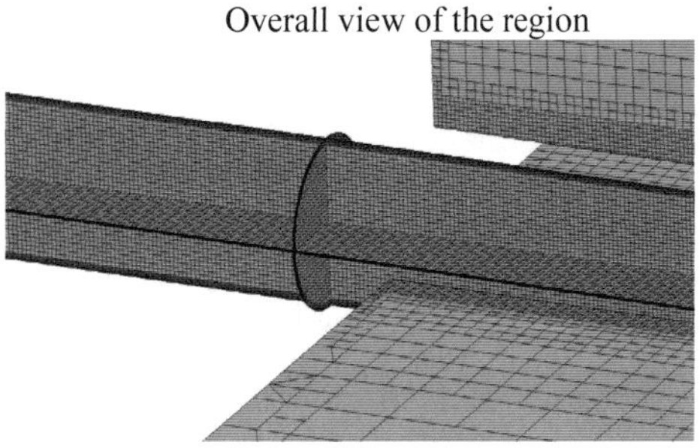

Zoom on the tube

Figure 9: Views of the background region mesh

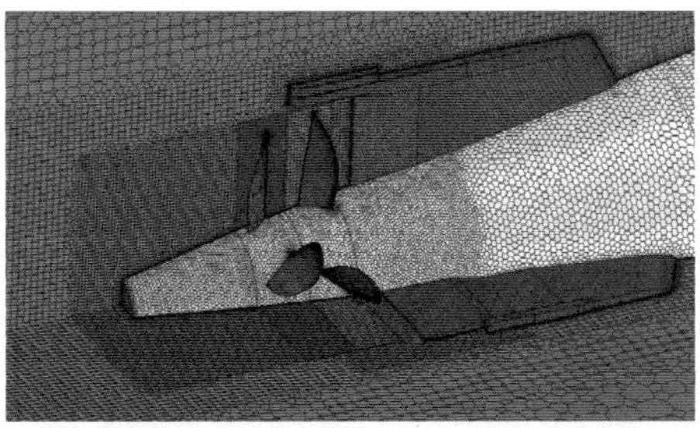

Figure 10: Surface mesh of the rear part of the drone and its counter-rotating propellers

3.4 PROPELLERS ROTATIONS TREATMENT

To take into account the rotation effects of the counter-rotating propellers on the flow, two methods were compared:

- the "MRF" method (Moving Reference Frame) consisting in solving the Navier Stokes equations in two rotating frames (one for each propeller) without any visible rotation of the propellers,
- the "sliding grid" method consisting in simulating the real rotations (in opposite directions) of the propellers during the launching of the drone. In that case, unlike the previous one, non conformal interfaces (which implies non coincident nodes on each side of the borders) are used between the three regions meshes.

For each studied configuration (different tube diameters), the global mesh included about 10 millions of cells. Meshes were made according to the experience acquired during previous studies of weapons ejection with overset approach..

3.5 MODELLING

To simulate the drone launchings from the tube with STAR-CCM+ code version 10.06, the assumptions taken into account were:

- the flow was unsteady, non compressible, turbulent, and in almost all cases monophasic. Nevertheless, some tests were carried out, by activating the Schnerr & Sauer [5] cavitation model. This model is close to those of Zwart & al [6] and Singhal & al [7, 8, 9, 10], since it introduces a dynamic equation for the volume vapor fraction with vaporization and condensation sources terms. But, unlike the second one, it does not take into account non condensable gases (N_2, O_2, CO_2...) contained inside nuclei, but only water vapor. Nevertheless, it has the advantage of only requiring, apart from the vapor pressure value (1700 Pa at 15°C), the nuclei concentration and the micro bubbles initial radius (in other words, the initial vapor volume fraction). Those values were both kept to their default ones of respectively 10^{12} m^{-3} and 1 μm.
- gravity was not activated,
- the rotating velocities laws of propellers obtained from the simulated experimental test were imposed,
- the drone moved forward thanks to its own propellers, without any solid friction on the guiding rails, according to one degree of freedom along the tube axis,
- the sea water density was equal to 1026 kg/m^3 and the dynamic viscosity to 1,219 $kg.m^{-1}.s^{-1}$.

The two equations k-ω SST RANS turbulent model was chosen, since it remains still valid in the vicinity of walls where turbulent Reynolds is low and viscous effects are dominating.

At the boundaries of the calculation domain, we imposed:

- no slip conditions on walls,
- zero relative pressure outlet far from the tube.

Moreover, at the beginning of the simulation, the fluid was assumed at rest.

All the computations were performed with the STAR-CCM+ segregated solver with implicit linearization of the flow equations, using a first order time scheme and second order space schemes.

For the 730 mm diameter tube, without cavitation, the time step value was about 5.10^{-3} s in "MRF" simulation, and 10 times smaller in "sliding grid" one. With cavitation, these time steps were reduced (time step between 10^{-4} s and 5.10^{-4} s for both approaches). For the reduced diameter tube, due to the strong confinement, the time step in "MRF" simulation without cavitation was of the same order of magnitude as for the large tube using the "sliding grid" simulation with cavitation. These time steps are issued from convergence studies.

4. COMPARISONS BETWEEN TRIALS AND SIMULATION RESULTS

4.1 TUBE OF DIAMETER Ø 730 MM

4.1 (a) Drone displacement, velocity and acceleration

Figure 11 and figure 12 show the drone velocity versus the drone aft position, for both cases of velocity command (Vmax and Vmin). On this figure, experimental and CFD results without cavitation are compared. Besides, like previously, two vertical lines indicate the exits of respectively the drone junction and the drone aft out of the tube.

This figure shows that the drone continuously accelerates inside and outside the tube. Moreover, it shows that, for each velocity command, the numerical results correlate quite well the experimental one. Globally, the "MRF" results are closer than the "sliding grid" ones, even if, up to an aft position of 2.5 m, the latter model leads to a better correlation. But, anyway, "MRF" computations are less time-consuming than the "sliding mesh" ones (15 to 20 hours versus 6-8 days, on 80 cores).

The maximum velocity deviations between numerical and experimental results at the tube exit does not exceed 0,2 m/s, which is of the same order of the magnitude of the velocities discrepancy at the tube exit, for the same configuration between two similar tests.

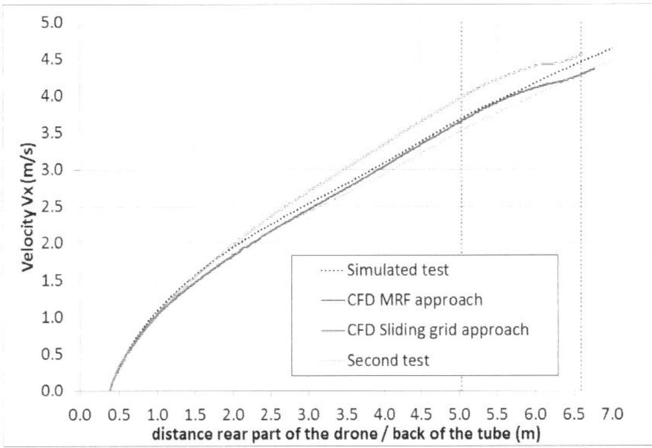

Figure 11: Drone velocity - 730 mm diameter tube - Vmax velocity command

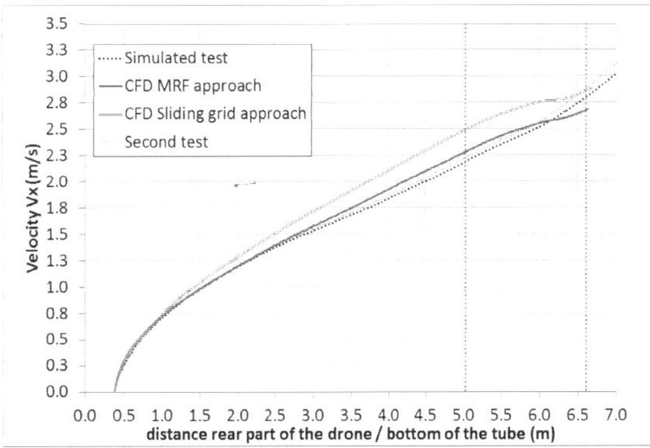

Figure 12: Drone velocity - 730 mm diameter tube - Vmin velocity command

In simulations, a significant and sudden drop of the drone acceleration (negative jerk) is observed, while the rear conical part is leaving the tube (figure 13). This acceleration drop, although existing, is far less important during tests and its effect is not visible on the experimental velocities curves. This difference of accelerations behaviors while the drone rear part is going out is coherent with the small velocity gap between tests and "MRF" calculations at the tube exit.

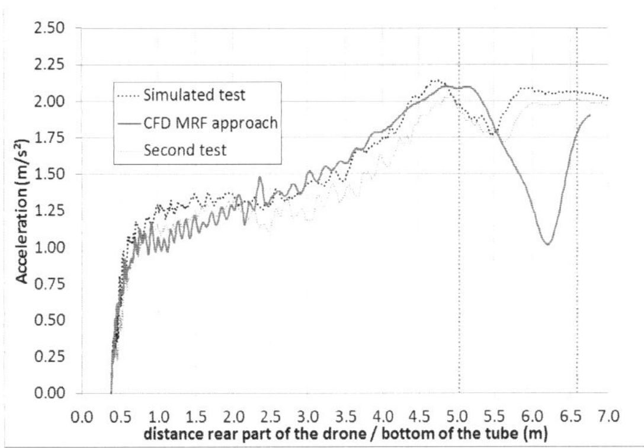

Figure 13: Drone acceleration – 730mm diameter tube - Vmax command velocity

This acceleration drop, generated by a thrust drop, is concomitant with an increase of the average velocity of the incoming flow on the propellers (especially at the blades feet). That is due to the progressive disappearance of a recirculation zone located on the drone upstream of the propellers, during the exit of the conical rear part of the mobile (figure 14).

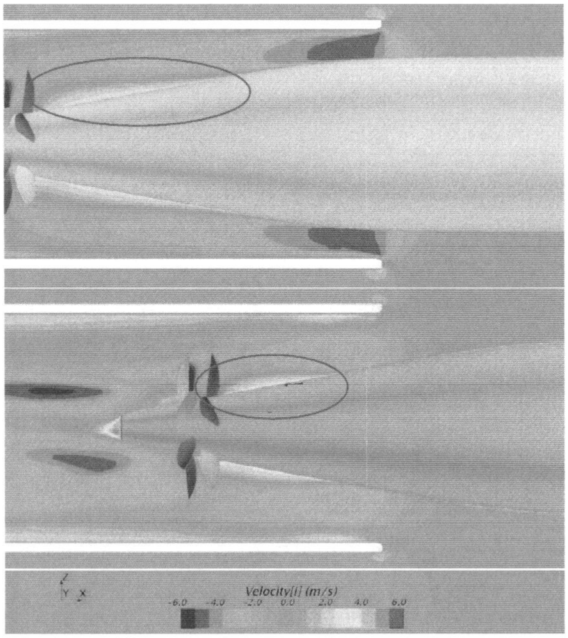

Figure 14: Disappearance of the recirculation zone upstream of the propellers – 730 mm diameter tube (remark: the drone rudders have been voluntary hidden)

To better understand the physics of this phenomenon, a simulation at Vmax velocity command, without any rudder on the drone was achieved. The obtained results showed that these rudders have no influence on the drone dynamics, including the exit phase of the mobile conical rear part. Therefore, it could be presumed that the tube modelling, in particular the removal of the guiding rails, altering the three dimensional local structure of the flow at the tube exit, is probably responsible for the drone dynamics discrepancies between simulations and experiments.

4.1 (b) Pressure and cavitation inception risk

Figure 15 allows to compare experimental and numerical ("MRF" approach without cavitation) pressure signals versus time, at three different points along the tube, for Vmax velocity command.

This figure shows a satisfying correlation obtained at each point between numerical and experimental signals, despite small offsets whose levels depend on the pressure sensors locations.

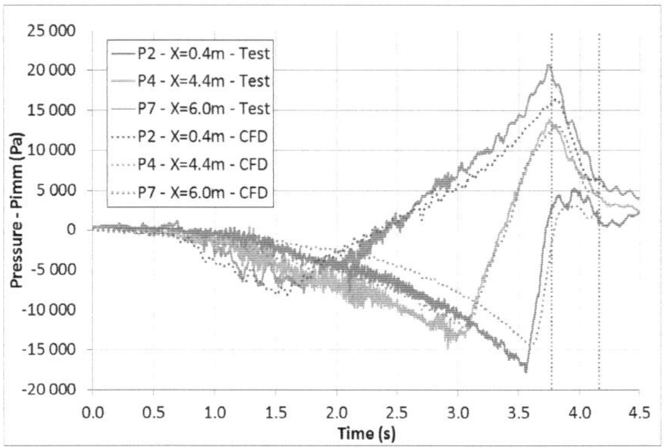

Figure 15: Relative pressures versus time inside the 730 diameter tube - Vmax velocity command

At each point, the pressure evolves in the same way in three stages:

- a first decrease stage, due to the drone motion, making the sea water flow into the tube. Except for P2 point, this phase lasts until the mobile junction overtakes the considered point.
- an increase stage, until the junction goes out, due to the compression of the flow between the drone conical part and the tube (diffuser-like annular space) and at the bottom of this one. It can be noticed that, at this time, the pressure is higher than outside.
- a second decrease stage, during which the pressure reaches the immersion one.

Although the cavitation was not simulated, the risk of cavitation inception was, nevertheless, assessed by comparing pressure levels on the blades with the vapor pressure value. If the pressure drops below this threshold value, this phenomenon should occur.

According to the "MRF" simulations results, there is no risk of cavitation inception in both cases of velocity command. But, according to the "sliding grid" ones, there are some small potential areas of cavitation inception on the leading edges of the blades extrados (in dark blue on figure 16) in the case of Vmax velocity command.

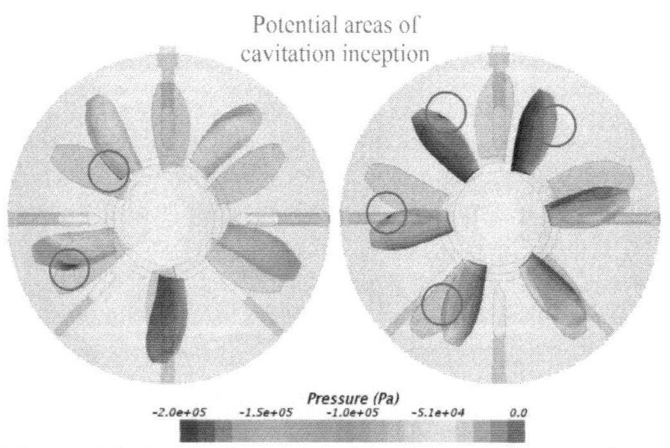

Figure 16: Cavitation inception risk - 730 mm diameter tube - "sliding grid" approach - Vmax velocity command

In order to understand, in that case, the impact of cavitation inception on the drone dynamics, an additional "sliding grid" simulation was carried out, by activating the Schnerr & Sauer [2] cavitation model, whose both parameters were kept to their default values. This simulation showed that cavitation actually appears on the blades (figure 17). Nevertheless, there was practically no impact on the dynamics (figure 18).

It can be noticed that this simulation was a bit time-consuming : it lasted more than 20 days on 80 cores.

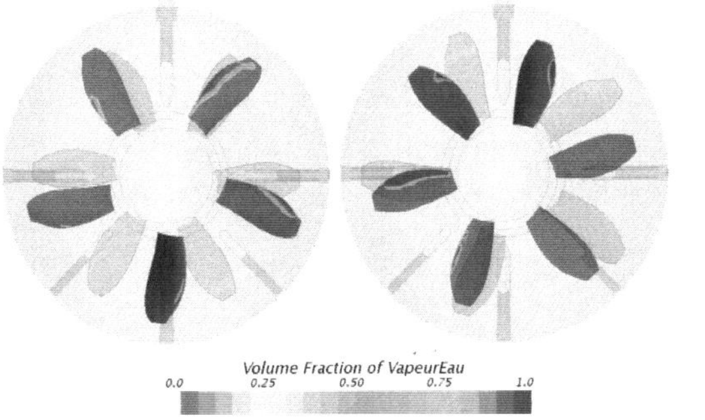

Figure 17: Vapour volume fraction - 730 mm diameter tube - "sliding grid" approach - Vmax velocity command

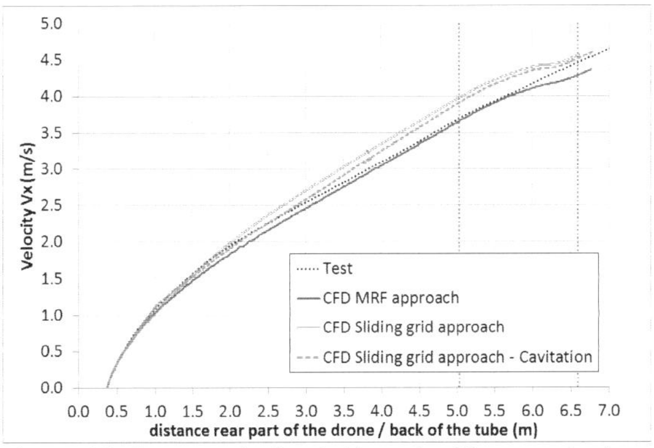

Figure 18: Drone velocity (with and without cavitation modelling) - 730 mm diameter tube - Vmax velocity command

4.2 TUBE OF DIAMETER Ø 580 MM

4.2 (a) Drone displacement, velocity and acceleration

For the 580 mm diameter tube, the drone confinement is so strong (annular gap about 23 mm between the drone cylinder part and the tube), that the fluid flow, in the case of Vmax velocity command, was particularly difficult to solve. The "MRF" simulation without cavitation modelling required a small time step (about 10^{-4} s), as indicated previously. Consequently, 1 month on 80 cores was needed to complete a 8 s physical time computation. The figure 19 and figure 20 show, for a Vmax velocity command, respectively the velocity and the acceleration of the drone versus the drone aft position. On these figures, experimental and "MRF" numerical results without cavitation modelling are compared.

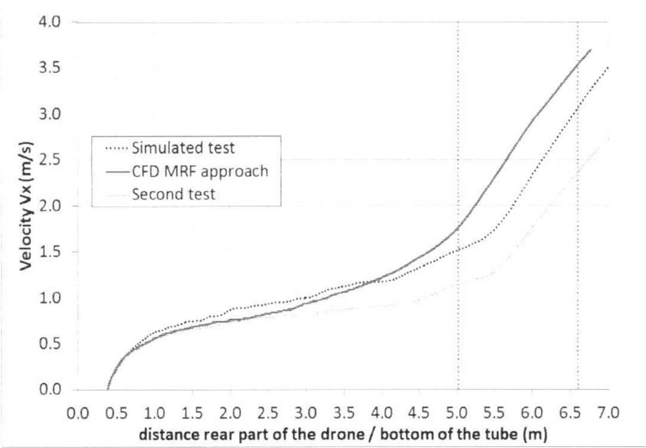

Figure 19: Drone velocity - 580 mm diameter tube - Vmax velocity command

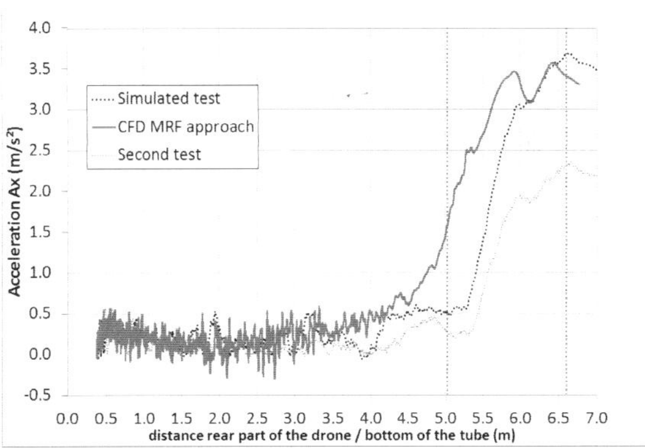

Figure 20: Drone acceleration - 580 mm diameter tube - Vmax velocity command

The numerical and experimental results are close to each other up to a drone position of 4.5 m. Then, the curves evolutions are similar in both cases but with a time offset, due to the significant rise of the time-imposed downstream propeller RPM as soon as the drone junction leaves the tube (figure 7). Thus, as in CFD simulations, at this moment, this junction is not arrived yet at the exit, the drone accelerates "earlier".

The velocity when the drone emerges from tube is overestimated of about 16 %, which is however less than the relative discrepancy between both similar tests.

The strong confinement tends to limit the mobile acceleration inside the tube until the drone rear conical part begins to go out. Unlike the largest diameter tube configuration, no acceleration drop is observed experimentally, while this conical part is leaving the tube. But, the code predicts, such a drop, even if its amplitude is twice lesser than in the 730 mm diameter tube configuration, for which there is a significant difference between real and equivalent diameters.

This observation seems to confirm the assumption according to which the tube modelling without rails is responsible for the drone dynamics differences between trials and simulations while the conical part leaves.

4.2 (b) Cavitation inception risk

According to the "MRF" simulations results, cavitation should appear on the leading edges of both propellers blades extrados. A new computation with cavitation modelling would probably lead to an exit drone velocity closer to the trial one, but would be prohibitive due to calculation time.

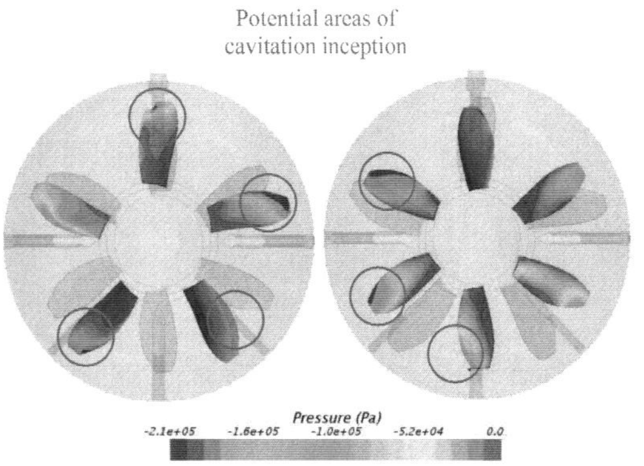

Figure 21: Cavitation inception risk - 580 mm diameter tube - "sliding grid" approach - Vmax velocity command

5. CONCLUSIONS & PERSPECTIVES

For both tube diameters, regardless of chosen numerical approach, the correlations between the different trials and simulations in terms of drone displacement, velocity and acceleration, and of pressure within the tube are globally quite satisfactory. The developed methodology on the basis of STAR-CCM+ software appears promising, and, without doubt, will be used, in the future, to predict the performances of swim-out launchings of torpedoes from submarines tubes.

However, the drone dynamics is not perfectly well captured, while the conical rear part of this mobile is leaving the tube. Indeed, the method implemented always predicts an acceleration drop which either does not exist in reality (when the tube confinement is strong) or is overestimated. This is probably due to the removal of the guiding rails and the replacement of the real tube by a fictitious equivalent one in the modelling.

The best compromise between computation time-consumption and results accuracy is reached when the "MRF" approach without cavitation modelling is used. If this calculation points out large areas where the pressure is lower than the vapour one on the propellers blades extrados, a second simulation using a cavitation model could be achieved.

To further improve the accuracy of the developed numerical CFD methodology, additional work could be carried out on both the real geometrical guiding rails modelling and also the cavitation one, by performing parametric studies on the nuclei concentration and the initial micro bubbles radius.

6. REFERENCES

1. STAR-CCM+, is a CFD solver edited by CD-adapco, http://www.cd-adapco.com

2. R. Ramaturi, W. Sandberg & R. Löhner, 'Simulation of a Torpedo Launch Using A 3-D Incompressible Finite Element Solver and Adpatative Remeshing', NRL/MR/6410--95-7797, November 30, 1995

3. J-M. Bolot, T. Chambenois, J-M. Laurens, L. Le Saint, X. Merlin, P. Perdon, J-M. Quenez, C. Rigoux, 'Numerical pre-qulification weapons launching from a Scorpène submarine', UDT Pacific 2006

4. A. Georges-Picot, B. Pengam, R. Luquet, P. Perdon, 'Numerical simulations of interacting multiple bodies launched from a submarine', UDT 2016

5. W. Yuan, J. Sauer & G. H. Schnerr, 'Modeling and computation of unsteady cavitation flows in injection nozzles', Med. Ind. (2001) 2, 383-394

6. P. J. Zwart, A. G. Gerber & T. Belamri, 'A two - Phase Flow Model for Predicting Cavitation Dynamics', ICMF 2004 International Conference on Multiphase Flow, Yokohama, Japan, May 30 - June 3, 2004, Paper No. 152

7. M. M. Athavale, H. Li, Y. Jiang & A. K. Singhal, 'Mathematical Basis and Validation of the Full Cavitation Model', Journal of Fluids Engineering, September 2002, Vol 124 ASME

8. H. Q. Yang, A. K. Singhal & M. Megahed, 'Industrial two-phase flow CFD - The full Cavitation Model', Von Karman Institute for Fluid Dynamics, Lecture series 2005-04, May 23-27, 2005

9. A. K. Singhal, M. M. Megahed & H. Q. Yang, 'Recent Advances (ventilated cavitation) with the Full Cavitation Model', European Congress on Computational Methods in Applied Sciences and Engineering (ECCOMAS 2004), Jyväskylä, Finland 24-28 July 2004

10. M. M. Athavale, H. Y. Li, Y. jiang & A. K. Singhal, 'Application of the Full Cavitation Model to Pumps and Inducers', International Journal of Rotating Machinery, Vol 8, pp. 45-56, 2002

7. AUTHORS BIOGRAPHY

Fabian PECOT has been holding the current position of R&D engineer at DCNS Research/Sirehna, for more than 15 years. He is responsible for hydrodynamic technical studies and research, particularly in the field of hydro-mechanical problems.

Thierry TAILLEFET holds a Ph. D in combustion and has been holding the current position of hydrodynamic Expert at DCNS for 5 years. He has more than 20 years of experience in many fields of fluid mechanics and heat transfer.

THE OPERATIONAL BENEFITS OF COVERTLY REFUELLING A CONVENTIONAL SUBMARINE WHILST AT SEA

S Harrison, H Diab, K Slater, Defence Science and Technology Group, Australia
M Renilson, University of Tasmania, Australia

SUMMARY

The Maritime Division of the Defence Science and Technology (DST) Group has developed the Integrated Performance System Model (IPSM) to assess platform level performance for Australia's future submarine fleet, and to enable trade-off studies to be conducted between concept of operations and design parameters.

This paper presents an example case study to investigate performance impacts of various technologies and sub-systems that could be potentially considered for further investment. For demonstration purposes, the IPSM was used to investigate the operational benefits of refuelling a conventional diesel electric submarine whilst at sea. Conventional submarines are significantly cheaper to operate than nuclear powered submarines but their capabilities can be constrained by the available fuel quantity.

Whilst the refuelling concept is introduced in this paper, a feasibility study has not yet been conducted. Additionally, for classification purposes, the platform model used for this study is a declassified generic submarine model and the mission configuration is an unclassified scenario.

1. INTRODUCTION

The Defence Science and Technology (DST) Group, which forms part of the Australian Department of Defence, have been applying integrated systems modelling techniques to provide whole of platform advice to the Australia's Future Submarine Program (FSP). Since 2010 DST has been developing a tool known as the Integrated Platform System Model (IPSM) [1-4] to model, simulate and analyse platform performance parameters. This capability forms part of a wider DST initiative to provide Defence stakeholders methods for clear and robust assessment of top level capability objectives, platform characteristics, and system requirements.

The IPSM enables trade-off studies to be conducted between platform performance; platform concepts and configurations; and operational parameters. The IPSM integrates system models of power and energy systems (including main storage batteries, diesel and air independent propulsion systems), propulsion and habitability systems (including atmosphere management, consumable food, waste stowage and discharge; and accommodation spaces), and can compare the platform's overall performance against various capability and performance requirements and constraints.

In this paper, a demonstration of the IPSM's adaptability to assess the impacts of novel technology concepts is presented. The IPSM has been used to consider the operational benefits of refuelling a conventional diesel-electric submarine at sea. The intent of replenishment at sea (RAS) of diesel is to increase the level of capability achievable by existing submarines without significantly compromising the stealthy nature of their operations.
The technical feasibility of the refuelling concept is not the subject of this paper. However the operational benefits and impacts are investigated with the use of DST Group's IPSM. A brief overview of the technology concept is given prior to the analysis and provides an excellent case study for the IPSM to examine a novel technology with low technology readiness level (TRL).

The advantage of using the IPSM at the conceptual phase is that it can be re-configured relatively quickly to assess operational impacts resulting from different technologies, concepts of operations, and scenarios. Therefore the IPSM has the potential to provide valuable inputs to business cases considering whether or not to develop the technology further. This makes the IPSM a powerful tool that could be used to support trade-off studies for future technology concepts considered as part of the FSP.

2. TECHNOLOGY CONCEPT

A concept for refuelling conventional submarines at sea is presented in [5]. This is based on the use of submerged fuel tanks that could be statically located at strategic positions at sea. For routine operations the fuel tanks could be transported to the designated area in advance by surface ships, and then moored to a float in the desired location. The engineering issues associated with transport of tanks into theatre are outside of the scope of this study. Figure 1 shows a diagram of the concept as described by Renilson [5].

It is proposed that the submarine will be equipped to allow it to approach a fuel tank which is moored in place and capture it. The connections between the submarine and the fuel tank may be similar to what is proposed in [6].

The submarine will deploy its catching equipment when approaching the moored fuel tank arrangement at a depth of around 50-60 m. At this depth motions imposed by the surface waves will be minimal, and the submarine will be able to be controlled easily [7].

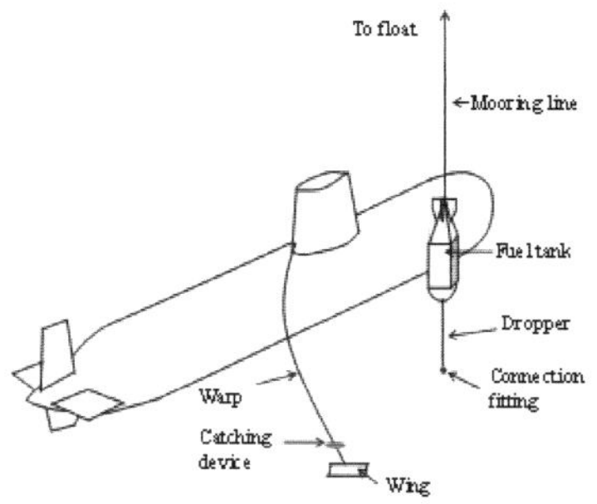

Figure 1: Schematic of Submarine Approaching Fuel Tank and Equipment (not to scale)

Once the fuel tank has been captured and appropriate connections between the submarine and fuel tank have been made refuelling will occur. Equipment will be required to ensure that the fuel tank can remain neutrally buoyant. Likewise the submarine will need to take appropriate action to maintain the desired buoyancy during the refuelling process.

It is acknowledged that this novel concept is far from reality at present and that there are many engineering difficulties to be overcome before being able to refuel a submarine at sea. However, assessing the technical feasibility of the technology concept is not the intent or priority of this paper. Rather, the refuelling concept provides a case study for DST to demonstrate the ability of IPSM to assess operational impacts of various technology concepts, irrespective of its technology readiness level.

3. ANALYSIS

In order to determine the potential benefits of replenishment at sea (RAS) of diesel fuel for a conventional diesel-electric submarine it is useful to demonstrate the conditions under which conventional submarines may become constrained in operation by the quantity of embarked diesel fuel. Constraining factors for conventional submarines can include: overall endurance, platform speed, power, energy storage, internal atmosphere, victuals, and diesel. The constraining factor may vary from mission to mission for any given submarine design but the overall range and endurance for a single deployment is usually constrained by either the submarine's overall endurance limit or by the quantity of usable diesel fuel. Typically overall endurance is characterised by the endurance of the submarine related to the embarked quantity of food and endurance of the crew.

To demonstrate these constraints an example submarine and mission are modelled within DST Group's IPSM. The demonstration submarine is described by the parameters in Table 1. These values are based on open literature sources [8-9] and are based on Australia's Collins Class submarine. Figure 2 shows the format of the example mission where the measure of success is time spent in the operational area and the transit range to the operational area has been set at 2500 nm.

Table 1: Parameters of example conventional submarine.

Parameter	Value
Submerged displacement	3353 tonnes
Diesel fuel capacity	268 m^3
Usable diesel	90 % of capacity
Overall endurance	70 days
Hotel load submerged (deep)	200 kW
Hotel load (snorting)	200 kW
Diesel installed power	4420 kW

3.1 DETERMINING WHEN THE SUBMARINE IS DIESEL LIMITED

In order to show how a submarine can become diesel constrained, a parametric analysis was conducted to show the variation of time spent in the operational area and diesel consumed with transit speed. In order to reduce the degrees of freedom of the problem the speed whilst snorting was fixed at a constant 10 knots. This was done for the purposes of this study only; the IPSM supports the specification of speed in snort and submerged deep states. Variation of transit speed was achieved through variation of the submarine's speed whilst submerged deep using energy from the main storage battery (MSB).

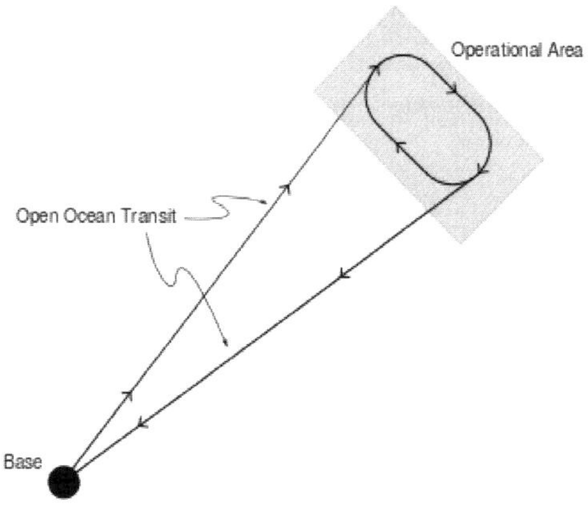

Figure 2: Example mission

Figure 3 shows the results of a parametric variation of transit speed, reported as speed of advance (SoA), and the resulting percentage of the diesel fuel and overall endurance resources consumed. The shading on the plot indicates the proportion of resource used in transit and in the operations area (OA). The figure demonstrates that the SoA to achieve the maximum time in the operations area occurs at approximately 8.5 knots and that at this SoA the diesel fuel and overall endurance resources are exhausted coincidentally. At a SoA less than 8.5 knots the entirety of the available overall endurance is exhausted whilst leaving some portion of diesel fuel unused. Conversely, at SoA higher than 8.5 knots all available diesel fuel is consumed whilst a portion of the overall endurance remains unused. The later indicates the case where the submarine is constrained by the quantity of embarked diesel fuel and thus the topic of this study.

The simulations conducted for this study placed no additional constraints on the submarine whilst in transit. In real world submarine operations it is likely that various constraining factors will interact and result in the constraining the speed of the submarine below the optimal speed for maximum OA endurance. Examples of such factors include varying threat levels and prevailing weather conditions. Therefore it is not unreasonable to assume that a submarine conducting a mission similar to that shown in Figure 2 may have difficulty in maintaining transit speed to attain the maximum OA endurance associated with the submarine design.

The SoA required to achieve maximum OA endurance is a function of the submarine design and the mission, specifically the range to the OA that is being conducted. Longer transit distances result in a reduced speed for maximum OA endurance. This in turn means that a larger range of transit speeds result in a diesel constrained operation. This is analogous to reducing the embarked quantity of diesel fuel, which may be the case for different submarine designs. Hence the degree to which a submarine can become diesel constrained increases with range to the OA and reduced diesel capacity. It is expected then that the benefit of RAS of diesel would be less for a submarine with large diesel capacity conducting a short range mission and greater for a submarine with a smaller quantity of diesel conducting a long range mission.

The key message is that if the submarine conducts a transit at any speed greater than the SoA to achieve maximum OA endurance, it will be constrained by the quantity of diesel fuel. This is clearly shown in Figure 3. Refuelling the submarine mid-deployment is a possible strategy to relieve this effect.

The conditions under which a conventional submarine may become constrained by the quantity of embarked diesel fuel have been shown; the analysis of the following sections will investigate the potential benefits of RAS of diesel fuel. The hypothesis of the analysis in the following sections is that RAS of diesel fuel will allow the submarine to operate at a transit SoA that would otherwise result in diesel constrained operation. With the key factor of the hypothesis being that faster transit reduces the proportion of overall endurance spent in transit legs and increasing the proportion spent in the operations area.

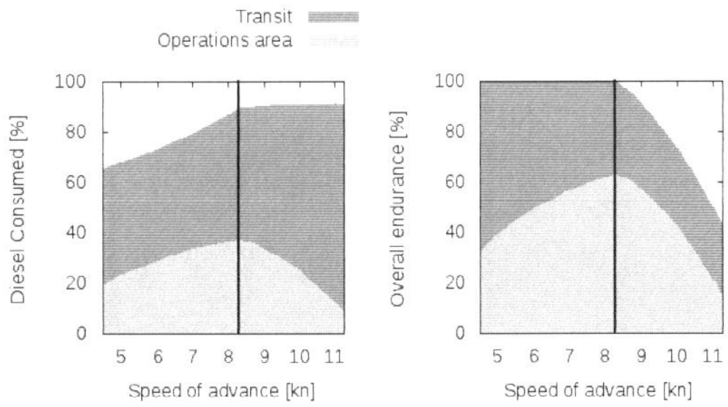

Figure 3: Diesel and overall endurance use variation with speed of advance, constant snort speed, range to OA of 2500nm, final 10 % of diesel held in reserve.

3.2 MAXIMISING OPERATIONAL AREA ENDURANCE

Given that faster transit will result in greater operational area endurance if the issue of increased diesel consumption can be resolved through RAS, the maximum transit SoA of the submarine needs to be determined. To achieve this, a parametric simulation of snort cycle transit was setup and run in the IPSM to show the resulting transit SoA for different submerged deep and snorting speed combinations. As previously, the snorting speed was held constant at 10 knots, usable diesel capacity was capped at 90% of total embarked diesel fuel, and the simulation was terminated at the exhaustion of the usable diesel capacity.

The results of the parametric simulation are shown in Figure 4 and illustrate the variation of specific diesel consumption with speed of advance. Speed of advance was varied by increasing the speed of the submarine when submerged deep under battery power whilst speed during snort was held at a constant 10 knots.

There are two points of interest, "A" and "B", marked on Figure 4. Point A on the figure indicates the minimum with respect to specific diesel consumption and occurs at SoA of approximately 5.5 knots. Concretely this SoA corresponds to the speed for achieving maximum range, any increase or reduction in speed will result in a reduced possible range.

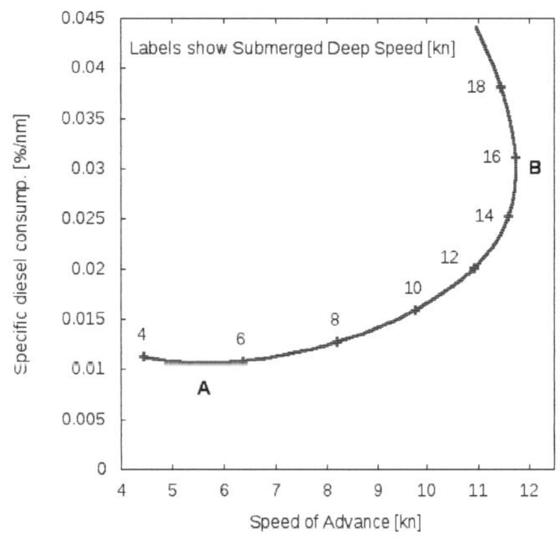

Figure 4: Specific diesel consumption variation with SoA. Labels on the figure show the Submerged Deep Speed. Snorting speed was held constant at 10 knots.

Point B in the figure shows the maximum achievable SoA of approximately 11.75 knots. Increases to the speed whilst deep beyond 11.75 knots will reduce SoA. This result is somewhat counter intuitive as it shows that increased speed deep utilising battery can result in a reduction of SoA. This effect occurs because as the speed of the submarine increases the endurance of a single battery discharge decreases. Eventually the submarine spends a greater amount of time snorting than submerged deep thus resulting in the lower SoA.

The maximum achievable SoA of the submarine occurs at point B in Figure 4. This therefore represents the fastest possible transit SoA the submarine can achieve. If we assume that the submarine has enough diesel fuel to transit at the maximum SoA whether through RAS or otherwise, the maximum possible OA endurance can be calculated. The reader should recall that for this study the speed during snorting has been held at a constant 10 knots for all simulations. A faster maximum SoA than what is shown in Figure 4 may be possible with variation of the snorting speed.

Equation 1 shows the maximum OA endurance as a function of overall submarine endurance, transit speed, and range to the operational area. Where E is endurance given in days, R is range in nautical miles, and S is SoA given in knots.

$$E_{oa,\max} = E_{overall} - \left(\frac{R_{oa}}{12S_{\max}} \right)$$

Equation 1: Maximum operational area endurance.

If the submarine can transit at its maximum transit SoA this will result in the greatest amount of time spent in the operations area. The increase in OA endurance will need to be traded off against the steep increase in diesel consumption. It is possible that some short range missions could be conducted with transit at maximum SoA but the achievable transit distances are likely to be very short. Similarly it may be possible for submarines with large diesel capacity relative to their displacement to complete missions at maximum SoA.

In the presence of the overall endurance constraint the maximum endurance in an operations area occurs when transit is conducted at the maximum achievable SoA. As previously mentioned this is not an achievable goal due to the high diesel consumption at maximum SoA. Furthermore the previous section demonstrated that the optimal transit SoA to maximise OA endurance is significantly lower than the maximum SoA which corresponds to a reduced OA endurance. Refuelling at sea provides one possible solution to allow the submarine to transit at maximum SoA and therefore achieve the maximum OA endurance.

3.3 TIMING OF RAS

Given that transit at maximum SoA requires a specific fuel consumption that very likely cannot be supported by the embarked quantity of diesel it is important to investigate the possible timings for RAS of diesel.

Consider the case where a single RAS event is conducted for the duration of a deployment at the completion of the first transit section prior to entry to the OA. If we assume that this event refuels the submarine to maximum diesel capacity the usable diesel quantity must be sufficient for both the required OA endurance and the return journey transit. Thus when the transit range to the OA becomes large enough such that the return journey and OA patrol together require more than the usable quantity of diesel embarked, two RAS events will be required. Nominally this could be one before and one after patrol in the OA.

Given that for the example submarine in this study (see Table 1) the diesel consumed in a 2500 nm transit is ≈75%. The return journey is assumed to require the same quantity of diesel but may vary in reality according to tasking, prevailing weather, and other factors. As such if a single RAS is conducted immediately prior to entering the OA, 75% of diesel must be reserved for the return transit. This leaves 15 % (allowing for 10 % reserve) for patrol in the OA, approximately 17 days of patrol. This brings the total duration of the deployment to 35 days, indicating that the submarine is still constrained by diesel.

This also shows that the required capacity of the refuelling tank is dependent on the quantity of fuel used in transit. The sizing of the refuelling tank is investigated in the following section.

If two RAS events were conducted, each refuelling the submarine to maximum capacity one before entry and one after exiting the OA, diesel sufficient to exhaust overall endurance is available (assuming refuelling back to maximum capacity) whilst still allowing for transit at maximum SoA. In this case the OA endurance will be limited to 52 days by the overall endurance and maximum transit SoA constraints.

There is a significant risk for the submarine when operating in a manner as has just been described. Allowing the available diesel to reduce to a point where the submarine may not have enough diesel to transit to a friendly port presents a situation with significantly increased risk. This risk needs to be traded against the operational benefit of the increased OA endurance. This trade-off is beyond the scope of this study and is a potential avenue of future work.

3.4 SIZING REFUELLING TANKS FOR MAXIMUM OPERATIONAL AREA ENDURANCE

The conditions under which diesel constrained operation have been shown, and the illustration of the maximum transit SoA that limits the maximum possible operational area endurance has been discussed. The benefits from refuelling a submarine mid-deployment are now investigated, specifically with respect to the mission shown in Figure 2.

Table 2 shows the increase in the achievable operational area endurance for three set points of transit SoA. The benefit to OA endurance from refuelling is significant. However there is a significant increase in the quantity of diesel consumed and the values shown represent what is possible if the submarine can be refuelled to maximum diesel capacity. Additionally the indiscretion ratio of the submarine in order to achieve high transit speed is greatly increased. This should be traded along with other considerations such as size of refuelling tanks, cost, logistics, stealth, and maintenance of a RAS system against the increased operational effect achieved by the increased OA endurance. Furthermore the signature profile and risk associated with RAS of diesel fuel is unknown and is a key factor in determining the difference in operational effect of the submarine. However this is beyond the scope of this study.

Table 2: Operational area endurance at selected transit SoA, assumes refuelling back to maximum diesel capacity.

Transit SoA [knots]	OA End. [days]
5.5 (min diesel consump)	32
8.5 (no RAS, baseline)	44
11.75 (max SoA with RAS)	52

Table 3 shows comparison of the OA endurance possible with a number of different combinations of transit SoA and refuel tank capacities.

Table 3: OA endurance for different transit SoA and refuel tank capacities.

Transit SoA [knots]	Refuel Tank(s) Diesel Capacity [tonnes]	OA Endurance [days]
8.5 (baseline)	Nil	44
11.75 (max SoA)	34	0
11.75 (max SoA)	221	44
11.75 (max SoA)	237	52

The first row of Table 3 shows the baseline case where no refuelling occurs. The best OA endurance possible in this case was illustrated in Figure 3 and is 44 days using a transit SoA of 8.5 knots.

The second row in Table 3 shows the original concept from [5] with a refuelling tank with capacity of approximately 34 tonnes of diesel. A refuelling station with 34 tonnes capacity would be sufficient to refuel only 15 % of the submarine's diesel tank and would not be sufficient to allow any patrol of the OA when using the maximum transit SoA, 11.75 knots. Further work to investigate other transit speeds and the resulting possible benefit over the baseline case should be done as follow-on work to this study.

A minimum capacity of the refuelling tank or tanks is proposed in the third row of Table 3 showing a required capacity of 221 tonnes of diesel. This is the quantity of diesel needed to match the OA endurance from the baseline case whilst transiting at maximum SoA. The quantity of diesel is such that two separate refuelling events will be required; each delivering half the required diesel. Nominally this could occur before and after patrol in the OA. The benefit of this style of operation is negligible; whilst the submarine is no longer constrained by overall endurance due to the fast transit it remains constrained by diesel quantity. No increase in OA endurance is achieved above what is possible with a baseline transit SoA and no refuelling. In order for further OA endurance to be possible a greater diesel capacity is required from refuelling.

A possible upper limit for the capacity of the refuelling diesel is shown in the final row of Table 4, the capacity in this case is sufficient to allow transit at maximum SoA and OA endurance such that the overall endurance is exhausted, 52 days in this case. The capacity of the required diesel for refuelling to satisfy this is 237 tonnes of diesel and will need to be delivered across two refuelling events. This will allow for the full use of overall endurance and maximum transit SoA. Diesel and overall endurance will be exhausted simultaneously.

To envisage the physical dimensions of tank sizes required, Table 4 shows the length and diameter for each of the tank capacity options shown in Table 3 allowing only for the volume of diesel required per refuelling event. The tank sizes are given as a length of 2 and 3 metre diameter cylinders. The values are very large indicating that refuelling tanks sized to maximise OA endurance and allow for the use of maximum transit speeds are infeasible. Additionally towing of tanks this size by the submarine during the refuelling process is likely to also be problematic.

Table 4: Example refuel tank sizings

Refuel Tank Diesel Capacity [tonnes]	2-metre dia. Cylinder Length [m]	3-metre dia. Cylinder Length [m]
34 (concept)	12.7	5.7
111 (minimum)	41.5	18.4
119 (maximum)	44.4	19.7

Large benefits from refuelling the submarine can only be realised when the submarine transits at maximum SoA and the transit distance is significant. Possible minimum and maximum tank sizes have been presented but are significantly larger than proposed in the original concept and sensitive to the quantity of fuel used in transit. Alternatively a succession of refuelling events from smaller tanks may be able to provide the required diesel quantity. This presents a number of feasibility issues that would need to be resolved for refuelling submarines at sea from tanks such as those presented in [5] to become worthwhile. Furthermore the duration of the refuelling event, which has not been investigated in this study, will reduce the benefit of refuelling significantly if the process is slow. This would be compounded in a concept where multiple refuelling occurs from a succession of smaller tanks. This may occur in order to embark the desired quantity of diesel whilst limiting refuel tank sizes to feasible dimensions. The technical aspects of this are beyond the scope of this study and should be investigated as future work.

4. CONCLUSIONS AND FUTURE WORK

In summary, platform performance impacts of the ability to refuel a conventional submarine whilst at sea have been investigated as a case study for the DST Group's IPSM. The IPSM has been designed with the intention of providing a capability to conduct trade-off studies across different technologies, concepts of operations, and scenarios in a robust and validated manner. This study provides an example of how the IPSM can be used as a test-bed to assess novel technology options in order to assist decision makers and stakeholders to determine which options may be worth pursuing further.

With regard to the specific findings of the case study, the following conclusions have been drawn:

- An optimal speed for transit exists specific to the mission and submarine performing the mission, where the measure of success is endurance in the operations area.
- A physical upper limit exists for transit SoA, placing a lower limit on the time transit stages can be completed.
- When operating at the maximum transit SoA it is very likely that a conventional submarine will become constrained by its diesel capacity, refuelling the submarine offers a potential relief to this effect if the required capacity of diesel can be embarked.
- A minimum refuel tank size such that transit can be conducted at maximum SoA has been discussed. This relieves the overall endurance constraint but not the diesel quantity constraint on OA endurance.
- A maximum refuel tank size has been discussed that allows diesel quantity sufficient to exhaust overall endurance and allow transit at maximum SoA. However, the tank size is very large.
- The quantity of diesel required for refuelling is sensitive to the submarine design and the transit range to the OA.
- The largest benefits from refuelling occur for submarines with small diesel capacity and in missions with long transit distances.
- If RAS of diesel can be shown to be practical it may have an impact on submarine design. Possibly resulting in designs with reduced diesel capacity, this may allow for increased food stores.

Whilst there are some clear benefits to the ability to refuel the submarine at sea, a range of areas for future work and other considerations have been identified:

- Further investigation and optimisation of transit speed across a range of mission types is required in order to fully determine the potential benefits of refuelling to OA endurance.
- Required refuel diesel capacity is large when transit range is large. This presents a significant engineering and logistical problem.
- The signature aspects and risk associated with refuelling at sea should be further investigated.
- The benefit to OA endurance from refuelling is significant. However there is a significant increase in the quantity of diesel consumed in order to achieve this increase and as a result the quantity of diesel for refuelling is very large. This should be traded along with cost, logistics, stealth, and maintenance of a RAS system against the increased operational effect achieved by the increased OA endurance.
- The risk associated with allowing the submarine to use the majority of embarked diesel whilst operating far from friendly ports should be assessed in terms of the operational effect that may be gained.

- If self compensating fuel tanks are used in the submarine, refuelling will require diesel contaminated seawater to be discharged from the submarine during refuelling. The potential environmental and legal impacts of this should be investigated further.

5. REFERENCES

1. SLATER, K. et al., 'Integrated Platform Modelling of Modern Conventional Submarines to Support Strategic Decision Making', RINA Warship Conference: Naval Submarines & UUV, UK - Bath, 2017.

2. HARRISON, S. et al., Analysis of Platform Configurations Using the Integrated Platform Systems Model', International Maritime Conference – Pacific 2012, Australia - Sydney, 2012.

3. HEPWORTH, G. et al., 'High Level Capability Analysis of Submarine Platforms', RINA Submarine Conference, UK - Bath, 2014

4. HARRISON S., Wharington, J. and Slater, K., 'An Investigation into the Penalties Incurred due to Operational Constraints Imposed on an SSK-type Submarine in an Operational Area', International Maritime Conference – Pacific 2015, Australia - Sydney, 2015.

5. RENILSON, M.R., 'Extending the Range of a Conventional Submarine by Autonomous Covert Refuelling when on Patrol', Proceedings of Pacific 2015 International Maritime Conference, Sydney, Australia, October, 2015.

6. RENILSON, M.R., 'A Simplified Concept for Recovering a UUV to a Submarine', Underwater Technology, Vol. 32, No. 3, pp. 193-197, 2014.

7. RENILSON, M.R., 'Submarine Hydrodynamics', Springer Briefs in Applied Sciences and Technology, ISSN: 2191-530X, ISBN: 978-3-319-16183-9.

8. GRAZEBROOK, A.W, 'Collins Class Comes Up Down Under', Jane's Navy International, Janes Information Group, 1998.

9. 'Jane's Fighting Ships 2012-2013', Jane's Information Group, 2013

6. AUTHORS BIOGRAPHY

Karl Slater works in the Maritime Division at DST on several tasks, including the development of methods and tools for undertaking integrated performance analysis of maritime platforms through modelling and simulation. Before joining DST in 2011, Karl spent 7 years with BAE Systems working primarily with surface warships, and 8 years with the UK Ministry of Defence working in submarine naval architecture. Karl obtained a Bachelor of Engineering (Honours) in Marine Technology (Naval Architecture) from the University of Newcastle upon Tyne in 1996 and a Master of Science in Naval Architecture from University College London in 1998.

Simon Harrison is a Defence scientist and analyst at DST developing and implementing platform system models for naval platform assessments, through simulation activities. Simon joined DST in 2010 and has a Bachelor of Science in Physics and a Bachelor of Engineering in Electronics from RMIT University Melbourne.

Hamid Diab is in the Maritime Division at DST leading a modelling and simulation team in whole of platform performance analysis. During his time at DST from 2003, Hamid worked on the F-111 Sole Operator Program as a structural analyst and fatigue test engineer; provided analysis and modelling support on the sustainment management for Defence's AIR 9000 Helicopter Program and air lift platforms; and became the lead researcher and advisor for the Air Force on strategic asset management.

Martin Renilson is an Adjunct Professor at the Australian Maritime College, an Institute of the University of Tasmania.

I DIDN'T SEE THAT COMING: THE PERILS OF UNDERWATER AUTOMATION

M Cook, M Thody and **D Garrett,** BAE SYSTEMS Maritime Submarines, UK

SUMMARY

This paper considers the potential influence of automation in accident sequence development in future submarines. Human factors analyses of accidents in other domains which involve automated systems have identified a number of recurring themes that contribute to accident development. From one perspective the residual accident occurrence in systems which are progressively and increasingly automated is attributable to the human operator(s). However, it can be argued that automation has largely dominated the performance envelope of systems in areas which are completely functional and well understood. Thus, the human operator is given the more difficult area of the performance envelope and the transition from automation to human operator occurs at a time when the operator is likely to face the greatest challenge in both understanding and managing the system upset.

GLOSSARY

Situation Awareness – A description of a mental model the operator(s) have, where better mental models are predictive of future events and pre-empt upsets. It can be very poor and result in reactive management of events.

1. INTRODUCTION

In a review of human-automation interaction failures the reasons for human-automation interaction issues were considered [32]. In Table 1 of Sheridan and Nadler's paper the majority of issues occurred in design of the interface or in management of the automation, although it was claimed that interface design and procedures were linked, and management of the organisation was linked to training and use of procedures. The most significant lesson was that "No amount of training can make up for poor design and procedures" [32]. Thus, the Human Factors Engineering of the interface and the procedures that direct the user are primae facie causes of the difficulties with automation. It had been previously noted that new technology, which includes automation, created new opportunities for erroneous operations, with negative consequential outcomes [18, 19]. Concerns about automation and the potential negative impact on performance have been a long-standing concern in human factors research. Failure especially of highly reliable automation may be catastrophic because the operator trusts the automation too much [25] and this may be viewed as automation-induced complacency [24].

Reliance on the automation to manage the task and disengagement from the task may result in reduced situation awareness [14] or automation may give rise to an automation induced decision bias [21, 22]. Long term reliability and low frequency unpredictable failure is especially problematic because operators adapt their use of cognitive resource to expectancies. This is not controversial because automation is intended to reduce operator workload and engagement in the task. It was suggested that increasing automation capability increased the risk of more significant problems at transition because operators adapt to the presence of automation [23]. There are reduced opportunities to practice skills, a point made in aviation with regards to manual approach to airports. The quality of the experience is reduced as well as the quantity, meaning the unusual events have more impact on performance. Many of these concerns were already identified by 1978 [17]. With over sixty years of research on automation the limitations have been reduced but automation is still implemented with a fixed closed-system model which cannot encompass all possible events.

As noted modern submarines spend the majority of their time operating under autopilot control [27], which means that operators have very limited opportunities to accrue experience, low familiarity with upsets and recovery procedures, and potentially all the attendant problems noted above, when transitioning from failed automation to manual control. Even in drone technology there is a strong supervisory element and discretionary involvement, suggesting that the boundary between automated and autonomous systems has not been crossed.

2. IMPACT OF AUTOMATION

2.1 EARLY PREDICTIONS OF ISSUES

In reviewing the impact of automation the impact on primary task automation may seem positive under normal operating conditions [32]. In a meta-analysis of 18 studies, many research papers applied automation to a primary task and then considered the impact on a secondary task, which became the operator's main task. It was found, perhaps unsurprisingly, that the performance on the primary task managed by automation became better. The transfer of primary task responsibility meant that the human operators had more cognitive resource to devote to the secondary task and performance on their task improved. However, there were a number of negative consequences of automating the primary task which included reduced situation awareness, which may have been a consequence of the reduced workload. In addition, returning to manual operation of the primary task was associated with poorer performance a point predicted as long ago as 1978/1983 [17, 2].

2.2 TRANSITIONS BACK TO OPERATORS

The problems with transitions at automation failure can be summarised in a single graph [23] and this has been adapted below into two separate graphs which summarise the observations below.

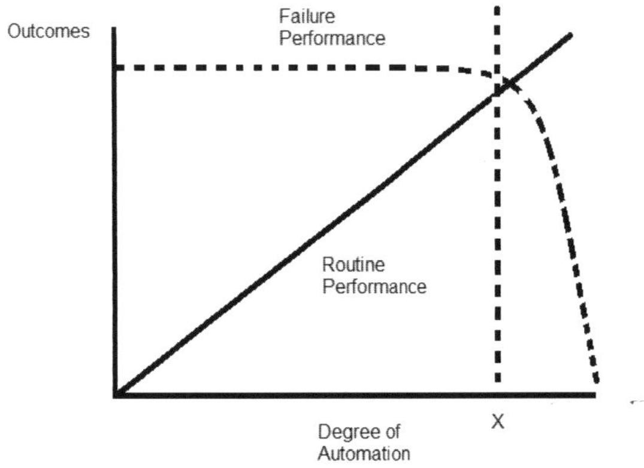

Figure 1: After Onnasch, Wickens, Li and Manzey (2014) routine and failure performance outcomes against degree of automation.

Figure 1 illustrates the general finding that the degree of automation (both in levels and stages) can increase and give rise to improvements in performance. However, in failures performance results in catastrophic collapse in performance, which is related to the impact of automation on other human performance characteristics illustrated below.

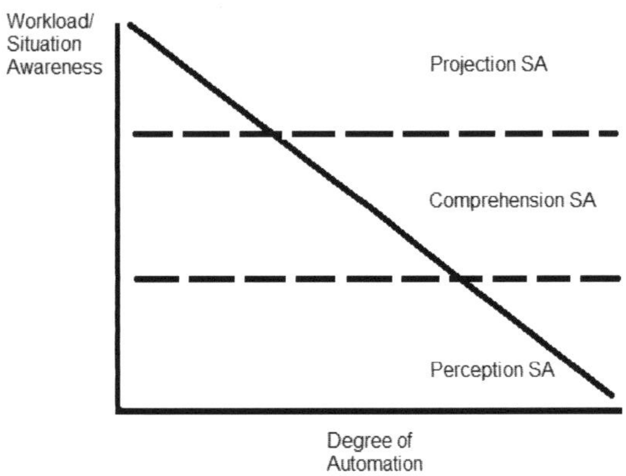

Figure 2: After Onnasch, Wickens, Li and Manzey (2014) workload and situation awareness against degree of automation.

Figure 2 illustrates the decreases in workload, which is potentially part of the design intent in making use of increasing automation because this accompanies increasingly effective performance outcomes. However, as workload falls the quality of operator situation awareness decreases because less operator effort is devoted to processing information and it may be this which eventually results in perceptual level situation awareness with its reactive response to events. This was compared to the aphorism the bigger they are the harder they fall, in that greater degrees of automation result in more catastrophic performance decline. It was suggested that there was a point at which increasing automation offered less benefit and created a latent risk. However, definition of this point was not adequately defined.

The evidence available thus far suggests that automation can have benefits in terms of allowing an operator to accomplish more under normal operating conditions because some of the resource demand and workload is transferred to the automated system. However, the reduced effort applied to the automated task means that at transitions where tasks are transferred from the automation to the human operator, the operator is poorer at task performance and this may in part be a result of reduced situation awareness.

Situation awareness (SA) is a term coined in military aviation where it was defined as knowing what is happening, where you are and what you may be required to do next [11, 12, 13]. There are three levels of SA, perception, comprehension and projection. If the operator has projection SA then they may not only know what they expect to do next but also how to manage a system upset. Sometimes operators are in the vernacular, behind the curve, when they comprehend what is happening but don't have a workable mental model of future events, they have comprehension SA. In the worst case operators may have perceived events but not understood their meaning, they are only able to react to on-going events.

The analysis a number of the events identified as human-automation problems can be clearly associated with different levels of SA [32]. Often operators appeared to be reacting to events and this suggested that they were only able to create perception level SA. The transition between perception and comprehension or projection is a reflection of the quality of the mental model the operator, which in turn may reflect the operator's effort (workload) on the task they are attempting. Automation is often designed and justified by the reductions in operator workload. The analysis of automation interaction is highly pertinent because it clearly indicates that automation has the potential to impair SA on a safety critical task managed by automation [23, 32]. In addition, infrequent abrupt and unpredictable transitions between automation and manual operations are areas of vulnerability and require being a focus for design and implementation of graceful degradation of automated systems.

2.3 MENTAL MODELS AND PERFORMANCE

Mental models have been a key focus for human factors designers involved in automated systems analysis. A mental model is defined as one might expect. A model is a predictive device or process which enables individuals to predict future outcomes and it is mental in the sense that it is held within the operator's mind, both in short-term information and long term accrued experience. Effective mental models require extended experience to develop and for that reason many junior operators are vulnerable, as their mental model may be incomplete or not sufficiently robust. It has been argued that automation interferes with the development of a system mental model because all the behaviour of the system in terms of inputs and outputs is mediated, via automation. In some cases the behaviour exhibited whilst in automation may be different from that in manual control. Thus, the operator's expectancies could be based on stereotypical expectations developed during automated control which do not apply in manual operation.

2.4 FIXED PROCESSES TO ALLOCATE FUNCTIONS

The divergence between automated and manual system behaviour is often justified on the typical Fitt's List [15] approach. Men Are Better At and Machines Are Better At (MABA-MABA). Automation can calculate complicated solutions and execute actions in parallel (or near parallel) which human operators cannot. At transition, a system which has been taken to an area of the performance envelope by automation that requires complex calculation and fast execution is an area of greater risk of compromising human performance. For this reason graceful degradation that leaves an automated system and transitions to human operation in a more benign area of the performance envelope is required. Very few designers consider that the automation frequently receives a benign transfer of control in a manner that human operators do not, there are often severe asymmetries in demand at transition to manual.

Some have argued that Fitt's list obstructs progress in improvement in human-machine interfaces because it derives the human component of the task as a residual after automation and shifts the focus away from coordination [8]. The same authors have been extremely critical of the use of folk models in analysis of human-automation interaction [7]. It has been argued that automation problems can be identified by the entropy (disorganisation) of behavioural performance observed, more accurately than through the use of characteristics inferred inside the operator. However, the same work fails to address the mapping of many internal states onto the same observable outcomes. Thus, decision to allocate cognitive resources, the rate of information flow in events, and the failure to direct attention to critical sources of information can all give rise to erroneous information process and poor performance, the outcome is the same but the causal mechanism and the remedy is different. It is perhaps in the nature of more complex tasks that human operators strategically make use of cognitive resources and disregard or not actively search for contradictory information [5], a form of the availability heuristic. This variable strategic use of cognitive resources may be the root cause of the concern about falsifiability criteria for genuine scientific constructs [7]. Without addressing this designers may not fully appreciate the impact of their automation interventions and this may explain the variability found in the meta-analysis of automated tasks [23].

It has been strongly argued that even though designers of automation identify the problems with performance as associated with the human operator it is in fact the joint coordination and lack of implicit coordination from the automation that is a more significant contributor to accidents sequences [16]. In illustrating the mapping of poor performance outcomes onto causes one can see that the communicative process is clearly part of the difficulties faced by operators. [26] have argued against [7, 8] who claim that concepts such as situation awareness, mental workload and trust are viable concepts but it is possible that despite their utility they need to be more formally equated with underlying psychological constructs such as working memory capacity, attention or executive function.

The concerns about the current design approaches which have a relatively poor "relationship" between the automation and the human operator are not new [4,16] have both identified the need for improved coordination between automation and the human operator dependent on its support. One of the components identified as problematic is the notification of impending failure by the automation [16] because often the failure is abrupt and occurs without warning. This has cognitive implications because automation has an advantage where the operator is released from tasks by automation. If the operator is required to maintain SA about the automated task then there is little or no benefit from the automation. This was the common problem identified in the meta-analysis of automation impact [23].

3. EVOLUTION CONTROL SYSTEMS

Control systems have evolved as [27] notes, with increasingly mediated actions and sensing. Early control systems communicated operator decisions directly to controls to create outcomes and operators viewed direct or nearly direct representations of parameters affected by the change in state. In modern systems the operators actions are frequently converted into potential control actions mediated by a high-level system and transformed into an acceptable output for a local controller before changing the status of the system. In aviation control systems may make decisions about the transformation applied and even the legitimacy of the request, denying the response in some cases. Similarly in modern control systems the change in the state of the system is communicated back via micro-processors and a central computing system to be displayed to the operator. These control systems, old and new illustrated below, represent different ends of the spectrum of possible technology solutions. In reality the sensing process in newer systems may not actually represent the actual change in state but the expected change in state or a consequence of the change in state. Thus, there is residual uncertainty about the accuracy of the sensing.

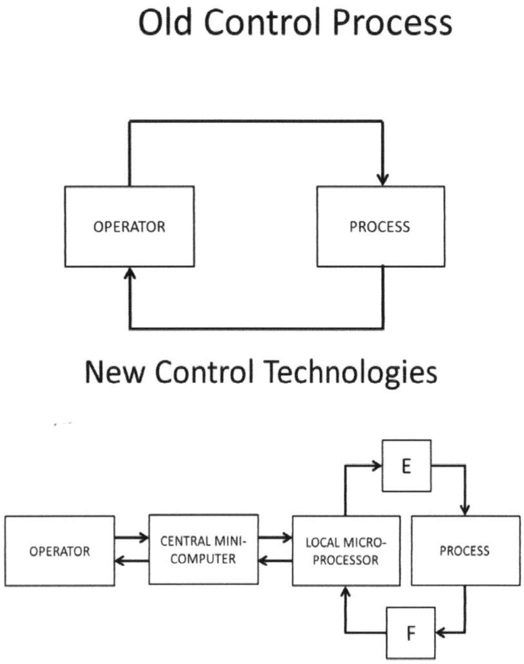

Figure 3: Illustration of the increasingly mediated experience of control operators in newer control systems.

In old technologies the manual force communicated to the control systems often gave proprioceptive and somaesthetic feedback and surprisingly with modern technology many designers are keen to make safety claims against the sensory feedback in operation of control systems, whether this is vestibular sensing of orientation and rotation, or feelings from the seat of the pants. However, such claims appear wholly inappropriate because of the dominance of visual and auditory senses in sense-making activities. One of the indirect consequences of modern control technologies is the insertion of a delay which creates hysteresis in modern control systems which are inherently delayed in submarine operations, at least relative to other vehicular systems. In larger vehicles, such as submarines, the momentum and inertia makes the immediate observation of actions difficult and this can make adjustments an iterative process which may under or over-compensate.

4. SUBMARINE CONTROL SYSTEMS

4.1 OVERVIEW

As noted [27] submarine control is delivered by three mechanisms hydrostatic control, hydrodynamic control and navigation control. Automation of any of these systems, as in aviation, creates the opportunity for new types of error specifically related to automation.

4.2 HYDROSTATIC CONTROL

Hydrostatic control nominally is associated the 'trim of the boat' but this may be linked to ballast in a wider sense. The trim is affected by the weight mass and centroid of the buoyant volume within an acceptable 'out of trim' margin. Submarines when submerged normally manage trim by moving masses of water between trim and compensation tanks through pumping. Bodily weight is managed as well in relation to consumables within the system or buoyancy in response to water density. Trim is a safety function in that when a submarine loses propulsion the forces influencing the boat behaviour will be relatively small, affording the crew time to recover. Time is a key issue because human operators are inherently slow to respond. When automation fails, transferring control to the operator, it may be at a location within the performance envelope that requires a speeded response and the operators are already less aware of the current system state, lacking situation awareness [23]. Hydrostatic control is complicated by changes in the boat with increasing depth. Essentially the increasing pressure on the boat with depth results in changes in density and that affects the buoyancy of the boat. Thus, the hydrostatic management process is not necessarily uniform across the full range of operating envelope.

4.3 HYDRODYNAMIC CONTROL

Hydrodynamic control makes use of directive and propulsive systems. The propulsive systems given the boat forward movement and as the speed of the boat increases the plane authority increases. Planes, both the fore and rear planes, provide directive control for the boat. Hydrodynamic control implies that at low speeds the boat is vulnerable because the planes cannot effectively control the vectored movement to the same degree. Thus, in different regions of the submarine performance envelope different systems dominate in recovery. At slow speeds hydrostatic control systems dominate and at higher speeds hydrodynamic systems become more dominant. The interaction noted above between the speed and plane authority creates non-uniformities in performance.

It is possible to imagine situations in which submarines have latent failure conditions. For example, during flooding events the mass of the boat is increasing as a result of water ingress and to some extent this can be offset by thrust from the planes at higher speed. Thus, boats at slower speeds or without propulsion are more vulnerable to flooding.

The delicate interplay between boat mass, speed and hydrodynamic lift means that failures of watertight integrity, failure to supply propulsive force and failure to control planes are potentially interlinked in accident sequence development. It is these inter-relationships that make recovery of the submarine in reversionary modes, after failure of automation, problematic. In many aviation accidents operators have either actively sought to regain control, failed to understand the restrictions imposed by automation modes or failed to foresee the consequences of their actions, resulting in realisation of the consequences of accident sequences.

5. SUBMARINE AUTOMATED SYSTEMS

5.1 DRIVERS FOR AUTOMATION

In 33 recent accidents over half of the accidents have human error as a potential root cause [27]. Many of the accidents are associated with navigation errors, an area in which there have been a large number of accidents. Of 18 incidents identified 8 involved collisions with another vessel. Navigation accidents occur in aviation, with incorrect data entry to the Control Display Unit (CDU) and some of these are documented in [32]. In terms of accident development the use of automated systems under time pressure has been a potential explanation of occurrence. Thus, using data entry into the on-board systems to re-plan a runway change on approach rather than flying the approach manually is a typical example of a significant contributor to accidents.

Submarine platforms are increasingly complex and both system complexity and capability growth have not been matched by evolution of automated electronic control systems [27]. As appealing as automated systems are they introduce significant up-front costs to the design lifecycle because the software and automated processes need verification and validation.

5.2 OPPORTUNITIES AND OBSTACLES TO AUTOMATION

Two areas of opportunity in submarine automated systems are in control of ballast and bodily weight, and steering and diving. In any simple control model involving either a human or automation, there are three steps that underpin control of any process. Events are detected, these are identified and then this is used to calculate a compensation response. Examination of human-automation interaction failures there appears to be opportunities for human and automated control to de-synchronise and even become antagonistic [32].

Perhaps surprisingly the events that influence hydrostatic and hydrodynamic control may not always be sensed directly or the information collected may not be accessible to automated systems. For example, a flood may be sensed and electronically communicated throughout a boat but the exact mass of water may not be known to the hydrostatic control system. Similarly for control surfaces which fail, the failure may or may not be directly sensed but the consequence of the failure may be observable. The indirect sensing of actual events may be problematic because this may impose limits on the sensitivity, speed or reliability of the detection process. The divergence of reported and actual observed events can be influential in shaping the human operator's trust of the automation. If the operator does not trust the automation they may devote more time to the supervision of the automated task, undermining the value of the automation. Thus, one can see that over-inclusive trust and under-inclusive trust of automation both create issues with realising the benefit.

5.3 AUTOMATION AND SUPERVISION

The fallacy that automation can reduce manning has frequently been used in a wide range of systems, where automation is promoted. It was this in part that was driven by the two-man cockpit in civil aviation, dispensing with the role of the flight engineer. The difficulty is that even after processes are partially automated they are not autonomous and they require human supervision. Even at an early stage in automation development this was recognised as ironic and problematic. Human operators are not effective at long-term monitoring and supervision roles. Less direct human involvement creates poorer performance at transition from automated control back to manual operation [23]. Indeed, recognition of this has meant that a more graceful degradation is proposed with additional interim automation, returning systems to an expected or benign state where operators can cope more effectively.

5.4 AUTOMATION'S INCOMPLETE KNOWLEDGE

Consider bodily weight and trim. As a boat floods sensors detect that water is building in the compartments within the boat. Thus, the total mass and the trim of the boat are affected by the inrush of water. An automated trim system has no easy method of interpreting this inrush of water and it will simply seek to compensate for the change in the boat. Water may be pumped forward or backward to balance the increasing mass at the other end, re-balancing the trim. The automation will attempt to fight the flood and move a mass of water around the boat.

Or, in some cases water could be pumped in to equilibrate the mass of the water at the alternate end of the boat. Either way, the automated system could respond to flood event in an inappropriate way because it does not incorporate a state transition model that reflects significant changes. Thus, the trim and ballast management system may need to be suspended at a critical time. The operator's performance on this previously automated task is likely to be poorer [23]. Automation results in skill fade, as manual operation is rarely used, operators cannot maintain the same quality of performance with training exercises alone and in some cases, as [32] found the interface to the system is poor in supporting the transition. An uncontrolled increase in bodily weight is one issue with an automated trim and ballast system, increases in buoyancy is another. Thus, presence of unexpected air within the ballast tanks may create unexpected buoyancy.

Both of these problems that reflect a general issue with many engineered automated systems in that they can only embrace a closed system model, where the unexpected or unplanned events do not occur. It was this that prompted commentators to suggest that automation had adopted the easier tasks and left the less tractable ones for the human operator. Thus, it was no surprise that human error was not only the residual part of factors contributing to accident development but it was responsible for a larger amount of the variance. The management of "normal" operational states by automation and the overwhelming presence of abnormal states in the residual tasks, allocated to human operators, results in an unfair bias of failures to human operators.

Trim and ballasting is a relatively simple system to automate but its simplicity belies its potential impact on successful operation of the boat. The control of steering and diving is a more complex system which is likely to have transverse representation in failures with electrical, pneumatic, hydraulic and mechanical failures contributing to changes in system state. The challenge that faces the operator of the system is that whilst the source of a failure may reflect the functional availability of elements of the system it is the latter that may be crucial in real-time responses.

5.5 MASKING OF FAULTED STATES

One possibility is that automation may mask the underlying fault and the operator is not notified or fails to notice a failure condition [32]. In one of the incidents identified it was only at transition from automatic to manual operation that the faulted condition became apparent, which resulted in a chain of events that may have led to loss of aircraft [32]. In aviation the automation compensated for the reduced engine performance but in submarines it may be compensation for steering, via control surface failures that could result in problems. At transition the control surface failures may result in changes in control over heading, depth and/or roll characteristics of the vessel.

Thus, masking of faulted states by automation is a possibility but equally the failure to foresee the consequences of an event at retaking manual control is another. The compensation by automation is intrinsic to its function but it is one that can compromise human performance at a forced abrupt transition. Again time to regain situation awareness at transition is an important issue. This should be a clear warning to those who seek to expedite transition performance by automatically dropping out of automated control into manual. The failure to notice state changes is another example of a latent accident development route which has occurred in aviation. Sometimes the incorrect belief that the system is under automated control creates failure conditions because operators stop monitoring events, whilst they concentrate on solving an unrelated problem. Or, they believe that they have assumed manual control of systems when they have in error not completed the necessary steps. This is the reason why instruction of submariners includes the SCAR (Select Check Action and Review) principle. However, as Sheridan and Nadler noted training and procedures are an ineffective mitigation for poor human-machine interface design.

5.6 MENTAL MODELS AND SURPRISES

In respect of accident sequences the mental model of the human operator has been identified as the locus for accident development. There can be mismatches between operator's mental model and the apparent behaviour of the automated system programmed by designers. This should perhaps not be surprising because the algorithmic approach to problem solution used in automation may not be the same as that used by the human operator. It may be that they appear to correlate for many situations which provide a false sense of confidence and it is surprising to the operator when there are divergences. At divergence the operator is in uncharted territory and is faced with the additional workload of trying to figure out what is going on.

Automation surprises [30] were described over 20 years ago and at that time they were linked to joint-working of automation and human operators, identified much later in [16]. Some observers suggest that checking the model will help to reduce but not necessarily eliminate automation surprises [28]. Going sour events or automation surprises were well established problems for automated systems [32]. It is clear that automation surprises are in part a consequence of interface design because the operator of any system relies upon the observable operator behaviour [1]. Where the mapping of observable interface events onto system behaviour is opaque or insufficient then automation surprises are more likely, echoing the observations of others [32], in that interface design is crucial to supporting the user.

Automation can create many different modes of operation which are an attempt to constrain the operator's input and to ensure that they do not produce erroneous behaviour. However, the restrictions imposed by different modes can result in operators being confused about what is possible. Mode errors are a frequent and persistent problem identified with automated systems [29]. Mode errors can reflect the imposition of the need to adapt onto the human operator because of the need to support automation. Whilst constraining opportunities for error may seem like an obvious design strategy it may result in a greater burden on the operator, in training to avoid mode errors, through error recovery during use and in adapting their preferred method of working.

This may result in conflict between the automation and the human operator. For this reason it may be inappropriate to make use of mixed modes of operation when introducing automation because indirectly this increases the operator's workload in tracking the mode, as well as increasing the potential for an automation surprise when the operator is trying to discern what the automated system is attempting to do. Prospective memory errors are another route by which human operators omit required responses, delaying or disqualifying responses. Prospective memory errors occur where the operator stores away a delayed response but fails to execute it later, as they think they have. System designers rarely consider that operators can forget in short time intervals even though multi-tasking results in interruption related errors and forgetting. Part of the problem is that operators organise their tasks to be self-supporting and structure them to stimulate information recall and prompt actions. The imposition of automation creates a temporal gap between actions and other information processing, which can result in disruption of prospective memories. Again there are insights that can be transferred from aviation where multi-tasking results in the possibility of prospective memory errors [9. 10]

6. AUTOMATION TRANSITION INDICES

There is guidance that can be given with regard to transitions between automated and manual control, to help identify areas of risk. First, transitions where the operator is required to act quickly are likely to present more significant safety problems because of the deterioration of situation awareness during automated operation. Second, transitions which require acquisition of five or more parameters to make and take effective decisions are problematic as they present a challenge to working memory. Such information acquisition activities may be necessary to re-build situation awareness. Third, complex control sequences that are executed following the decision making are problematic because the preparation time will be greater. Fourth, the need for a speeded response often excludes the possibility of genuine supervision because there is insufficient time to execute recovery actions, in the event of an error. Time and available cognitive resource can be seen as a thread which runs throughout these warning indices and it is part of the Fitt's list, frequently used to guide function allocation. Thus, it would be somewhat ironic to allocate a recovery action requiring a speeded response to a human operator in the event of automation failure. The problems with function allocation, in developing automation, have encouraged some to suggest that designers and users should have the opportunity to refuse new technology in its current form [31].

Some have suggested that Fitt's List has significant value because it is still in use [6]. The same authors have suggested that the ironies of automation were already implied by Fitt's List [15] and for that reason designers could have foreseen the problems before they were realised in actual systems. However, the authors failed to appreciate that increasing automation would result in amplification of the automation failure related problems, a point now accepted to some extent by those reviewing automated systems [23]. In that increasing degrees of automation result in increased risk at transitions back to operator control.

The degree of automation can be described either as increased automation within a specific level or by increased automation to later stages within an overall task [20]. The stages of processing have been divided into four, with information acquisition, information analysis, action-selection and action implementation. It is predicted that routine performance will continuously improve with automation but that non-routine performance, which includes automation failure will decline. The evidence supports the notion that as the degree of automation increases the risk associated with non-routine performance and automation failure increases. As with information processing it is recognised that automation which offers actions as choices may be problematic because the effort of processing action choices helps to better inform decisions. Thus, substitution of human processing with automation has potentially deleterious effects on performance. Collectively the evidence presented raises serious concerns that the risk from automation induced human performance failures may not have peaked. Until a tipping point is reached where automation errors are greater contributors than human actions, or when human contributions to systems are so small, care should be taken in assessing automated system designs. The tipping point is illustrated in Figure 2 where the degree of automation directly contributes to collapse of performance when the switch to manual operation occurs.

7. CONCLUSIONS

Automation is often identified by some as a way of reducing error by excluding the human operator from the control process. It can also be identified with reduced workload and by implication lower manning requirements for systems. However, historically automated systems are not completely autonomous and they are not universal in their functions. Automated systems require human inputs to guide their behaviour and this can introduce error into performance. The absence of autonomy also requires that the human operators, that automation is in part replacing, are required to supervise the automation. However, it is generally agreed that the human operators may not be able to understand the way in which the automation is achieving its goals and human operators may not elect to operate in the same vulnerable areas of the performance envelope which the automation operates within. Thus, whilst automation increases capability by operating in a broader envelope it increases risk which is in part shared with the human operator, especially at times of unexpected events. To some extent there is always a balance between capability and risk because demands for greater capability often incur greater risks and the issue is then managing the risk effectively.

The greatest irony is that some people did foresee these issues with automation. Lisanne Bainbridge [2] was perhaps one of most significant in appreciating that automated systems are not completely autonomous and human operators are required to operate them. Bainbridge's work, like Nancy Leveson's almost two decades later predicted that automation creates new opportunities for human errors. That it has been claimed that the ironies are still going strong after 30 years is perhaps an indication that the problems associated with human-automation interaction have not been successfully resolved [3]. For environments which do not share the momentum and breadth of input that aviation engineering has it is important to learn the lesson before implementing them in systems. That many of the problems are rooted in design supports the view that this is where the greatest change will result in the greatest improvement [32]. Now it seems impossible and unacceptable to say we didn't see these problems coming out of implementing automation.

8. ACKNOWLEDGEMENTS

The authors would like to acknowledge the generous support of BAE Systems in preparing this paper. However, the authors take full responsibility for the contents of the paper and the opinions expressed. The opinions expressed in no way reflect those of BAE Systems.

9. REFERENCES

References should be numbered in the text where they occur, with the number in square brackets [1] and a complete list given at the end of the paper.

1. Adachi, M., Ushio, T., & Ukawa, Y. (2006). Design of user-interface without automation surprises for discrete event systems. *Control Engineering Practice*, *14*(10), 1249-1258.

2. Bainbridge, L. (1983). Ironies of automation. *Automatica*, *19*(6), 775-779.

3. Baxter, G., Rooksby, J., Wang, Y., & Khajeh-Hosseini, A. (2012, August). The ironies of automation: still going strong at 30?. In *Proceedings of the 30th European Conference on Cognitive Ergonomics* (pp. 65-71). ACM.

4. Christoffersen, K., & Woods, D. D. (2002). How to make automated systems team players. In *Advances in human performance and cognitive engineering research* (pp. 1-12). Emerald Group Publishing Limited.

5. Cummings, M. (2004, September). Automation bias in intelligent time critical decision support systems. In *AIAA 1st Intelligent Systems Technical Conference* (p. 6313).

6. de Winter, J. C., & Dodou, D. (2014). Why the Fitts list has persisted throughout the history of function allocation. *Cognition, Technology & Work*, *16*(1), 1-11.

7. Dekker, S., & Hollnagel, E. (2004). Human factors and folk models. *Cognition, Technology & Work*, *6*(2), 79-86.

8. Dekker, S. W., & Woods, D. D. (2002). MABA-MABA or abracadabra? Progress on human–automation co-ordination. *Cognition, Technology & Work*, *4*(4), 240-244.

9. Dismukes, R. K., & Nowinski, J. (2007). Prospective memory, concurrent task management, and pilot error. *Attention: From theory to practice*, 225-236.

10. Dismukes, K. (2006, October). Concurrent task management and prospective memory: pilot error as a model for the vulnerability of experts. In *Proceedings of the Human Factors and Ergonomics Society Annual Meeting* (Vol. 50, No. 9, pp. 909-913). Sage CA: Los Angeles, CA: Sage Publications.

11. Endsley, M. R. (1995). Toward a theory of situation awareness in dynamic systems. *Human Factors: The Journal of the Human Factors and Ergonomics Society*, *37*(1), 32-64.

12. Endsley, M. R., & Garland, D. J. (2000). Theoretical underpinnings of situation awareness: A critical review. *Situation awareness analysis and measurement*, 3-32.

13. Endsley, M. R. (1996). Automation and situation awareness. *Automation and human performance: Theory and applications*, 163-181.

14. Endsley, M. R., & Kiris, E. O. (1995). The out-of-the-loop performance problem and level of control in automation. *Human factors*, *37*(2), 381-394.

15. Fitts PM (ed) (1951) Human engineering for an effective air navigation and traffic control system. National Research Council,

Washington, DC.

16. Klien, G., Woods, D. D., Bradshaw, J. M., Hoffman, R. R., & Feltovich, P. J. (2004). Ten challenges for making automation a" team player" in joint human-agent activity. *IEEE Intelligent Systems*, *19*(6), 91-95.

17. Kragt, H. (1978). Human reliability engineering. *IEEE Transactions on Reliability*, *27*(3), 195-201.

18. Leveson, N. G. (2002). System safety engineering: Back to the future. *Massachusetts Institute of Technology*.

19. Leveson, N. (2004). A new accident model for engineering safer systems. *Safety science*, *42*(4), 237-270.

20. Manzey, D., Reichenbach, J., & Onnasch, L. (2012). Human performance consequences of automated decision aids: The impact of degree of automation and system experience. *Journal of Cognitive Engineering and Decision Making*, *6*(1), 57-87.

21. Mosier, K. L., & Skitka, L. J. (1996). Human decision makers and automated decision aids: Made for each other. *Automation and human performance: Theory and applications*, 201-220.

22. Mosier, K. L., Skitka, L. J., Burdick, M. D., & Heers, S. T. (1996). Automation bias, accountability, and verification behaviors. In *Proceedings of the Human Factors and Ergonomics Society Annual Meeting* (Vol. 40, No. 4, pp. 204-208). Sage CA: Los Angeles, CA: SAGE Publications.

23. Onnasch, L., Wickens, C. D., Li, H., & Manzey, D. (2014). Human performance consequences of stages and levels of automation: An integrated meta-analysis. *Human Factors*, *56*(3), 476-488.

24. Parasuraman, R., Molloy, R., & Singh, I. L. (1993). Performance consequences of automation-induced 'complacency'. *The International Journal of Aviation Psychology*, *3*(1), 1-23.

25. Parasuraman, R., & Riley, V. (1997). Humans and automation: Use, misuse, disuse, abuse. *Human Factors: The Journal of the Human Factors and Ergonomics Society*, *39*(2), 230-253.

26. Parasuraman, R., Sheridan, T. B., & Wickens, C. D. (2008). Situation awareness, mental workload, and trust in automation: Viable, empirically supported cognitive engineering constructs. *Journal of Cognitive Engineering and Decision Making*, *2*(2), 140-160.

27. Ritchie, D. (2017). Past, Present and Future Submarine Manoeuvring Control. Paper presented at RINA Warship 2017: Naval submarines and UUVs, 14th-15th June 2017.

28. Rushby, J. (2002). Using model checking to help discover mode confusions and other automation surprises. *Reliability Engineering & System Safety*, *75*(2), 167-177.

29. Sarter, N. (2008). Investigating mode errors on automated flight decks: Illustrating the problem-driven, cumulative, and interdisciplinary nature of human factors research. *Human Factors: The Journal of the Human Factors and Ergonomics Society*, *50*(3), 506-510.

30. Sarter, N. B., Woods, D. D., & Billings, C. E. (1997). Automation surprises. *Handbook of human factors and ergonomics*, *2*, 1926-1943.

31. Sheridan, T. B. (2000). Function allocation: algorithm, alchemy or apostasy?. *International Journal of Human-Computer Studies*, *52*(2), 203-216.

32. Sheridan, T. B., & Nadler, E. D. (2006). A review of human-automation interaction failures and lessons learned. *Volpe National Transportation System Center, Cambridge, MA*.

33. Woods, D. D., & Sarter, N. B. (2000). Learning from automation surprises and "going sour" accidents. *Cognitive engineering in the aviation domain*, 327-353.

LLOYD'S REGISTER CODE FOR UNMANNED MARINE SYSTEMS

Matthew Palmer Lloyds Register EMEA, Marine & Offshore, UK

SUMMARY

Lloyd's Register has created a goal-based code for the certification of Unmanned Marine Systems (UMS) – including Unmanned Underwater Vehicles (UUVs). This has come about through Lloyd's Registers work with the marine industry to assess the applicability of existing regulatory regimes to unmanned vessels and to determine appropriate alternatives. As a result the LR Unmanned Marine Systems Code (the 'Code') was created to provide a safety framework to capture the novel design aspects of these vessels, in a goal-based format.

The use of a goal-based format allows the Code to capture the rapid advances in technology of unmanned vessels in an environment where prescriptive requirements do not exist or are unsuitable. Compliance is achieved through the use of appropriate standards or where these do not exist, by risk-based analysis methods.

This paper explains how the Code was developed, outlines its application and how it can be applied to the certification of UUV's. The paper will also show how the Code is designed to assist Naval Administrations in supporting their safety case arguments in order to certify these vessels.

NOMENCLATURE

ANEP	Allied Naval Engineering Publications
COLREGS	International Regulations for Preventing Collisions at Sea
ConOps	Concept of Operations
IMO	International Maritime Organisation
LR	Lloyd's Register
MASRWG	Marine Autonomous Systems Regulatory Working Group
MARPOL	Marine Pollution Convention
MCA	Maritime and Coastguard Agency
SOLAS	Safety of Life at Sea
TQ	Technology Qualification
TRL	Technology Readiness Level
UMS	Unmanned Marine Systems
UUV	Unmanned Underwater Vessel

1. INTRODUCTION

The autonomous vessel's sector has developed rapidly over the past few years, and it is expected that Unmanned Marine Systems (UMS) will soon be in widespread use in many aspects of the maritime industry. Lloyd's Register has been active in this area in understanding how to best serve this new and expanding technology in the marine area through the certification of these UMS.

The current maritime regulatory framework is developed on the basis of mitigating the risks for vessels that are manned. The risks for the operation of UMS differ materially from those for a manned vessel and so any equivalent regulatory framework developed for UMS needs to take this into account.

It was concluded that a simple adaption of Classification Society Rules within the existing regulatory framework would not meet the requirements of the Regulator, nor Owners and Operators, for the range of vessels envisaged, and that a different approach would be needed for the assurance of UMS. Due to the rapid advances in technology prescriptive rules would not be able to keep up with industry and their limited scope would be too narrow to cover most UMS.

Lloyd's Register determined that the best way to provide assurance to this sector was through a goal-based code and consequently has developed the LR Code for Unmanned Marine Systems (the 'Code') [1] to provide a framework for the assurance of safety, and mitigation of risk for UMS. This LR Code for Unmanned Marine Systems is written to work alongside evolving requirements for Safe Operation and Environmental Protection, which would complete a Regulatory framework for UMS that provides equivalence to that for manned vessels.

The LR Code for Unmanned Marine Systems is written to be applicable to all autonomous and remote controlled UMS operating above and below the surface. The application of the Code is dependent upon the owner defining their operational requirements and risk, which will then determine the verification requirements through the 'Level of Integrity' process which Lloyds Register has developed as one of the features of this Code. The scope is wider than that currently covered by Classification Rules as it is written to cover all areas of SOLAS that are applicable to UMS.

As a part of the development process Lloyd's Register has worked closely with industry and participants in the UK Marine Autonomous Systems Regulatory Working Group (MASRWG). This group is made up of stakeholders from industry working together to create a regulatory framework for UMS equivalent to the current regulatory framework for manned vessels. The UMS Regulatory Framework is shown in Figure 1; the Lloyd's Register Code for Unmanned Marine Systems fits into the 'Safe Design' category.

2. CURRENT REGULATORY FRAMEWORK

Looking at the regulatory framework that has evolved for manned vessels, and with reference to Figure 1; regulatory requirements may be broadly broken down into those covering; safe design, environmental protection, and safe operation.

The main source of requirements for the safe design of commercial ships is the International Maritime Organisation (IMO). The IMO is in turn dependent upon the application of predominantly prescriptive requirements defined within International Conventions, Classification Society Rules and Regulations, and International and National Codes and Standards. Similarly the main requirements for environmental protection are defined within MARPOL and requirements for safe operation within COLREGS, STCW and the ISM Code - also IMO Instruments. All of these elements work together to provide a 'safe system'.

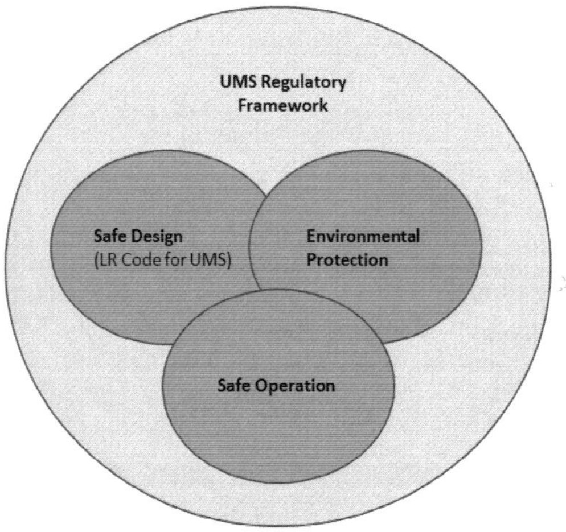

Fig. 1: UMS Regulatory Framework

The traditional role for a Classification Society is to provide third-party independent verification to give assurance to the Owner/Regulator that the vessel as designed and built is safe. Whether this is for hull and machinery as covered by Classification; or stability, fire, escape and evacuation which are carried out against SOLAS requirements under delegation from a Flag State.

When looking at the regulatory regime for commercial ships, it was agreed that for UMS the whole of safe design needed to be incorporated to show equivalence to the existing statutory regime. In the absence of a UMS 'friendly' regulatory regime, 'equivalence' only can be achieved as existing requirements are predicated towards both protecting persons onboard but also requiring their presence or action. For example the provision of life saving appliances; and that some requirements within SOLAS are based upon the availability of crew to mitigate risk.

The Lloyd's Register Code for Unmanned Marine Systems is intended to cover safe design in its entirety; therefore we must establish what the risks are from the operation of UMS and include these in a framework which can then ensure they are adequately mitigated. The ultimate goal in writing the Lloyd's Register Code for Unmanned Marine Systems is to be able to assure the safe design, build and maintenance of UMS against an established framework that minimises the effort required by an owner or operator to achieve certification.

Safe operation and environmental protection are outside the scope of the Code. The requirements for safe operation of UMS are currently being reviewed by the UK Marine Industry Alliance, which is a group of key UK industry representatives established to develop the UK position on regulation and which has the support of the MCA. Environmental protection has not yet been reviewed in detail but it is considered that the requirements of MARPOL should be broadly applicable as currently written. Some interpretation may be required which can be addressed by a separate environmental risk assessment.

By studying the current regulatory regime and looking at what industry wanted, it was decided that the Lloyd's Register Code for Unmanned Marine Systems would have to be equivalent to Statutory Codes and sit above traditional Classification Rules as shown in Figure 2.

3. STRUCTURE OF THE CODE

3.1 WHY GOAL-BASED?

As previously described the Lloyd's Register Code for Unmanned Marine Systems is a goal-based code. This is due to the rapid advances in autonomous and remote control technology for which a prescriptive set of rules would not have the scope to cover the wide range of vessels that are currently out there; and which would become outdated in a very short space of time.

Lloyd's Register has experience in both developing and applying goal-based codes as it is one of the founding members of the International Naval Safety Association (INSA) authors of the Naval Ship Code - ANEP 77 [4] and the Naval Submarine Code. It was found that SOLAS is not appropriate for many naval ships because their military role demands design and operating solutions that are not compatible with the provisions of SOLAS. The Naval Ship Code was produced to show equivalence for IMO conventions by providing Naval Administrations a similar to structure to what is found in the commercial world to demonstrate that their ships and submarines are safe to operate.

The Naval Ship Code is being used by navies worldwide and the Naval Submarine Code is currently in the process of being adopted by navies. This experience of creating goal-based codes was used in the creation of the Lloyd's Register Code for Unmanned Marine Systems.

3.2 STRUCTURE OF THE LLOYD'S REGISTER CODE FOR UNMANNED MARINE SYSTEMS

The Lloyd's Register Code for Unmanned Marine Systems has been written using a goal-based structure, providing a minimum level of safety for UMS and a method to define and assess against the owner's or operator's declared tolerability of risk for loss or damage to its asset. The benefits to the designer of using the goal-based structure are that it defines an ultimate safety objective whilst allowing for the consideration of alternative designs and solutions that meet the safety objective; thereby supporting innovation in an area that is developing rapidly.

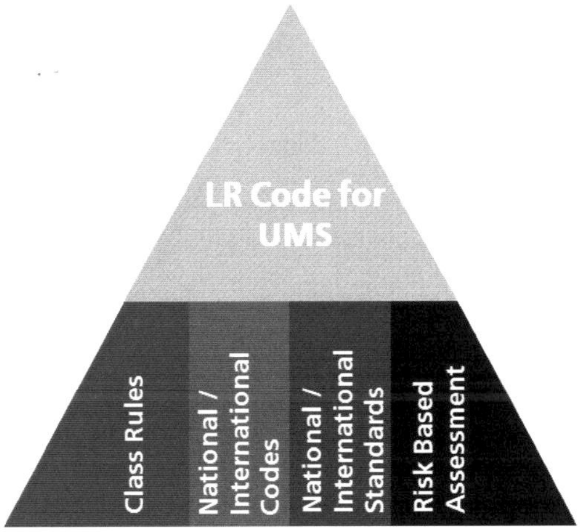

Fig. 2 How different solutions can satisfy the Codes Performance Requirements

The structure of the Code follows the goal-based triangle structure which is also seen in the Naval Ship and Submarine Codes. With reference to Figure 3; the tiers of the Lloyd's Register Code for Unmanned Marine Systems are as follows, with the increasing width of lower tiers representing increased level of detail:

- Tier 0 is the overall Aim of the Code which is: "The Unmanned Marine System (UMS) shall be safe, dependable, capable and resilient in all Reasonably Foreseeable Operating Conditions."

- Tier 1 defines high level Goals for each hazard area, and these form the chapters, of the Code. These Goals align to support the overall Aim.

- Tier 2 Functional Objectives identify the functions to be satisfied in order to meet the Tier 1 Goals for specific functions.

- Tier 3 Performance Requirements define the detailed requirements to meet Tier 2 Functional Objectives and are qualitative to allow them to be met by a range of solutions.

- Tier 4 Solutions will describe the design solution, how it is to be verified and will need to be justified against the Tier 3 Performance Requirements to show that they are adequate

Tier 4 solutions are not defined within the Code. This is because of the potential diversity in design and operation of UMS. The Code expects an appropriate solution be selected from a combination of: Classification Rules, National or International Codes and Standards or where standards do not exist or are not appropriate, through the application of risk-based assessment methodologies that demonstrate the performance requirements are achieved. This *tailoring* of solutions is shown in Figure 2.

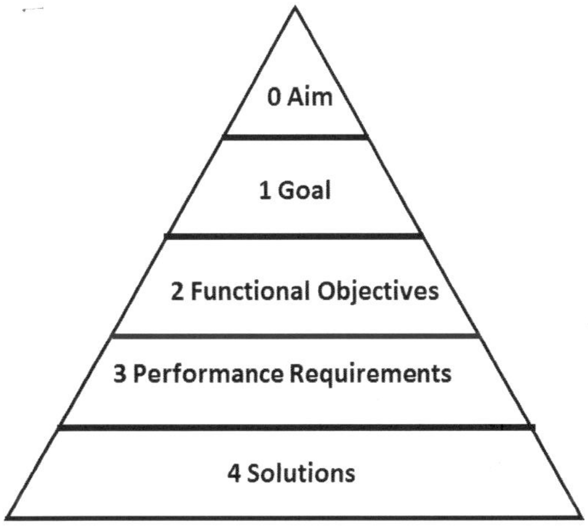

Fig.3 Lloyd's Register Code for Unmanned Marine Systems Structure

3.3 SCOPE OF THE LLOYD'S REGISTER CODE FOR UNMANNED MARINE SYSTEMS

The Lloyd's Register Code for Unmanned Marine Systems will cover all aspects of safe design for unnamed vessels and remote controlled vessels operated on or below the surface. It is not applicable to remotely operated underwater vehicles tethered to a mothership (ROVs). This limitation is due to the presence of well-established regulatory regimes covering the safe design, operation and environmental requirements for these vessels.

The Code does not cover the risks resulting from embarked cargo or mission specific equipment; and whilst it is designed to support safe operation and maintenance it does not address Operator training and qualification (Safe Operation).

One of the dangers of focussing on the autonomous elements of unmanned vessels is that the human element in the systems can be overlooked, just because the vessel is autonomous doesn't mean that the vessel does not interact with humans. In this regard where the vessel will also be manned then it must also comply with the relevant manned Regulations. The human element is discussed in more detail in Section 7 of this paper.

The Goals, Functional Objectives and Performance requirements have been divided into the following key hazard areas forming the main technical chapters of the Code:

- Structure
- Stability
- Control System
- Electrical System
- Navigation System
- Propulsion and Manoeurving
- Fire
- Auxiliary Systems

The above hazard areas are common to all types of UMS. The scope and boundary of each hazard area is defined together with its interaction with other hazard areas to provide clarity of what is to be covered when carrying out an assessment.

The Functional Objectives and Performance Requirements are written to cover identified hazards relating to the operation of UMS. To ensure completeness, these have been validated against existing Conventions, Codes and Standards including; Classification Rules and Regulations, SOLAS, and ANEP77 and also by engagement with existing UMS projects.

As previously mentioned the Lloyd's Register Code for Unmanned Marine Systems only covers the UMS and does not cover any safety, environmental and commercial risks resulting from embarked cargo or mission specific equipment. These can be covered through separate hazard identification and risk assessment techniques.

3.4 LEVEL OF INTEGRITY

In order to apply the Code, a minimum level of integrity for safety is defined for each of the hazard areas. The required level of integrity is determined for each of the UMS systems. The level of integrity is to be determined by assessing the effect on the UMS, considered as a system of systems, of all reasonably foreseeable system failures and by considering their consequence on the ability of the UMS to achieve the Aim of the Code.

The level of integrity is broken down into the following areas:

System Safety Consequences which consider reasonably foreseeable system failures and the consequences of failure in the following areas:

- To people onboard
- To people/objects in the vicinity
- To the environment

System Operational Consequence which considers the effects of failure has on the UMS in the following areas.

- To capability
- To resilience

The levels of integrity associated with the system operational consequences are determined with reference to the Concept of Operations and the design requirements for the UMS and need to be agreed between the Owner and Lloyd's Register.

This process is conducted for each system of the UMS and these results in a Safety Levels of Integrity (SLoI) and Operational Levels of Integrity (OLoI) which shall be categorised as:

(a) High; a LOI for which the consequence of system failure on the achievement of the aim is not acceptable
(b) Medium; a LOI for which the consequence of system failure on the achievement of the aim is acceptable subject to the presence of mitigating factors; or
(c) Low; a LOI for which the consequence of system failure on the achievement of the aim is acceptable

The highest Level of Integrity for each system is then used to define the verification activities for that system

If it is not possible for a designed system to achieve the required Level of Integrity, or as an alternative to the above method for the determination of Level of Integrity, a full hazard analysis shall be carried out. The details of the risk assessment process are discussed in Section 5 of this paper.

3.5 APPLICATION OF THE UMS CODE

The application of the Lloyd's Register Code for Unmanned Marine Systems will be dependent upon how the UMS is operated and the configuration of the design solution. As an example, the hazards to other vessels will not be as significant for smaller vessels as for larger vessels. In order to understand this a Concept of Operations (ConOps) has been included as a key criterion of the Code. The ConOps is the key to the whole certification process as it bounds the UMS and provides a clear description of the roles, environment and capabilities expected of the UMS. Also the ConOps supports Lloyd's Register as the process only takes into account the bounding requirements in the ConOps and if the UMS is used differently to what is laid down in the ConOps this can invalidate the Certification of the UMS.

An example of this would be small coastal UMS being used in the middle of the Atlantic Ocean and suffering a failure in an area in which it was not designed to operate in.

As the Lloyd's Register Code for Unmanned Marine Systems is designed to be use for a wide variety of surface and sub-surface UMS, it is expected that solutions will be different for each type of UMS. Therefore the solutions for smaller UMS may be derived from tried and tested equipment manufactured to recognised standards, whereas for larger vessels it would be expected to make greater use of the requirements and processes defined in Classification Rules and international and national standards. Likewise the maintenance and survey regime for equipment will be proportional to the threat posed by the failure of the equipment.

For areas of the design where there are no recognised standards to apply, which will be more prevalent for UMS due to the fact they encompass a lot of novel technologies. Or where equipment is being used in an environment for which it has not been previously demonstrated, it is expected that recognised risk assessment methods are used to justify acceptance of the design solution. This practice is already applied in the Technical Qualification process which is explained in a Section 5.

4. LLOYD'S REGISTER CODE FOR UNMANNED MARINE SYSTEMS APPLICATION FOR UUVS

4.1 LLOYD'S REGISTER CODE FOR UNMANNED MARINE SYSTEMS TRIAL

Part of the development process for the Lloyd's Register Code for Unmanned Marine Systems was to apply it to a current UMS project. The aim being to both calibrate the Code against an actual UMS and also to assist the project in highlighting any areas that may need to be covered in more detail to meet the Code requirements.

The UMS in the study was, intended for use in either a manned mode or an unmanned mode for military operations. During the assessment over 160 Performance Requirements were tested, covering all chapters of the Code; this was run as a desk-top exercise designed to stress-test the content of the UMS but also the Code.

The study required the creation of a body of evidence by the project team; including Manufacturers Declarations, Class Certificates and construction information; which was examined as a part of the process to show compliance against the Lloyd's Register Code for Unmanned Marine Systems Performance Requirements.

The outcome of this assessment was ultimately successful for both parties, however did lead to the identification of several key findings which are presented below.

Currently the Lloyd's Register Code for Unmanned Marine Systems is being applied to an UMS which has just completed the trials phase and it is expected that that the process will be complete later in 2017.

4.2 OUTCOMES

Following the assessment minor edits were made to the Lloyd's Register Code for Unmanned Marine Systems requirements, however in particular several key actions were identified:

The human element should be considered early in the creation of the ConOps and then throughout the system design process. Lloyd's Register has already commenced work on an additional Chapter to the Code looking specifically at the interaction of UMS with humans. Whilst not intended to replace 'manned' regulation this will ensure that human element requirements specific to UMS operation are considered in the design.

As an extension of the assessment the next stage of development for the Code is to generate a Verification Procedure which supports the application of the Performance Requirements via Solutions against each of the Code Chapters. This will be developed through Lloyd's Register's experience with certification of small boats and the In-service Classification cycle which is used for commercial ships.

For areas that cannot be satisfied through the evidence basis approach taken above a different approach is required. A risk based approach in the form of the Technology Qualification could be used and this is discussed in the next section.

5. RISK ASSESSMENT METHODS

5.1 TECHNICAL QUALIFICATION PROCESS

The Technology Qualification process is key tool in evaluating the maturity of the systems and components of the UMS and informs the verification activities that need to be undertaken to assure those system and components.

The Technology Qualification process (TQ) [3] has been developed by Lloyd's Register to provide assurance of novel technologies for which standards do not exist. It is a goal risk based process that uses a readiness level framework, total system perspective and whole lifecycle approach to qualify new technologies, unconventional designs and new ways of applying existing technology. Arguments and evidence are based where possible by the application of existing standards which for the case of a UMS may include Classification Rules. For elements of systems that are novel, a risk assessment is carried out and qualification based upon a combination of analysis, simulation and testing.

The TQ methodology is developed with specific acceptance criteria for the technology under review. This is focused on, but not limited to, the following areas:

- Safety
- Environmental
- Functionality
- Performance
- Reliability
- Availability

The TQ process covers the following activities:

- Identifying technology elements and their readiness levels
- Assessing risks posed by the elements
- Establishing acceptance criteria and goals
- Evidence Collection
- Further analysis and testing including test failure analysis
- Performance review

5.2 Technical Qualification Stages

The TQ process is divided into the three stages shown in Figure 3.

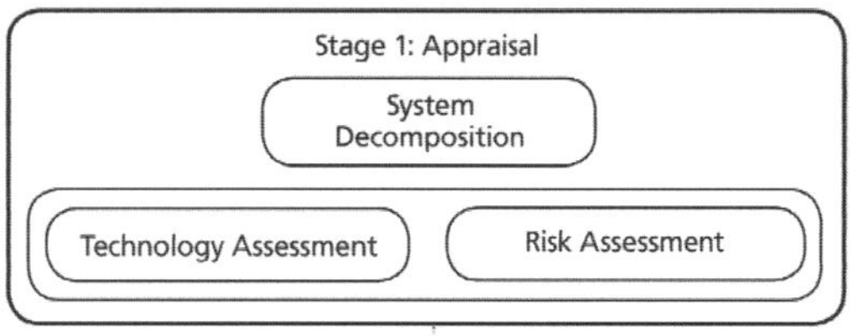

Fig. 4 Technical Qualification Process Stages

Appraisal comprises of three steps and these are carried out as separate staged activities.

- System Decomposition
- Technology Assessment
- Risk Assessment

5.2(a) System Decomposition

A system is defined as a group of interacting elements or technologies having a functional relationship that when grouped or integrated, provide some processes or service. System performance is often as strong as its weakest link so all technology elements within the systems of the UMS are broken out and goals are assigned to the sub-system. The resulting System Decomposition is the structural and functional breakdown of the overall system into subsystems, equipment, and components down to a level at which the component or element can be assessed. The decomposition process should start with the functions of the UMS before breaking them down into further detail.

The outcome of the System Decomposition provides the basis for review and assessment of the UMS. It defines the initial scope for the Technology Qualification review, and also identifies the document system, operating conditions, functions and sub-functions requirements, process sequences and operations and capabilities with respect to qualitative and quantitative claims and goals. It is important to note the System Decomposition process is an initial step and should not limit the review when issues and risks emerge later on.

The following examples provide guidance to what level the decomposition is carried out:

Example 1: Fire System

System	Subsystem	Elements
Fire System	Engine room self-contained extinguisher	extinguisher
		piping
		detector

For the fire system the extinguisher is provided by the supplier as a complete packaged system that is again built and tested to a recognised standard and therefore does not need to be broken down any further.

Example 2: Fuel System

System	Subsystem	Elements
Fuel System	Wing tanks	Tank
		Level sensor
	Forward tank	Tank
		Level sensor
	Fuel distribution	Pipes
		Transfer pump
		Solenoid valves
		Fuel flow meters
		Fuel filters
		Water detection modules
		Pipe clips/mounts
		Pipe fittings

Looking at the fuel distribution system, this has been further broken down to components such as pipes and transfer pump. These components are supplied by a manufacturer which has designed and built the component to a standard or carried out testing to demonstrate its performance therefore it is not necessary to decompose any further.

5.2(b) Technology Assessment

The next step within the Appraisal stage of the Technology Assessment, considers the Technology Readiness Level (TRL), the experience of use in the defined operating environment and the integration of the technology or items of equipment.

All elements of subsystems identified at the decomposition stage are screened by their TRL and experience in the defined operating environment and assigned an uncertainty rating from the screening matrix which is colour coded Red, Green or Yellow, see Figure 5.

This initial screening enables the quick elimination of proven technology elements in that elements in the Green category are considered to not be introducing any new technical uncertainties being subject to established codes, standards and manufacturing processes.

The elements in the Red category are considered to be introducing a high level of new technical uncertainty and are therefore subject to a high level of qualification review which may include; testing, risk mitigation, and design improvements. Elements in the Yellow category, presenting a lower level of uncertainty will be subject to a subset of qualifications expected for the red category.

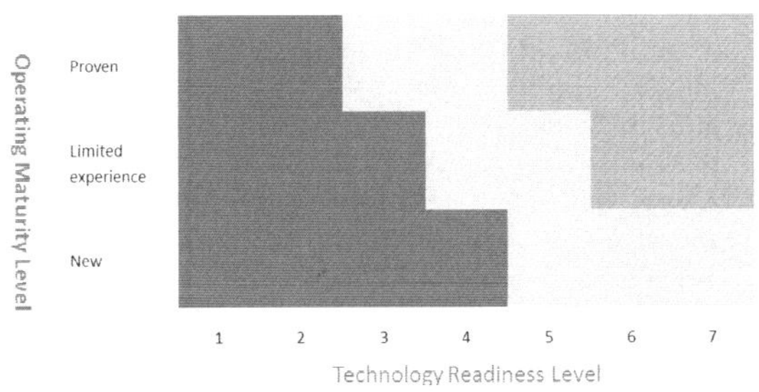

Fig. 5 Initial Screening Matrix

Looking at the system decomposition example for the fuel system given previously, the initial screening is as follows:

Example 1; The transfer pump for the fuel distribution system may be from a manufacturer of pumps designed and manufactured specifically for operation in fuel systems in the marine environment, with a number of years of operational experience. This item of equipment would therefore be assigned a TRL of 7, and its level of experience proven. On this basis it has an uncertainty level of green and further assessment is not required.

Example 2; The fuel flow meters may be from a manufacturer of fuel flow meters that have extensive non-marine operational experience but have not been used in the marine environment before. This item of equipment would therefore again be assigned a TRL of 7, but its level of experience New. On this basis it has an uncertainty of Yellow and further assessment is required.

5.2(c) Risk Assessment

For each of the technology elements that have not been eliminated in the screening process because they are mature and proven in the marine environment (green), a risk assessment is required. This process is carried out to ensure that all reasonably foreseeable hazards associated with a particular technology are identified and adequately controlled. This risk assessment is carried out using standard methods such as those defined in ISO and IEC standards.

Each technology element is identified and reviewed individually together with its interactions. Recommendations for qualifications or mitigations are identified, recorded in a manner which is to be traceable throughout the qualification process. It is an iterative process and continues until all mitigation and qualification activities are satisfactorily identified. It is possible that the list of technology elements and/or their screening assessments identified at earlier qualification steps might be adjusted during the detailed risk assessment steps. The assessment process aims to ensure that all reasonably foreseeable hazards associated with a particular technology are identified and adequately controlled, such as hazards related to novel aspects of system integrity failure, and control/isolation failure.

Returning to the example case for the fuel flow meters, which are part of the fuel distribution subsystem; if the failure of the fuel flow meter were to be assessed to result in the ultimate failure of the fuel system and so loss of power, this will result in a loss of control of the UMS. Depending on the acceptance criteria this may be deemed unacceptable and so action required to mitigate against this hazard.

5.3. VERIFICATION PLAN

At the end of the TQ process a Verification Plan is created to prescribe a method by which the Tier 3 Performance Requirements will be met, including the details and justification of the Tier 4 Solutions, the method of assessment of these and the means of demonstrating conformance.

The verification method is determined by relevant Level of Integrity for each system, as discussed previously in this paper. Verification Methods may include but restricted to the following:

- ConOps
- design review,
- independent calculation,
- equipment and materials certification,
- Details of software integrity testing and cyber-security audits;
- audit,
- inspection,
- survey,
- testing and trials.

6. SIMULATOR TRIALS AND SEA TRIALS

One of the key areas of the certification process is the trials phase. This is essential to provide confidence that the UMS will perform in its intended environment for all reasonable foreseeable conditions as defined in the ConOps.

Lloyd's Register has been involved in the MAXMAS Project. The goal of this project was to develop a method to interpret COLREGs through a compliant path planning system for autonomous vessels. This involved simulator testing to prove that the software worked for a variety of scenarios which was repeated many times to provide assurance that the software worked.

Simulators trials are a very good way of proving that a system does what it says it will do in a stable environment but real-life operations add extra variables that cannot be accounted for in a simulator – most specifically environmental conditions.

One of the keys areas of the sea trials is if the UMS has a problem how does it fail and does it fail to a safe state. In certain circumstances you do not want the UMS to fail in the same way every time. An example of this would be that one failure state would lead the UMS to stop and turn off all propulsion systems. In certain operating areas is this the safest way to fail, but there is a danger that the UMS can drift and the chance of causing increased harm, to human, environment or property. Therefore the failure states of the UMS are captured in the ConOps and allow the correct level of integrity to be assigned to each of the UMS safety systems.

The Lloyd's Register Code for Unmanned Marine Systems requires the UMS to undertake simulated trials and sea trials to see how the system performs.

7. THE HUMAN ELEMENT

The reason why Lloyd's Register uses the term UMS is because the human is one of the most import elements and cannot be ignored in the certification of the system. Humans appear frequently in the design of the UMS even though by definition they are unmanned or autonomous. During the assessment the role of the human in the safe design and in particular safe operation of the UMS became more prominent.

Even if the UMS is not designed to periodically carry people onboard, people will still be required to interact with the UMS on an occasional basis. For example: to carry out maintenance; repairs; data recovery and recovery of the UUV; or, in the event of failure, to retrieve or restart the UMS at sea. As such UMS design still needs to consider interactions with humans at certain points.

The interaction between persons onboard operating the vessel and person's in-the-loop acting as the remote supervisor. A chain of command process needs to ensure that one operator does not cancel out the actions of another operator.

The presentation of information to the operator is another key feature that is covered by the Code. The Code requires the clear presentation of all information for the safety of the operations of the UMS. There have been many instances of human operators being overloaded with information and not being able to process the information that is being presented to them. The human in the loop needs to be able know when the UMS is not responding correctly without getting swamped with information that cannot be processed in time to make a safe decision.

A lot of UMS are operated from consoles which can be tailored to show the screens that Operator wants to see; however it is important that the operator is always provided with information to operate/supervise the UMS safely.

One of the possible solutions to this is the introduction of the 'S-Mode' or Standard Mode. The reason for a common user interface is as different manufacturers are creating different user interfaces and this can become confusing for operators which in turn could lead to accidents. The human factors area of the Code will be used to assess the control system layout for the information that is presented to the operator.

8. CONCLUSIONS

This paper discusses the issues with trying to assure a rapidly evolving technology in which a Marine industry Regulatory Framework has been designed around human not onboard the ships and submarines which they are controlling or supervising. Lloyd's Register has found that prescriptive solutions are at a disadvantage in trying to apply assurance to UMS - by the time a new rule set has been approved industry has moved on. To enable Lloyd's Register to provide assurance to UMS we have developed a goal-based Code which is broken down into functional areas which are applicable to all types of UMS - surface or sub surface.

The key feature of the Code is its flexibility to be tailored to different types of UMS through bounding the scope with a ConOps and identifying the correct solutions to satisfy the Performance Requirements of the Code with various risk based techniques including the TQ process.

As the UMS industry evolves the Lloyd's Register Code for Unmanned Marine Systems will develop with it.

The Lloyd's Register Code for Unmanned Marine Systems provides a structured approach to assessing the design of UMS against operational and safety related goals and to provide assurance for the safe, dependable, capable and resilient design of UMS in all reasonably foreseeable operating conditions. The use of a goal-based format allows the Code to capture the rapid advances of technology in unmanned vessels in an environment where prescriptive requirements do not exist or are unsuitable. Compliance is achieved through the use of appropriate standards or where these do not exist, by risk-based methods.

A copy of the Lloyd's Register Code for Unmanned Marine Systems can be made available on request by contacting the authors at:

Email: mod-office@lr.org

Website: www.lr.org/marine

Tel: +44 (0)1275 515000

9. REFERENCES

1. Lloyd's Register's Code for Unmanned Marine Systems, February 2017

2. Jesus Mediavilla Varas, Spyros Hirdaris, Renny Smith, Paolo Scialla, Walter Caharija MAXCMAS project. Autonomous COLREGs compliant ship navigation

3. Lloyd's Register's Guidance Notes for Technology Qualifications, December 2014.

4. NATO, Naval Ship Code, ANEP 77, Edition F Version 1, dated Aug 2014.

10. AUTHORS BIOGRAPHY

Matthew Palmer BEng CEng MIMechE holds the current position of Senior Specialist in Lloyd's Register's Naval Liaison Office. Matthew works on developing Goal-based Standards and Rules and is contributing to the development of the Lloyd's Register Code for Unmanned Marine Systems for UMS. Matthew is also the part of the Secretariat for the INSA Naval Ship and Submarine Code. Matthew has been leading the industry engagement, particularly in the coordination and delivery of the two trial projects run as a part of the Lloyd's Register Code for Unmanned Marine Systems development.

PAST, PRESENT AND FUTURE IN SUBMARINE MANOEUVRING CONTROL

D Ritchie, **A Cowley** BAE Systems, UK and **S Russell**, MOD, UK

SUMMARY

The evolution of Submarine manoeuvring control systems is driven by a need for improving the performance and safety of platforms.

Since the end of the Second World War the aircraft industry has largely been seen as the 'early adopter' of complex vehicle control systems driven by its own needs to drive for performance, safety and the necessity of automation to increase safety and reliability whilst reducing crew numbers and operating cost. Submarines have typically lagged the aircraft industry in their adoption of new technologies in this area, and so the aircraft industry therefore provides an interesting portal into the future.

This paper aims to examine the trends of functionality of UK Submarine manoeuvering control systems in terms of performance, safety and cost, by looking back at legacy mechanical systems, the progression into digital control and its evolution to the current 'state of the art'. It will conclude by presenting a vision for an idealised future Submarine Manoeuvring Control System and outline the technological enablers and challenges required to realise it.

NOMENCLATURE

GPS	Global Positioning System
HMI	Human Machine Interface
SSN	Nuclear powered Attack Submarine
SSK	Diesel powered Attack Submarine
SIL	Safety Integrity Level
SOE	Safe Operating Envelope
UUV	Unmanned Underwater Vehicle
WECDIS	Warship Electronic Chart Display Information System

1. INTRODUCTION

From a high level viewpoint the Command responsibilities of a modern submarine have changed little in the last century. Intrinsically the Submarine command Team use sensors and other information sources to develop and maintain a picture (the plot) of their physical and tactical surrounding. This plot allows the Submarine to be effectively manoeuvred to a position where it can affect the situation with its mission capability.

Over the last century, the technology used to sense its the environment and control submarines has evolved. Whilst some of these functions have changed beyond all recognition, others have changed surprising little.

The development of air vehicles over the last century provides a useful benchmark to compare and contrast the relative development of submarine technology. Military aircraft like submarines are vehicles required to work in a hostile environment, manoeuvring with six degrees of freedom whilst analysing large amounts of sensor data to achieve tactical advantage.

2. SUBMARINE CONTROL MECHANISMS

2.1 OVERVIEW

Submarine control is affected through three key mechanisms; the requirement to manage these three functions over the history of Submarines will be investigated later in this paper

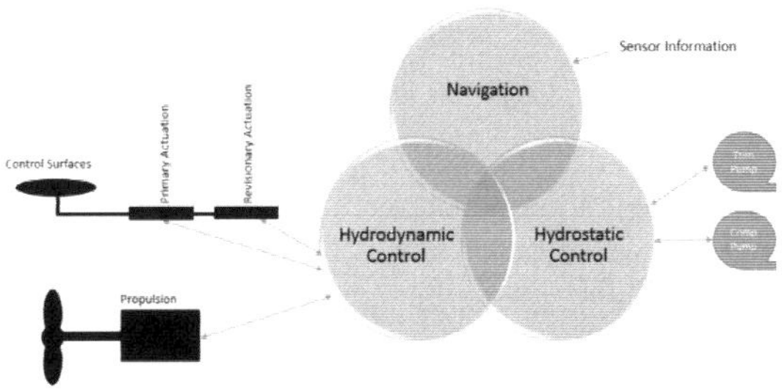

Figure 1 Control System Overview

2.2 HYDROSTATIC CONTROL

Hydrostatic control is undertaken by maintaining the ‘trim of the boat’, managing the ‘balance’ between the mass centroid and the centroid of the buoyant volume within an acceptable ‘out of trim’ margin. Virtually all military Submarines ‘trim’ and ‘compensate’ by using pumping systems to effect a moment change using seawater within an internal tank system (trim) or effect a Mass compensation by ballasting or deballasting water, exactly the same principles as in use on Holland 1 in 1903

Hydrostatic control is generally undertaken in response to bodily weight or buoyancy changes for example use of consumables or changes in buoyancy of the Submarine with depth (compressibility / seawater density).

Historically the increase in density with depth has been broadly proportional to the compressibility with depth, therefore little compensation was required. The advent of more compressible hull coatings and sonar suites has in recent times led to a greater imbalance, requiring compensation to be an essential part of any depth change.

Hydrostatic balance can be used as a fundamental safety argument. If a perfectly 'in trim' Submarine loses propulsion the forces acting on it will be small allowing time for the crew to regain the situation.

2.3 HYDRODYNAMIC CONTROL

Hydrodynamic control is undertaken using forces linked to the movement of the platform, typically main propulsion and hydroplane lift, and occasionally vectored thrusting. The submarine's main propulsion is capable of producing significant force to drive the platform forward, the hydroplane system utilises the flow generated over the control surfaces to generate forces proportional to both the speed of the boat squared and the angle of plane deflection. Again little has changed with the principle of hydrodynamic control since Holland 1 in 1903.

Hydrodynamic forces have the advantage of being able to be rapidly altered; this is beneficial when forces need to be generated to effect manoeuvres such as heading and depth changes.

It is important to note that the effectiveness of hydrodynamic control surfaces will diminish, as speed tends to zero. The use of thrust generating dynamic control systems is therefore necessary if effective hydrodynamic control is required at very low speeds – e.g. UUV Systems.

Hydrodynamic control can again be used as a fundamental safety argument. Forces generated by the control surfaces are capable of indicating out of trim. If continuity of propulsion can be maintained hydrodynamic forces are capable of counteracting extreme hydrostatic failures such as flooding

There is a symbiotic relationship between hydrostatic and hydrodynamic control on naval submarines. When correctly managed hydrostatic control provides a safeguard against failure of hydrodynamic control mechanisms conversely hydrodynamic control provides both indication of and a safeguard against failure in the hydrostatic balance of the platform.

When control of a Submarine is lost the ultimate safety response is defined by the mantra 'Blow and Go' i.e. 'blow' air into the main ballast tank system (hydrostatic control) and 'go' by opening the throttle to gain speed and control force through the hydroplanes. There is significant advantage from achieving a design which uses the main ballast tank system as defence in depth against manoeuvring control hazards rather than a primary mitigation.

UK Submarines generally observe the key principle of compensating for static forces (e.g. variable load items and compressibility) on the Submarine within the hydrostatic trim and compensation system. Conversely short acting dynamic forces e.g. manoeuvring forces, and high frequency seakeeping environmental forces at PD are compensated with dynamically generated lift from the control surfaces. All rules have their exceptions and surface suction effects because of their large size and long duration are often hydrostatically corrected for to provide a greater operating envelope of lift forces to counter high sea state forces.

2.4 NAVIGATION SYSTEMS

Navigation systems are required to sense and track both the geographical position of the Submarine as well as classifying and tracking vessels and objects in its vicinity both for military and collision avoidance purposes.

Navigating systems can be split into the following categories:

2.4 (a) Guidance Systems

Guidance systems present the Submarine crew with information to support ascertaining their location, typically on a modern submarine an inertial guidance system is used and periodically corrected for drift using fixes from alternative sources like GPS. Traditional navigation data sources such as compass, log and depth sounders and when on the surface Pelorus bearings complement the more modern systems.

2.4 (b) Sensor Systems

Submarines carry numerous powerful sonar systems for the primary purpose of detecting and tracking contacts in the marine environment. From a navigation perspective classification of the range, bearing and vessel type requires virtually the same level of effort as to target it with a weapons system.

Above water sensors include visual masts (through which navigation bearing and vessel classifications can be made) radar and radar warning receivers (which can be used to provide a fix) and GPS antenna.

2.4 (c) Information Management systems

Assembly of all the sources of navigation data into a user friendly form requires an information management system, traditionally this was a paper chart marked up with the relevant information by the Command team - the Plot - however in recent years digital systems such as

WECDIS have started supplementing paper charts to improve functionality.

3. TRENDS IN THE AUTOMATION OF SUBMARINE CONTROL SYSTEMS

As time progresses, Submarine platforms have become increasingly complex both in the number of systems that are being operated but also in the operation of the platform itself to ensure both safety and combat performance.

Cook, Thody and Garrett [5] present the case that automation is often seen as a panacea for handling the control and optimisation of the operation of systems but does not come without significant implementation challenges.

The design of an effective automated control system that meets the SIL required for the safe operation of a nuclear Submarine is a difficult task.

3.1 DRIVERS FOR AUTOMATION

3.1 (a) Submarine Safety

Analysing a list of recent significant publicly reported Submarine accidents [2], and making high level assumptions about the root cause of failure that led to the incident makes for some interesting observations.

Table 1 Submarine Incident since 2000

2000 - Kursk – system failure (release of stored energy)
2001 - Ehime Maru and USS Greeneville collision – Navigation error (human)
2002 - USS Dolphin major flooding and fire – system failure (WTI)
2002 - USS Oklahoma City collision with tanker - Navigation error (human)
2002 - HMS Trafalgar Skye grounding - Navigation error (human)
2003 - HMAS Dechaineux flooding - system failure (WTI)
2003 - Ming 361 accident – system failure (asphyxiation)
2003 - HMS Tireless – Collison with iceberg - Navigation error (human)
2003 - K-159 sinking – system failure (WTI)
2003 - USS Hartford grounding - Navigation error (human)
2004 - HMCS Chicoutimi fire – system failure (WTI)
2005 - USS San Francisco collision with undersea terrain - Navigation error (human)
2005 - AS-28 emergency – entanglement
2005 - USS Philadelphia collision with MV Yasa Aysen - Navigation error (human)
2006 - Daniil Moskovsky fire – system failure (electrical)
2006 - USS Minneapolis-Saint Paul incident – Operational error (human)
2007 - USS Newport News collision with tanker Mogamigawa Navigation error (human)
2007 - HMS Tireless-O2 Candle explosion – system failure (release of stored energy)
2008 - HMS Superb – underwater collision - Navigation error (human)
2008 - Russian K-152 Nerpa gas leak – system failure (asphyxiation)
2008 - HMS Vanguard and Le Triomphant collision - Navigation error
2008 - USS Hartford and USS New Orleans collision - Navigation error (human)
2010 - INS Sindhurakshak fire - system failure (battery)
2010 - INS Shankush incident – Operational error (human)
2010 - HMS Astute grounding - Navigation error (human)
2011 - HMCS Corner Brook grounding - Navigation error / or hydrostatic control (human)
2012 - USS Miami arson – sabotage (human)
2012 - USS Montpelier collision with USS San Jacinto - Navigation error (human)
2013 - INS Sindhurakshak explosion – system failure (release of stored energy)
2013 - Russian K-150 Tomsk fire – fire
2013 - USS Jacksonville Mishap - Navigation error (human)
2015 - HM Submarine and MV Karen - Navigation error (human)
2016 - HMS Ambush collision - Navigation error

Eighteen of these incidents (over 50%) are attributable to navigation or control errors, fortunately none of these incidents led to loss of the submarine involved.

Of these 18 incidents 8 involved collision with another vessel – failure of Navigation function.

All the reported navigation incidents involve US (9), UK (7), Canadian (1) or French (1) platforms, given the high ratio of submarine accidents to operational boats it is probable a great number of navigational or control issues go unreported.

This evidence indicates that advances are still required in the area of Submarine manoeuvring control and navigation and that legacy control systems (and associated integrity) for both hydrodynamic and hydrostatic control will struggle to meet modern safety standards.

3.1 (b) Combat effectiveness

As Submarine design evolves systems will become more complex as functionality is developed. To be effective in combat, greater numbers of and more complex systems are required over time. Complexity of operation can be overcome by the development of automation or the addition of more crew.

It therefore follows that measurement of crew number and platform size over time could be considered as analogous to platform complexity and therefore combat effectiveness.

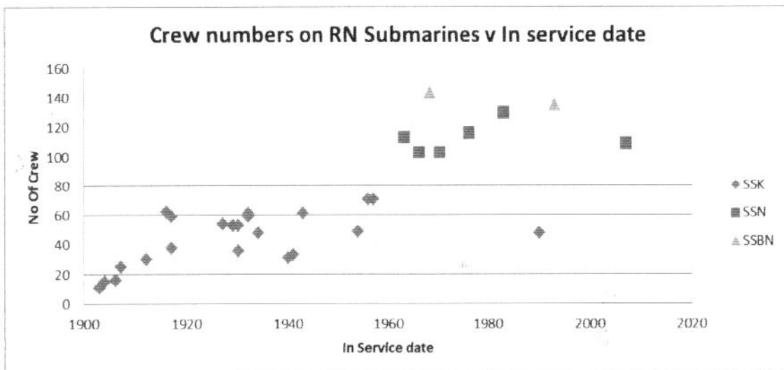

Figure 2 Crew numbers on RN Submarines against in service date

From examination of crew numbers on UK platforms it can be seen that after an initial 20 year development period a trend of approximately 60 crew on SSKs and 115 on SSNs has become a virtually linear trend. This supports the conclusion that automated systems have been developed to maintain similar crew numbers on more complex boats.

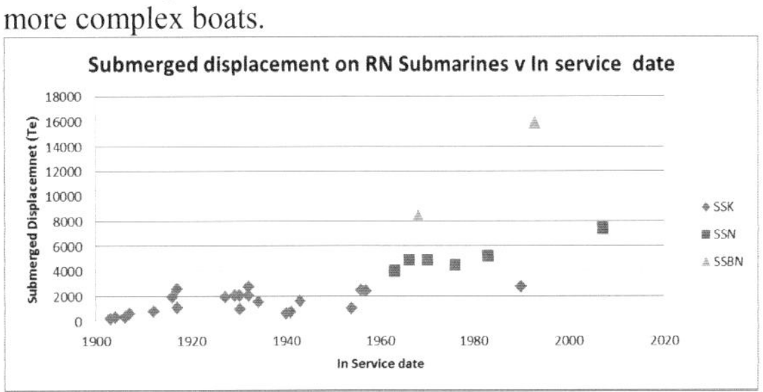

Figure 3 Displacement of RN Submarines against in service date

Analysis of displacement [9] over time shows that since about 1930 the SSK design has matured to a design of approximately 2000 tons, this is in marked contrast to the development of SSN platforms whose displacement has been steadily increasing as capability is added.

The measurement and analysis of the trends in automation of functionality over time could also be used to allude what functionality will be automated in the future. The question is how do we subjectively classify the level of automation being applied to a system?

The design of aircraft control systems – both civil and military has been heavily researched and documented over the time.

As part of the UK DERA Cognitive Cockpit project a policy for pilot authorisation and control of tasks (PACT) was developed by Bonner and Taylor [3] this system has read across to a Submarine manoeuvring and control system. Later in this paper the PACT system is used to classify the capability of past submarine and aircraft designs in order to track the progression of automation into the manoeuvring and control system design over time.

Table 2 PACT automation classification system

Primary Modes	Levels	Operational Relationship	Computer Autonomy	Pilot Authority	Adaptation	Information on Performance
AUTOMATIC	5	**Automatic**	Full	Interrupt	Computer monitored by pilot	On/Off Failure warnings Performance only if
ASSISTED	4	**Direct Support**	Advised action unless revoked	Revoking action	Computer backed by pilot	Feedback on action Alerts and warnings on failure of action
	3	**In Support**	Advice, and if authorised action	Acceptance of advice and authorising action	Pilot backed up by the computer	Feed-forward advice and feedback on action. Alerts and warnings on failure of authorised action.
	2	**Advisory**	Advice, and if authorised action	Acceptance of advice	Pilot assisted by computer	Feed-forward advice
	1	**At Call**	Advice only if requested	Full	Pilot, assisted by computer only when requested	Feed-forward advice, only on request
COMANDED	0	**Under Command**	None	Full	Pilot	None performance is transparent

3.1 (c) Cost Reductions

Automation has been muted as a driver for cost reduction through an associated reduction in crew.

Examining the numbers of personnel on board a Submarine associated with the manoeuvring and control function limits the opportunity for crew reduction to approximately 10% of the compliment. The potential crewing savings have to be offset against the significant cost of developing the solution. The payback period for a highly capable integrated solution would likely be measured in decades. Therefore automation of manoeuvring and control is not considered as a viable cost reduction measure.

3.2 ADOPTION OF AUTOMATION

In general automated systems are highly effective at tasks that can be constrained (i.e. there are a fixed number of permutations of the outcome). Modern data analysis techniques are able to reliably monitor and analyse the vast amounts of the 'big data' available in the platform management and environmental sensors and react at speeds that are unachievable to humans.

Up until very recently humans were believed to have an edge in analysing large scale unconstrained problems (e.g. Military Tactics), as computers struggle with the 'brute force' necessary to compute the number of permutations in the short timescales required. However this situation is rapidly changing.

By way of example, In 1997 a breakthrough in artificial intelligence was widely hailed when the IBM computer 'Deep Blue' defeated Gary Kasparov in the relatively constrained tactical game of chess (around 20 possible moves in each stage of the game) in January 2017 [6] the Google 'AlphaGo' beat expert predictions by ten years by becoming the first computer to defeat a grand champion of the Chinese game of 'Go'. 'Go' is considered to be the most complex game that can be played by humans with approximately 200 possible moves for each stage in the game and more configurations of the board than atoms in the universe.

It is therefore logical to assume that artificial intelligence will become increasingly common in complex automated systems in the near future.

Humans, on the other hand, are highly adaptable to their operating environment and the behaviours of 'human intuition' and of being able to think 'out of the box' are still very much in demand in combat Submarines.

In 1951 Paul Fitts [4] published a highly regarded set of criteria detailing where automated functionality was better employed and where humans would achieve better results. The introduction of AI technology and big data analysis techniques potentially change this balance.

Table 3 Fitts 1951 criteria comparison with recent automation advances

Fitts List (1951) **What People are Better at:**	**What Machines are Better at:**
Detecting small amounts of visual, auditory or chemical energy Perceiving patterns of light or sound Improvising and using flexible procedures Storing information for long periods of time and recalling Reasoning Deductively Exercising judgement	Responding quickly to control signals Applying great force smoothly and precisely Storing information briefly, erasing it completely Reasoning deductively Doing many complex operations at once
Fitts list adjusted for 2017 technology **What People are Better at:**	**What Machines are Better at:**
Improvising and using flexible procedures Exercising judgement	Responding quickly to control signals Applying great force smoothly and precisely Storing information briefly, erasing it completely Reasoning deductively Doing many complex operations at once Detecting small amounts of visual, auditory or chemical energy Perceiving patterns of light or sound Storing information for long periods of time and recalling Reasoning Deductively

For an automated manoeuvring and control system to be effective it requires to have access to all the information it's human counterpart has in order to make fully informed decisions, Sensor data, Principles of Operation, Signatures data, Environmental data, understanding of the material state of the platform etc. Even with modern whole boat data networks this represents a substantial integration challenge.

3.3 HUMAN MACHINE INTERFACE (HMI)

Due to the number of parameters under system control an efficiently designed Human Machine Interface becomes increasingly important as more automated functionality is added. HMI enables effective communication of intent from the human to the system, however feedback of 'edited' information to the operator becomes more difficult as the computer takes on increasingly complex responsibilities.

There are two main philosophies for operating an automated vehicle control system, 'Real time control' and 'Communicated intent'.

Real time system are typified by joystick type control interfaces which communicate 'rate of change' and direction, these systems are difficult to apply effective safety limits to as the system has no understanding of the desired end state. The real time HMI however is able to rapidly communicate helm commands into the system.

With a 'Communicated Intent' system the end state of the manoeuvre is input to the system and it optimises the manoeuvre being conducted and can ensure that it is carried out within the safety limits. E.g. depth changes manoeuvres respecting both SOE and trim limit constraints.

As complexity of the system increases it becomes increasingly difficult for humans to interact with the machine through 'Communicated Intent', as the number of inputs required when conducting a depth change manoeuvre for example are vast - there are multiple subtle ways in which the manoeuvre can be adjusted depending on the situational need.

BAE SYSTEMS have been exploring the analogy of VERBS and ADJECTIVES in the order chain to the system to address this issue, the VERB is to 'change depth by 100m' the ADJECTIVES describes subtlety within the manoeuvre e.g. 'quickly' 'without altering pitch' 'quietly' 'over the next 5 minutes'.

The submariner will be reluctant to adopt automated systems that they perceive will limit their available options, hard coded manoeuvring safety limits can sometimes present a safety hazard in themselves. (In a tactical environment is it better to dive past deep dive depth or be hit by torpedo?)

3.4 PRINCIPLES OF OPERATION

Submarines conduct dozens of manoeuvring operations, some operations are routine whilst others are rarely conducted.

K BESSEL, N STANTONs paper [1] provides a structured insight into the personnel, systems, processes, and interactions associated with a 'return to PD' manoeuvre on a T-class Submarine, this single manoeuvre involves over twenty people and has nuances dependent on the exact nature of the situation and the command team involved.

In order to effectively automate the functionality of manoeuvring and control these wider operations and interactions need to be both understood and standardised as far as practically possible for every manoeuvre the submarine is likely to undertake.

Once all the manoeuvres have been analysed in depth it becomes apparent that the 'command intent' can largely be embodied in a set of principles of operation, these can be used to form a basis for developing automated system functionality.

Contrasting this with Aircraft operation, Aircraft manoeuvres are conducted in phases that are highly defined, standardised and procedurally controlled – crucially they are also monitored and policed by air traffic controllers who manage the interrelationship with other air vehicles offline.

4. COMPARITIVE DEVELOMENT OF SUBMARINE AND AIRCRAFT PROGRAMMES

The three control principles used on Submarines discussed above are applicable to aircraft, In general aircraft are dominated by dynamic force control systems although they do possess 'static' weight control systems such as the fuel systems and jettisonable payload items. Airships are notable for having almost exactly the same control principles as Submarines all be it with poorer manoeuvring characteristics owing too there operation in a less dense medium.

4.1 1903 HOLLAND VS WRIGHT FLYER

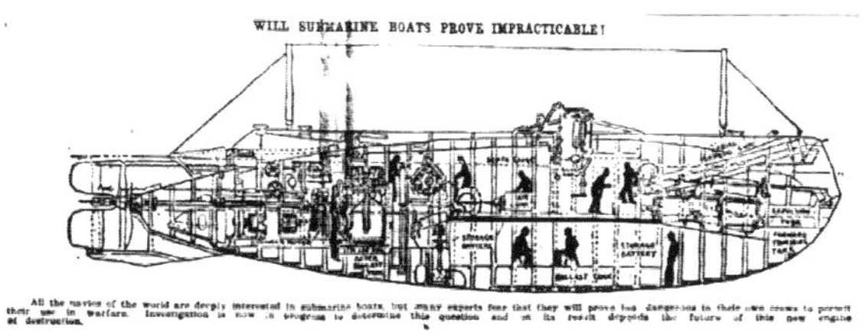

Figure 3 Holland submarine

Holland
Crew 6-8
Navigation / environmental sensors: Conning tower & basic navigational instruments
Fully Manual Mechanical control gear interfaced via hand wheels
Combat system: Torpedo

Range 250 miles surfaced 20nm submerged [7]

Wright Flyer 1
Fully Manual Mechanical control system – wire linkages to flexing wings
Range 852 feet [9]

The blueprints for the modern combat Submarine and fixed wing aviation were set in 1903, both breakthroughs were enabled by both technological advances in the key components and an understanding of the supporting physics associated with each vehicle. The Submarine at this stage was already a complex integrated system requiring a number of crew to operate manoeuvring functionality.

Both Holland and the Wright Flyer have been classified as PACT Level 0 (Under Command) across there Hydro dynamic, Hydrostatic and Navigation functions.

4.2 1920S E CLASS VS VICKERS VIMY

E-Class
Crew 30
Hydrostatic control – electric trim and ballast pump system – manual valve line ups
Hydrodynamic control –hydroplane actuation via direct acting rod gears interfaced with hand wheels
Navigation / environmental sensors: periscopes, compass, sextant
Range: 3000nm surfaces 65 submerged (12-15 days) [7]
Combat system: 5 Torpedo tubes

Vickers Vimy
Crew 3
Navigation / environmental sensors - Sextant
Fully Manual Mechanical control system – wire linkages operated by collective columns
Range: Transatlantic speed 100mph
Combat systems 12 250lb bombs defensive machine guns [9]

In a timeframe of twenty years the designs have evolved into capable combat systems, their military advantage has been proven (three Victoria crosses were awarded to E-class captains, and the genesis of submarine warfare tactics were being developed by the likes of Max Horton in E9) and large numbers are being procured even after the cessation of WW1. The notable advances are both in the area of platform range and the capability of the combat system being delivered, both have required larger crew numbers to be carried on what are essentially still manually operated platforms.

Both the E-Class and the Vickers Vimy have been classified as PACT Level 0 (Under Command) across their Hydro dynamic, Hydrostatic and Navigation functions.

4.3 1940s TYPE XXI U-BOAT VS B29,

Type XXI U-Boat
Crew 57
Hydrostatic control – electric trim and ballast pump system – manual valve line ups – system complexity is driving a HF Nightmare!
Hydrodynamic control – twin screws, hydroplane actuation via hydraulic systems actuated via tapper bars
Navigation / environmental sensors: passive and active sonar systems, Radar, periscopes, compass, sextant
Range: 15500nm surfaces 340nm submerged (3 days at 5 knots)
Combat system: 6 Torpedo tubes, Deck guns
Introduction of (mechanical) Torpedo Computer systems [9]

Figure 4 Type XXI Ballast Control Station

B29 Superfortress
Crew 11
Static control – complex fuel tank and distribution system – flight engineer required to manage fuel and engines –Dynamic control – servo assisted electrical control system, C1 electronic gyroscope driven Autopilot system
Navigation / environmental sensors: flux gate compasses, sextant and ground radar
Range: 3250m Speed 310 knots
Combat system: bomb payload, electronic computer operated gun systems (later radar guided)

The Type XXI and B29 represent the most expensive military programmes carried out by the Axis and Allied powers during WW2. [8]

By the end of WW2 the white heat of wartime development had injected huge amounts of new technology in aircraft design in order to improve the combat effectiveness of the platforms. Advanced electronic autopilot systems on the B29 were flying the aircraft and co-ordinating with the combat system to deliver improved bombing accuracy. It is notable that combat systems on the type XXI and B29 are the first to adopt computer systems to improve performance.

The necessity of war and the infancy of safety centred design, meant that whilst advanced in function many of these new systems lacked reliability redundancy, and integration

4.4 1960s DREADNOUGHT VS TSR2

Dreadnought
Crew 113
Hydrostatic control – electric trim and ballast pump system – remote system operation from a ballast control panel
Hydrodynamic control – two man interface (split control of bow and stern planes) electro - mechanical depth control autopilot
Navigation / environmental sensors: passive and active sonar systems, Radar, periscopes, compass, sextant, inertial guidance system
Range: unlimited nm
Combat system: 6 Torpedo tubes, [9]

TSR2
Crew 2
Static control –unknown possibly computer controlled– flight engineer required to manage fuel
Dynamic control –Advanced terrain following autopilot system driven via digital computer system (VERDAN) electro-hydraulic control system, pilot head up display, mission pre-planned and uploaded via digital waypoints (on punched tape!)
Navigation / environmental sensors: inertial platform, Doppler, radio altimeter, fwd looking radar, sideways looking radar, moving map display
Range: 3250m Speed Mach 2.35
Combat system: bomb (nuclear) payload, computer calculated release points [9]

Both Dreadnought and TSR 2 were at the height of sixties innovation, there are some notable similarities in that both platforms were pushing the known performance envelopes in terms of speed and capability and using numerous innovative control systems. The manoeuvring and control system on TSR 2 is of particular note as it both introduced digital computing and integrated the navigation and dynamic control functions, one of the key reasons the programme was cancelled during its flight testing programme was the cost overruns being occurred on the avionics development. It is important to note that the dynamic characteristics of both platform were such that they could be effectively controlled manually (nearly always on dreadnought), automation of platform control was purely to reduce crew workload and improve combat effectiveness.

4.5 1980s TRAFALGAR VS EUROFIGHTER

Trafalgar
Crew 130
Hydrostatic control – electric trim and ballast pump system – remote system operation from a ballast control panel
Hydrodynamic control – single man dynamic control interface (developed on Oberon program) electro mechanical autopilot
Navigation / environmental sensors: passive and active sonar systems, Radar, periscopes, compass, sextant, inertial guidance system
Range: unlimited nm
Combat system: 5 Torpedo tubes, [9]

Eurofighter
Crew 1
Static control – fully automated management of aircraft CofG
Dynamic control –quadruplex fly by wire system, dynamically unstable, automated implementation of safety limits, HMI systems include a configurable Glass cockpit, and Hemet based 'augmented reality' projection system
Navigation / environmental sensors: inertial platform, complex radars, Network enabled via LINK system
Range: 1800m Speed: Mach 2+
Combat system: variety of ordinance and electronic warfare on external hard points [9]

The functionality of the Trafalgar class manoeuvring and control system changed little since Dreadnought, the pioneering work on single person dynamic platform control was completed on the Oberon class SSKs to de-risk the integration onto Swiftsure class submarines. Whilst the functionality has change little the system performance and reliability have improved markedly, particular advances in inertial navigation and sonar performance have improved both the navigation and combat effectiveness.

System integration defines the Eurofighter programme, to have such a complex machine operated effectively by a single pilot has necessitated a vast expense of money and time to realise, the advent of the glass cockpit, complex HMI and a software based architecture has allowed continued development and upgrade of the avionics at regular intervals throughout the aircraft life. The fly by wire approach has required an onus on system reliability and despite it complexity the aircraft has enjoyed a very low accident rate for a front line fighter.

It is of note that the reliance, trust and safety integrity of computer assisted control systems on aircraft had evolved to the point that a dynamically unstable aircraft was acceptable and there is a virtual cessation of 'manual' control. This contrasts with the Submarine design which remains dynamically stable and operating at speeds which direct control of the planes with the 'man in the loop' remains perfectly acceptable. The autopilot is used to reduce workload as opposed to a primary control method.

4.6 2000s ASTUTE VS PREDATOR DRONE

Astute
Crew 109
Hydrostatic control – electric trim and ballast pump system – remote system operation via 'behind glass' ballast control workstation.
Hydrodynamic control – single man 'behind glass' interface, digital autopilot system, hardwired reversionary control system.
Navigation / environmental sensors: passive and active sonar systems, Radar, periscopes, compass, sextant, inertial guidance system, WECDIS chart table
Range: unlimited nm
Combat system: 6 Torpedo tubes, [9]

Predator
Crew 0 – 2 operators at remote from platform.
Static control –unknown possibly computer controlled– flight engineer required to manage fuel
Dynamic control –Autonomous autopilots updated via command from a ground station through satellite or microwave communication, reversionary recovery modes on loss of link
Navigation GPS
Range: 675m Speed 117 knots
Combat system: external weapon hardpoint [9]

Astute's control system was designed in compliance with UK Safety case legislation, although delivered on the same network; the system retains the separation of control of the hydrostatic and hydrodynamic functionality. The HMI is improved with a high integrity behind glass control system, basic reversionary instrumentation and rate control actuation is hardwired onto the control console to mitigate the potential risk of electronic control failure.

Compared with legacy UK Submarines Astute class have improved autopilot robustness and functionality, the autopilot is therefore the primary hydrodynamic control method for the first time. Although hydrostatic control is enabled through under glass technology it is still very much 'manual' with the crew calculating, initialising and executing ballasting and trimming operations.

The Predator system utilised the catalyst technologies of high integrity secure communication and digital sensor technology to introduce autonomy into combat aircraft. The unmanned architecture of the platform was exploited to improve range and reduce cost, at the expense of platform survivability and other attributes necessary for a manned vehicle.

By removing the control of the platform to a remote location the operators are able to access numerous data sources that would not be directly available to an organically manned platform, this is an effective enabler of the 'Network Centric' design approach where battlefield information is effectively shared between assets.

Of the 360 Predator systems delivered worldwide 66 have been lost in air accidents and 4 to enemy action. This can be viewed from the perspective that autonomous control technology isn't fully mature or that the 4 million dollar unit price tag represents a 'bargain' when compared to the 120 million dollar cost of an F35 fighter - in essence the Predator is so inexpensive it's viewed as disposable.

Table 4 PACT Level assessment of platforms

1900s	Static control	Dynamic control	Navigation
Holland	0 (Under Command)	0(Under Command)	0 (Under Command)
Wright Flyer	0 (Under Command)	0(Under Command)	0 (Under Command)
1920s			
E-Class	0 (Under Command)	0(Under Command)	0 (Under Command)
Vickers Vimy	0 (Under Command)	0(Under Command)	0 (Under Command)
1940s			
Type XXI	0 (Under Command)	0(Under Command)	0 (Under Command)
B29	1 (At Call)	4 (pilot assisted)	1 (At Call)
1960s			
Dreadnought	1 (At Call)	4 (pilot assisted)	2 (Advisory)
TSR2	1 (At Call)	5 (automatic)	5 (automatic)
1980s			
Trafalgar	1 (At Call)	4 (pilot assisted)	2 (Advisory)
Eurofighter	5 (automatic)	5 (automatic)	5 (automatic)
2000s			
Astute	1 (At Call)	4 (pilot assisted)	2 (Advisory)
Predator	5 (autonomous)	5 (autonomous)	5 (autonomous)

Tabulating the PACT levels of the examined platforms together enables the following trends to be drawn:

There is also correlation in the aircraft industry between platform speed and the level of automation, the reaction times required to operate safely operate platforms like TSR2 (flying fast and low) and Eurofighter (dynamically unstable) require safety critical automation to prevent catastrophic failure. Submarines by contrast remain relatively slow moving, are generally dynamically stable making them relatively easy to manually control.

Another strong correlation in the aircraft industry is to fully integrate control functions into a single user interface. There are many reasons to reduce the crew numbers on aircraft, for example reduction in aircraft size, increase in performance, a reduction in number of personnel exposed to danger and reduced training costs. Submarines by contrast have long patrol times requiring large crews to enable watchkeeping to be maintained. They are also required to operate autonomously, conducting defect-rectification and maintenance during for long periods away from base support.

Because of their large size, high value and different operating profile Submarine platforms tend to have higher levels of redundancy built into their control architectures reducing the impact of potential system failures.

5. FUTURE ADVANCES IN MANOEUVRING AND CONTROL

5.1 IDEALISED SUBMARINE MANOEUVRING CONTROL SYSTEM

Functional analysis of the PACT data has shown that UK Submarine control has changed little since the 1960s, although the technology has improved and become more robust from the perspective of usable HMI and safety.

It is evident that further automation of functionality will be a focus to improve both safety and combat effectiveness of the Submarine.

Currently the most challenging aspects of Submarine control exist at either end of the operating speed range; at high speeds the large hydrodynamic forces generated by the control surfaces can effect rapid depth excursions in the event of a failure. Conversely control at very slow speeds (and transition into a Hover mode if fitted) is complicated as the platform transitions from a rapidly effected dynamic stability dominated regime into a docile hydrostatic stability dominated regime.

The next generation of Submarines should take the key next step of integrating the dynamic and static control functions into a single system. This will reap the benefits provided by an integrated solution, for example safely managing the transition between hydrodynamic and hydrostatic depth control or maintaining trim within safe limits throughout a depth change without a man in the loop. Automated control of the platform during these complex manoeuvres will enable them to be conducted more accurately, reliably and safely.

Navigation and wider situational awareness will remain separate but should take advantage of improved HMI technologies and data analysis techniques to effectively communicate the environment and platform state to Command.

Like many surface ship system architectures, speed orders have historically been telegraphed to the manoeuvring room. This arrangement introduces unpredictability into the implementation of an automated manoeuvring control system. However, the option of electric propulsion makes the prospect of an integrated solution conceivable if safe bounds can be derived.

5.2 CHALLENGES TO AN IDEALISED MANOEUVRING CONTROL SYSTEM

The aircraft industry has shown that it is incredibly expensive to implement higher levels of automation effectively – for Submarines (with a smaller number of vehicles to recoup the investment over) this effectively a barrier to entry. The high SIL approach required drives significant cost into the design.

However the aircraft industry has also shown that once the control architecture has been matured there are significant performance and safety gains to be exploited. It is apparent that once the investment in new control architecture has been made it is relatively straight forward to port into new platforms.

There are challenges associated with capturing and documenting the full extent of tasks currently carried out by the crew, coupled with the nuances that are required dependent on the command intent. Increased automation also places a greater reliance (and integrity requirements) on sensor suites to enable action without going via the crew for validation.

High SIL design also tends to drive out functionally to reduce the number of failure modes, and the associated cost of developing the extra functions – this can be a double edged sword as the system becomes open to criticism of being inflexible.

5.3 FULL AUTONOMY?

UUVs exist today that operate in an semi-autonomous way, in general these vehicles perform well at isolated tasks, but lack endurance and have high discretion ratios brought about by their need to frequently communicate back to a command authority.

The two stalwarts of Submarine warfare are stealth and endurance. Stealth currently complicates the use of secure communications and endurance drives the need for a costly reactor plant with its associated a large crew requirement through the need to maintain watchkeeping oversight and the inevitable platform maintenance tasking's over a long period. Unless a 'game changing' technology emerges in these areas the practicality of fully autonomous fleet submarine seems some way off.

5.4 ENABLING TECHNOLOGY TO REDUCE THE COST OF IMPLEMENTING AUTOMATION?

A systematic phased approach needs to be applied to the development of manoeuvring and control automation.

The glass cockpit design approach can be viewed as a game changer as it not only allows the system to be refreshed more often than the 30 year design life of the submarine platform but also allows a common manoeuvring and control interface to be standardised across a fleet of Submarines. This could allow a reduced training burden and interchangeable crews across a platform types.

The concept of a 'Common Combat System' across the UKs submarine fleet has been around for some time now where a development team is constantly working on new technology. This could be a model for a 'Common Control System' being developed to avoid the cyclic nature of submarine control system design being compressed into the short timescale of the design and build phase and the associated risk of design skill fade between programmes.

6. ACKNOWLEDGEMENTS

The author would like to acknowledge the generous support of BAE Systems in preparing this paper. The opinions expressed in no way reflect those of BAE Systems Maritime - Submarines.

For the purposes of this paper platform data has been sourced from publicly available sources and may not reflect actual platform capabilities.

7. REFERENCES

1. K BESSEL, N STANTON, 'How a Submarine returns to periscope: Analysing the complex socio-technical systems using cognitive work analysis', Journal of Applied Ergonomics, 2014.

2. List of submarine incidents since 2000, WIKIPEDIA, extracted 12/4/2017.

3. R TAYLOR, 'Capability, Cognition and Autonomy', DSTL, 2003.

4. PM FITTS, Human engineering for an effective air navigation and traffic control system.' National Research Council Washington DC, 1951.

5. M COOK, M THODY, D GARRET, 'The Perils of Underwater Automation', BAE SYSTEMS, 2017.

6. https://en.wikipedia.org/wiki/AlphaGo, WIKIPEDIA, extracted 14/4/2017.

7. A PRESTON, 'The Royal Navy Submarine Service, A Centennial History'.

8. P O'BRIEN, 'How the War Was Won (First ed)', Cambridge University Press, 2015.

9. VARIOUS, ' data sources for publically available platform data for comparison', WIKIPEDIA 2016, extracted 14/4/2017.

8. AUTHOR'S BIOGRAPHY

David Ritchie, MEng, CEng, MRINA, Currently holds the position of Naval Architecture Department Manager at BAE Systems Maritime - Submarines

Adam Cowley, CEng, MRINA, Currently supports the Hydrodynamics function within the Naval Architecture Department of BAE Systems Maritime - Submarines

Seth Russell, MEng, MSc, CEng, MIMechE, MRINA, RCNC, holds the current position of Naval Architecture Technical Lead at the Ministry of Defense.

THE DYNAMIC STABILITY OF SUBMARINES ON THE SURFACE

P Crossland, C K Pope, QinetiQ Ltd, UK
S Machin, Naval Authority Group, Submarine Delivery Agency, UK

SUMMARY

In the UK, the stability of a submarine on the surface continues to be evaluated according to hydrostatic stability criteria that, for a surfaced submarine, are based on the examination of the metacentric height and the righting lever. The aims of these criteria are to provide sufficient stability to prevent capsize and to limit submarine motions to reasonable levels whilst on the surface. However, the methodology, based on the work of Sarchin and Goldberg (1962), was developed over a period of time to provide guidance in the design and operation of naval ships; the methodology is empirical in nature derived from data obtained from USN destroyers whose hull forms are quite unlike submarines. Furthermore, there has been significant amount of research undertaken to investigate suitable alternative measures of stability that incorporate the hydrodynamic response of the vessel to the environment and in the use of performance based measures rather than the traditional static stability criteria. This paper describes a programme of work aimed at applying some of the tools and techniques developed for surface ships to quantify the feasibility of using such techniques for the dynamic stability of submarines on the surface, primarily at zero speed. The tools have been validated using a series of model experiments and a proposal for a potential amendment to the current static stability criteria is proposed.

NOMENCLATURE

Symbol	Definition	(Unit)
a_ϕ	Total inertia	kgm
B	Beam	m
b_{lin}	Linear damping	kgm^2s^{-1}
b_{qua}	Quadratic damping	kgm^2
c_ϕ	Restoring term	kgm^2s^{-2}
erf	Error function	-
Fn	Froude number	-
G	Gravity	m/s^2
GM_{min}	Minimum GM	m
$GZ_{\phi m1}$	GZ value at ϕ_{m1}	m
GZ_{area}	Area under GZ curve	m.deg
$GZ_{area\phi m2}$	Area under GZ curve to ϕ_{m2}	m.deg
$H_{1/3}$	Significant wave height	m
k	wave number	m^{-1}
k_{xx}	Roll radius of gyration	m
l	length	m
m_0	Variance of roll	$rads^2$
m_2	Variance of roll rate	$(rads/s)^2$
N_β	No of headings	
N_{Hs}	No of wave heights	
N_{T1}	No of periods	
T_ϕ	Natural roll period	secs
T_0	Wave modal period	secs
V_{wind}	Wind velocity	m/s
Z	Heave amplitude	m
ζ	Wave amplitude	m
ϕ_{m1}	Heel angle 1	degs
ϕ_{m2}	Heel angle 2	degs
ϕ_{max}	Maximum roll ange	degs
ϕ	Roll angle	degs
$\dot{\phi}$	Roll rate	degs/s
$\ddot{\phi}$	Roll acceleration	degs/s/s
ϕ	Heel angle 2	degs
ϕ_{max}	Maximum roll angle	degs
θ	Pitch angle	degs
τ	Duration of exceedance	secs
σ	Standard deviation of roll	rads
ψ	Yaw angle	degs
ω	Wave frequency	rads/s

1. INTRODUCTION

1.1 STATIC STABILITY CRITERIA

Whilst the design of a submarine is typically optimised for the case where it is deeply submerged, there are times when the submarine may find itself on the surface; such as a failure of the main propulsion system. Therefore, to aid the submarine designer there are a number of stability criteria, specified as standards that are to be met by design. These criteria are mandatory for all new and existing submarines and are applied throughout its life. The stability criteria are based on the examination of the values of BG, metacentric height, GM and righting lever, GZ achieved in the submerged, surfaced and damage conditions.

The aims of the criteria are:

a) To provide sufficient stability to prevent capsize induced by environmental hazards which the submarine might encounter during its life.
b) To limit submarine motions to reasonable levels during normal operational evolutions.
c) To provide sufficient reserves of stability to enable the submarine to recover safely from flooding

The criteria for an intact surfaced stability are based on consideration of the shape of the GZ curve, and on the ability of the submarine to withstand beam winds and icing without experience large angles of heel.

The structure of the criteria set is:

a) GM is to be not less than GM_{min} metres.
b) GZ at a heel angle of ϕ_{m1} is to be not less than $GZ_{\phi m1}$ metres.

c) The area under the GZ curve up to ϕ_{m1} is to be not less than GZ_{area} metre degrees.
d) The area under the GZ curve up to ϕ_{m2} is to be not less than $GZ_{area\phi m2}$ metre degrees.

These intact static stability criteria stem from the work of Sarchin and Goldberg (1962) which was based on the analysis of data from US Naval ships. Various criteria were developed to establish the ability of the ship to survive the effects of wind, sea, special operating conditions and flooding due to damage. The acceptable shape of the GZ curve and the minimum GM value are essentially empirical in nature, being based on what has been achieved and found to be satisfactory in previous designs.

These intact static stability criteria were then modified and adapted to suit modern designs of submarines when they were adopted in the UK in the 1970s.

1.2 PERFORMANCE BASED CRITERIA

Although these static based criteria are still used today, it has been recognised, in the surface ship community, that they should be updated by using performance based approaches that include the effects of hydrodynamics. Fundamentally, capsizing of an intact ship is a statistically rare event that is a consequence of the behaviour of that vessel in extreme wind and wave environments and so requires an approach that can account for the forces due to the waves and wind.

In the surface ship community, the IMO have defined three distinct physical phenomena responsible for failures in the stability of a ship, Belenky *et al,* (2008):

- Variation of the restoring moment in waves
- Instability in the deadship condition
- Manoeuvring related failures.

The physics of these phenomena should be modelled to a high level of fidelity to include such mechanisms as:

- Static loss of stability – the quasi-static loss of transverse stability (associated with an excessive righting arm reduction) on the wave crest.
- Parametric rolling – resulting from a time-varying roll restoring characteristic of a ship.
- Resonant excitation - when a ship is excited at or close to its natural roll frequency.
- Excessive acceleration – excessive lateral accelerations will impact on the performance and safety of the crew.
- Broaching - represents a wave-induced large amplitude change in heading angle. This mode is usually characterised by surf-riding at the wave phase speed and steadily increasing yaw angle.

Second Generation Intact Stability Criteria have been under development by the IMO SLF Sub-Committee, Peters *et al.*, (2011) for a number of years. A three tiered approach to assessment, and associated criteria, has been developed. A ship is required to comply at one of the three levels for each of the failure modes. The lower the level the less significant the amount of effort is required to demonstrate it passes.

In the case of the deadship condition, when there has been total loss of propulsion and the ship is in beam wind and waves, the first level criterion has been defined by IMO as the current weather criterion but with a modified wave steepness that is reflective of the specific geographical area. For the second level, a weighted average representing a failure index is evaluated considering different combinations of possible environmental conditions. Calculations of a possible critical roll angle are undertaken for a number of sea states that are representative of the geographical area.

The calculations are based on a single degree of freedom method relying on seakeeping codes or model tests. The next level is by a direct stability assessment that would use numerical simulations of nonlinear roll motion. For level 2 and above, validated numerical tools and or model tests, coupled with a prescribed procedure for applying these techniques, is required to ensure that the approaches are robust.

There has been a significant amount of research to understand the dynamic stability of naval ships and in the development of performance-based criteria, de Kat *et al.* (1994), and this work is still ongoing. Significant progress has been made in this area of ship hydrodynamics including the development of non-linear ship motions prediction capabilities and advanced physical model testing techniques, Belenky *et al,* (2008). Much of this research stems from the work of the Naval Stability Standards Working Group (NSSWG), a collaborative effort between a number of navies, with the joint aim of developing rational, physics based, criteria for intact and damaged surface ships; their approach is, primarily, using a numerical simulation tool (FREDYN) developed by the Cooperative Research Navies (CRNAV), validated by model tests, to understand the capsize phenomena.

The work described here looks to evaluate the suitability of the tools and techniques developed for surface ships to quantify the dynamic stability of submarines on the surface. Whilst there may be operational reasons why the submarine is on the surface during rough weather, this work is limited to the case where the submarine has suffered some failure of propulsion is forced to surface; effectively equivalent to the IMO deadship condition. Therefore, the main stability failure in this context is related to resonant excitation.

2. FAILURE MODES AND CRITERIA

2.1 IMO DEADSHIP

The IMO deadship stability failure mode is a case of the vessel, at zero speed, encountering excessive beam seas with the stability failure mode occurring as a consequence of resonant roll excitation. The assumption of beam seas is derived from the case of a ship experiencing power loss in a storm, aerodynamic wind forces and wave forces tend to turn the vessel into beam seas. The beam sea assumption may not be applicable for submarines whereby the above water hull form is significantly different to that of most surface ships. As a result, the deadship condition for a submarine maybe more complex than just assuming beam seas; meaning it may be necessary to predict the steady drift velocity and angle of the submarine rather than just assume beam seas.

The relevance of excessive roll to a surface ship is that at these extreme angles, which are below the angle of vanishing stability, the ship may experience significant quantities of green water on the deck that has the same effect of any free surface and reduces the GM making capsize a possibility. Furthermore, it is possible that green water on the deck could be the source of consequential flooding due to water ingress into non-watertight access routes.

This would not be the case for a submarine that is in a state of preparedness for being on the surface in rough weather. Therefore, any criteria must be representative of actual failure phenomena. A suitable mathematical model on this type of failure mode could be used to develop suitable criteria for the case of submarines. In this case, a dynamic stability criterion could be regarded as a relative measure whereby a new submarine design would be only assessed against an existing design.

But what is "suitable" in the context of a mathematical model to evaluate the means of quantifying the failure mode of a submarine on the surface? It appears appropriate to adopt a mathematical model that can capture the correct hydrodynamic (and to a lesser extent aerodynamic) forces on the body in realistic environmental conditions. Depending on the kind of mathematical model, the criterion could be either probabilistic or deterministic in nature. According to Belenky *et* al (2008), the principal advantage of a probabilistic approach is that information on the statistical description of wind and waves can be considered; a resultant probability can then be combined with other aspects such as the probability of the submarine suffering a propulsion failure in the first instance, to provide an overall probability of failure.

2.2 STOCHASTIC PROCESSES

If the sea state, course and speed of the submarine do not change, then the stochastic process of roll motions can be considered to be statistically stationary; this simplifies the derivation of the probability of stability failure.

A property of a stationary stochastic process is that the statistical characteristics of the process (mean, standard deviation, skewness etc) can be derived from a single record if it is sufficiently long; waves are regarded as a stationary stochastic process or ergodic, Belenky (2004).

It is also the case that a linear system will produce a statistically stationary response to ergodic excitation. However, roll response is generally considered to be weakly non-linear and hence may not be ergodic in nature. This means that longer simulations for time-based analysis may be required for roll limiting applications; for obtaining statistical parameters for roll response it may be necessary to simulate several records for post-processing Belenky (2012).

If the assumption of ergodicity is applied to the scenario of a disabled submarine on the surface in rough weather, the probability distribution of the roll amplitudes would be independent of time and can be assumed to follow a Rayleigh distribution, Lloyd (1989):

$$P(\phi > \phi_{max}) = e^{-\frac{1}{2}\left(\frac{\phi_{max}}{\sigma}\right)^2}$$

The assumption of linear roll motion may become an issue at the more extreme roll angles, i.e., the ones of most interest in stability failures since roll is inherently non-linear in nature. In the development of a methodology to investigate the risk of capsize of a surface vessel a nominal critical roll angle of 70° was considered the point of capsize for surface ships, which, in itself is a very rare event, and so involves to some extent a degree of extrapolation for the tails of these probability distributions using a Generalised Pareto Distribution.

2.3 EXCEEDANCE AND SEVERITY

As discussed earlier the case of the roll angle of a submarine exceeding a given level may not be appropriate as a criterion for quantifying failure. There may be systems on board a submarine that fail if the roll motion exceeds a defined amplitude for a certain length of time. In this case it is effectively a joint probability of exceeding an amplitude and duration of exceedance that should be considered.

This is somewhat similar to Graham *et al* (1991) in defining the frequency and severity of people sliding on deck. A probability distribution is derived for the motion phenomenon, in this case, lateral acceleration in the plane of the deck (so called Lateral Force Estimator (LFE)) and for the duration of threshold exceedance.

In the case of LFE, the time history is deemed to be a Gaussian stochastic process and hence the amplitudes will be Rayleigh distributed. Then, Graham *et al* (1991) assumed that if the duration of a given threshold exceedance is τ, the duration probability is $p(\tau)$ and high threshold exceedances the probability density function is given as:

$$P(\tau) = \lambda e^{-\frac{\lambda\tau^2}{2}}$$

Where

$$\lambda = \frac{\pi M_S^2}{2\left[f\left(\frac{\phi_{max}}{\sigma}\right)\right]^2}$$

$$f\left(\frac{\phi_{max}}{\sigma}\right) = \frac{1}{2}\left[1 - \mathrm{erf}\left(\frac{\phi_{max}}{\sigma\sqrt{2}}\right)\right]$$

$$M_s = \frac{1}{2\pi}\sqrt{\frac{m_2}{m_0}}e^{-\frac{1}{2}\left(\frac{\phi_{max}}{\sigma}\right)^2}$$

m_0 is the variance of the roll angle

m_2 is the variance of the roll rate

The probability of exceeding a maximum roll angle for a given duration is then the joint probability formula:

$$P(\phi > \phi_{max} | \tau > \tau_{max}) = e^{-\frac{1}{2}\left(\frac{\phi_{max}}{\sigma}\right)^2} . e^{-\frac{\pi}{4}\left(\frac{M_S\tau_{max}}{f\left(\frac{\phi_{max}}{\sigma}\right)}\right)^2}$$

Which is best explained through an arbitrary example; Figure 1 shows an example of the probability of exceedance a roll angle for any specific time which is just the simple Rayleigh distribution compared with the joint probability of exceedance a roll angle for a period of time, in this case for 5 seconds and 10 seconds.

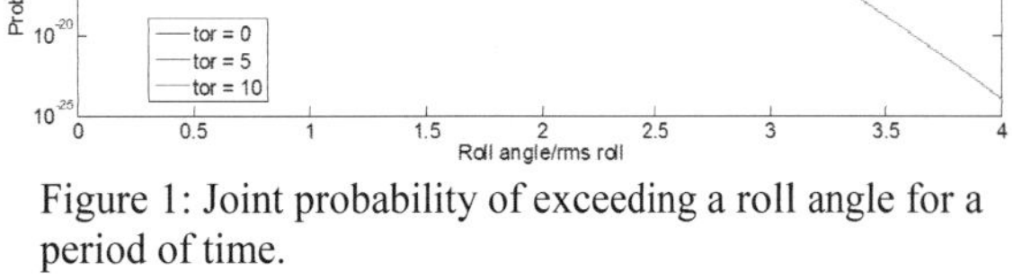

Figure 1: Joint probability of exceeding a roll angle for a period of time.

The joint probability has the effect of significant reducing the overall probability of occurrences compared with the simple Rayleigh distribution.

Having defined the joint probability distribution of the failure mode (probability of exceeding a roll angle for a period of time), the hourly probability of failure can be determined from the following:

$$p(\phi_{hourly} > \phi_{max} | \tau > \tau_{max}) = 1 - [1 - p(\phi_{amp} > \phi_{max} | \tau > \tau_{max})]^{No\ of\ peaks\ /hour}$$

This overall probability of hourly failure will be a function of both heading and sea state (recall that this is the deadship condition so speed is zero) and the probability of these occurring. So, the probability of failure is given as:

$$P(fail) = \sum_{j=1}^{N_\beta}\sum_{k=1}^{N_{Hs}}\sum_{l=1}^{N_{T1}} p_\beta(\beta_j)p((fail|\beta_j, Hs_k, T1_l)p_{Hs,T1}(Hs_k, T1_l)$$

Where p_β is the probability of being at the heading β

$p((fail|\beta_j, Hs_k, T1_l)$ is the probability of failure at a single heading for a given significant wave height and period

$p_{Hs,T1}(Hs_k, T1_l)$ is the probability of waves occurring with a given significant wave height and period

The probability of a particular vessel being on a p_β particular heading for, the deadship condition according to IMO, would representative of only beam seas.

The probability of a particular wave condition $p_{Hs,T1}(Hs_k, T1_l)$ will be dependent upon the sea area of concern; the data can usually be found in wave statistics databases such as Bales *et al* (1982).

Finally, the annual risk of failure can be determined as follows:

$$P(annual\ failure) = 1 - [1 - P(fail)]^{No\ of\ hours\ on\ the\ surface\ /year}$$

2.4 ACCEPTABLE RISK OF FAILURE

In general a level of risk is defined by considering the likelihood of an event occurring and the level of severity if that incident occurred. These can be defined in many ways, and in the context of the risk of capsize of a surface ship have been summarised by Jong *et al* (2005) on risk evaluation criteria. However, there are other considerations required when assessing risk, such as whether an incident would occur in isolation, the possible outcomes and the public perspective of the effects of that incidence. In many cases historical data are used to look at the annual number of fatalities/injuries for an activity, event or circumstance and this is used as the basis to define what is an acceptable or tolerable risk level. The main aim of this approach to risk assessment is to ensure that the risk is as low as reasonably practicable (ALARP). Peters (2010) proposed an annual capsize risk of $1x10^{-4}$ as a suitable level for the tolerable risk boundary for the loss of a naval frigate at sea. The work proposing new stability criterion for surface ships used an hourly relative risk of $3x10^{-3}$. This was determined from considering a number of scenarios for surface ship and the probability of exceeding a roll angle of 70° to find the relative risk level.

The overall risk level used for this work was the same as that adopted in the surface ship community, i.e., an hourly relative risk of $3x10^{-3}$ of exceeding a defined critical roll angle for a defined period of time.

However, in the overall context of assessing the risk of failure of a submarine on the surface, consideration must also be given to other factors that would impact on the overall calculation of probability. For example,

- the likelihood of the submarine requiring to undertake an emergency surface
- whether the submarine has lost main shaft power and any shaft independent propulsion system cannot be used to control heading

To understand the validity of this approach to the problem of submarines on the surface, this process was applied to a generic submarine as described in the following sections.

3. NUMERICAL MODEL DEVELOPEMENT

3.1 FREDYN

The CRNAV WG have been collectively developing the time domain simulation code FREDYN for a number of years and has been become the tool of choice, in that community, for intact and damaged stability related studies. FREDYN is a time domain potential flow code based originally developed by de Kat & Paulling (1989).

In FREDYN, the total forces and moments are divided into the following contributions:

- Froude-Krylov forces
- Wave radiation forces
- Wave diffraction forces
- Quasi-static hull forces
- Appendage forces
- Propulsor forces
- Wind forces
- Forces due to internal liquid or damage fluid (flood water).

The Froude-Krylov forces are determined by direct integration of the hydrodynamic pressure in the undisturbed wave over the instantaneous underwater geometry. The wave radiation forces are the time domain equivalents of frequency domain added mass and damping forces, and they are expressed in terms of added mass and damping at infinite frequency and linear convolution integrals. Wave diffraction forces are derived before initiating the time domain calculation using either 2D linear strip theory or 3D frequency domain panel methods; a database of linear transfer functions of the diffraction forces and moments as a function of wave frequency, heading angle and forward speed is generated and accessed during the time domain procedure.

The Quasi-static forces represent the manoeuvring forces and are generated using empirical formulations. The manoeuvring coefficients are based on model tests for frigates hull forms.

The control surface forces are determined at each time step as a function of the velocity at the control surface and the mechanical angle of attack. Although not relevant to the problem of the submarine at zero speed, propulsor forces are determined from the calm water thrust coefficients (specified by the user) and the instantaneous flow velocity. Wind forces are represented by empirical relationships; the user can specify a drag coefficient, or use a default value.

Confidence in the performance predictions using FREDYN has been enhanced as a consequence of a number of correlation exercises involving the comparison of data from physical model tests and simulations. However, these tests have not been focussing on the evaluation of the quality of FREDYN predictions of a submarine at zero speed.

3.2 MODEL GENERATION

FREDYN requires a mesh to be generated from a simple geometry of the submarine in order to generate the basic hydrostatic and hydrodynamic properties of the hull form. For convenience a CAD model from PARAMARINE was used as a basis to create the geometry for the FREDYN pre-processor to generate a mesh on the external surface.

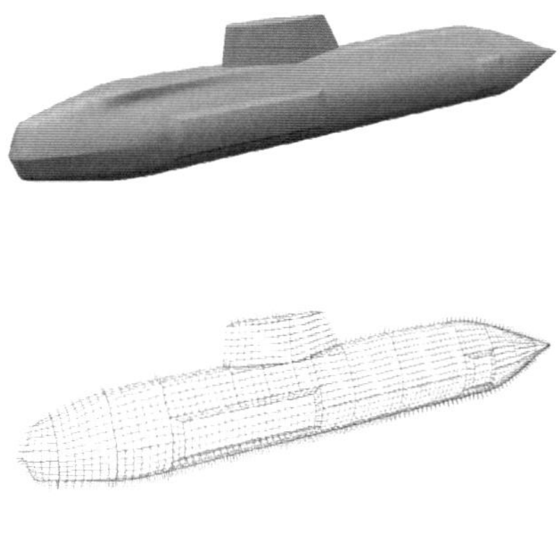

Figure 2: PARAMARINE model and FREDYN external mesh.

FREDYN only uses the basic shell of the hull since it treats any appendages differently within the code, therefore any rudders, stern planes and bow planes had to be removed from the PARAMARINE model; the bridge fin remained in the mesh model but was simplified as part of the meshing process.

It is clear from Figure 2 that panel method produces a surface mesh using quadrilateral panels. Using such panels may mean that in locations of significant curvature some loss of geometric fidelity may occur.

The geometry model was further developed to increase fidelity by including the free flood areas under the casing.

As mentioned earlier, FREDYN had been developed to evaluate the stability of damaged ships in waves; it was this functionality that was used to include the free flood spaces in the submarine. This was undertaken for the scale model configuration in preparation for comparison with model tests.

These free flood areas in the model were first defined within the PARAMARINE model and subsequently exported into FREDYN. To aid this process the PARAMARINE model of the free flood compartments were simplified to a number of separate zones in the areas of the bow, casing along the length, bridge fin and stern, Figure 3.

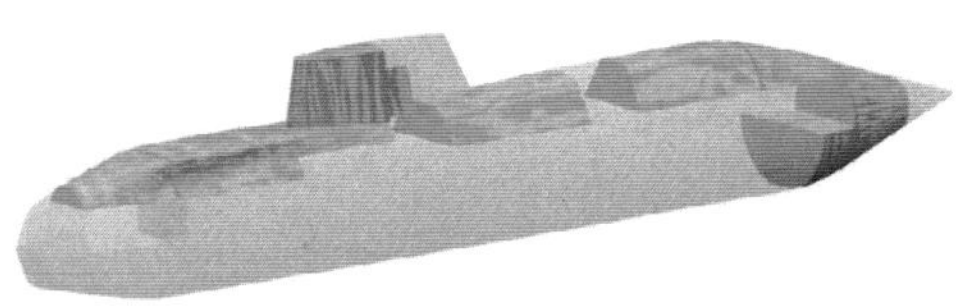

Figure 3: Simplified free flood compartments in scale model.

Drainage openings are defined in FREDYN to allow fluid to move between the free flood compartments and the external environment; to prevent any vacuums forming as the fluid drains from these compartments, vents are required to effectively connect the compartments to the atmosphere.

3.3 MODEL VERIFICATION

Both PARAMARINE and FREDYN are able to calculate the basic hydrostatics but, when including free flood space, they use different methods. PARAMARINE determines it then applies a loading condition based on the geometry and draught marks (which includes any free flood compartments modelled as tanks) from which the hydrostatics are determined. FREDYN calculates the hydrostatics from the initial conditions (draught marks prescribed within the input file) and the integration of user defined sections of the geometry. However, it does not include any differences in the loading condition as a consequence of the free flood compartments; the "intact" hydrostatics are determined before the simulation begins and then the free flood compartments are allowed to flood as part of the simulation. This means that if a set of draught marks are provided in FREDYN, since the free flood compartments are not modelled initially by FREDYN, the final hydrostatics will not be representative of the required condition since there will be additional mass of the water taken into the free flood compartments during the early stages of the simulation.

Therefore, the initial draughts in FREDYN are set so that one the free flood spaces have reached equilibrium then the final draught marks will be reflective of those required and match the condition and hydrostatics calculated in PARAMARINE.

As part of the verification process the hydrostatics from FREDYN and compared with PARAMARINE. Figure 6 shows the comparison of predicted displacement and LCG position for a range of draughts; also shows are the predicted GZ curves.

In general, the displacements and LCB position show very good correlation throughout the draught range. The small differences are probably due to the way the appendages are modelled in PARAMARINE and FREDYN. PARAMARINE includes the buoyancy of the appendages, where, as mentioned earlier, FREDYN deals with appendages in a different way to focus on obtaining the correct hydrodynamic rather than hydrostatic properties; therefore the do not contribute to buoyancy in FREDYN. The appendages are included in FREDYN using the input files, rather than the geometry file, and are modelled only as lifting surfaces.

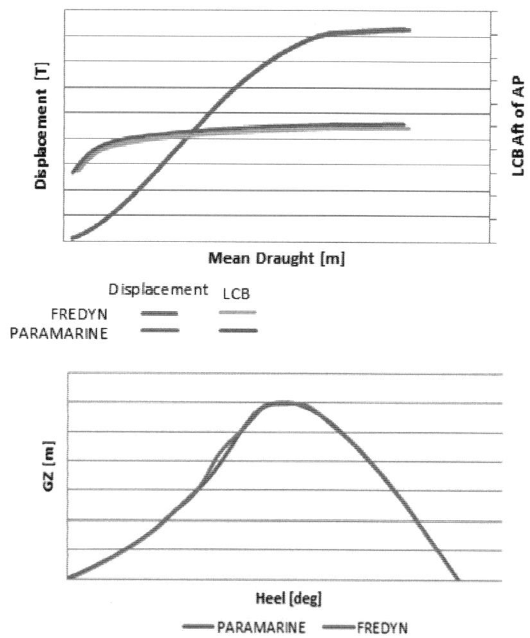

Figure 4: Comparison of hydrostatics from PARAMARINE and FREDYN.

The comparison of the GZ curve shows that FREDYN agrees quite well with PARAMARINE up to the point where the bridge fin becomes immersed; the differences here could be due to the way the mesh was simplified to accommodate the bridge fin in the geometry rather than as part of the input file like the other appendages. However, the impact of this difference in the restoring force is deemed to be minimal since it is not anticipated that the bridge will become submerged during the simulations.

3.4 ROLL DAMPING

Roll damping is a significant effect in the behaviour of any type of vessel in surface waves. For the case of the disabled submarine at zero speed there are effective two methods available in FREDYN; the method developed by Ikeda *et al* (1978) which stems from model tests on surface ships with non-frigate type hull forms at low to moderate speeds: and the option to provide specific roll damping coefficients for the hull form in question – this data most likely coming from model tests.

Considering the case of providing specific roll damping coefficients obtained from roll decay tests on a fully appended scale model.

The free roll decay of the submarine, at zero speed, can be described by:

$$a_\phi \ddot{\phi} + b_{lin} \dot{\phi} + b_{qua} \dot{\phi}|\dot{\phi}| + c_\phi \phi = 0$$

where

a_ϕ is the moment of inertia and added inertia
b_{lin} is the zero-speed linear damping coefficient
b_{qua} is the zero-speed quadratic damping coefficient
c_ϕ is the restoring coefficient

The linear and non-linear decay coefficients can be obtained by plotting the data from the peak roll angles as in Figure 5.

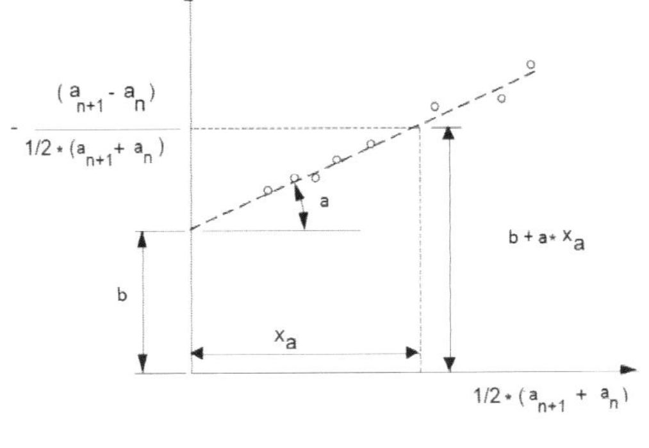

Figure 5: Linear and non-linear damping terms.

The resulting data can be approximated by a straight line with slope *a* and offset *b* and according to the ITTC (2012).

$$b_{lin} = a\frac{mgGM}{2\pi^2}T_\phi$$
$$b_{qua} = \frac{3}{8}b\frac{mgGM}{2\pi^2}T_\phi$$

The above method of extracting the linear and quadratic damping coefficients was applied to results from model scale roll decay tests; initial comparisons showed that even though the comparisons between the predicted roll decay curve and models tests were very favourable FREDYN over predicted the roll response in regular and irregular waves by a factor of 2. An alternative method of obtaining the roll damping terms that is available in later version of FREDYN is to choose a combination of using the default Ikeda *et al* (1978) method and inputting additional linear and quadratic terms to match the experimental roll decay curves. Early analysis suggests this provides a better estimate of roll damping across the frequency range; the analysis is not sufficient complete in time for this paper.

4. FREDYN SIMULATIONS

4.1 ENVIRONMENT

For the purpose of this exercise, the North Atlantic was chosen as a representative sea area and Bales *et al* (1982) was used to obtain the wave statistics for that particular sea area. Then, for each combination of significant waveheight and modal period the Bretschneider 2-parameter spectral formulation was used in the FREDYN simulation:

$$S_{BS}(\omega) = \frac{A}{\omega^5}e^{\left(-\frac{B}{\omega^4}\right)}$$

Where, for a wave with significant wave height $H_{1/3}$ and modal period T_0

$$A = 487.3\frac{H_{1/3}^2}{T_0^4}$$

and

$$B = \frac{1949}{T_0^4}$$

The effect of wind was also considered in the FREDYN simulations using the default relationship between the wind speed and significant wave height:

$$V_{wind} = a_0 + a_1 H_{1/3}$$

$a_0 = 2.88$ is the wind speed offset
$a_1 = 1.652$ is the linear coefficient

In order to understand the variability in probability of failure within each sea state, 10 different wave realisations, each of one hour duration, were considered in the simulations and the results averaged across these different time histories.

In simulation, as a consequence of the total loss of power the submarine is assumed to be stationary, with heading fixed, but free to drift in the waves. In this investigation of the probability of failure, all headings were considered to be equally likely to occur. The results, therefore, focus solely on the probability of exceeding the critical roll angle for the given time, over a period of an hour.

4.2 LOAD CONDITIONS

To help understand how consideration of the dynamic stability of a submarine, in the context of the probability of failure, compares with the current static stability criteria a number of load conditions were considered. These were all at a single representative displacement but with four different VCG positions:

- a representative surfaced condition (VCG1),
- a GM greater than the representative surfaced condition (VCG2).
- just passing the current static criterion for GM (VCG3)
- just failing the current static criterion for GM (VCG4)

Therefore, for each condition the GM will be different since the waterline remains constant and hence the natural roll period of each condition will be different (the roll inertia is assumed to be the same).

4.3 INDICATIVE RESULTS

For each combination for heading, significant wave height and modal period, 10, one hour simulations from FREDYN were obtained. From each of these simulations a probability of failure was derived (failure again being defined as when the roll angle exceeds a limits for a specific period of time), this probability was then averaged across the 10 different time histories to provide an average value for each of the headings and sea conditions; probabilities were then averaged across headings to give an unweighted mean probability of failure for each sea condition. The weighted probabilities were then obtained by applying the probability of each sea condition occurring. This final weighted mean was them summed to obtain an overall hourly probability of failure for that particular sea area. The probability of failure for each of the VCG conditions is shown in Figure 6.

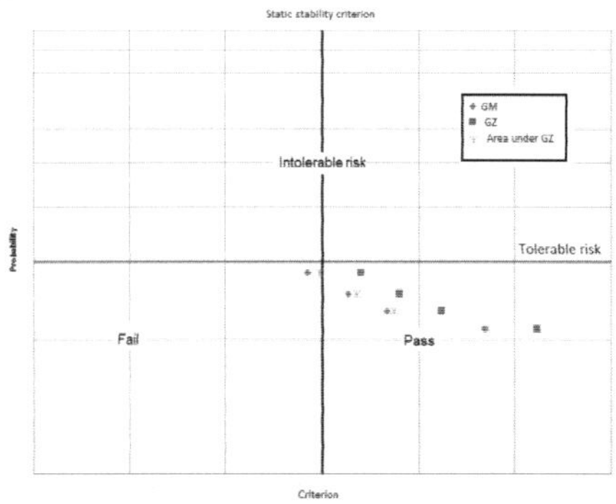

Figure 6: Probability of failure.

5. VALIDATION

5.1 PHYSICAL MODEL TESTS

An existing submarine model was chosen for this experiments; which being a deadship condition required minimal preparation. The load condition was taken to be a representative surfaced condition; the displacement, draughts and GM were obtained from stability statements and the radius of gyration assumed to be as recommended by the ITTC (2008). The natural roll period was calculated using the following:

$$Roll\ Period\ (s) = \frac{2\pi\ k_{xx}}{\sqrt{g\ GM}}$$

$$k_{xx} = 0.35\ B$$

Where B is the beam, k_{xx} roll radius of gyration and GM is the transverse metacentric height.

The experiment was conducted in both regular and irregular waves for a range of initial headings from head seas to following seas. To achieve these initial wave headings, the model was held in position until it was in a fully develop sea condition then allowed to move in full 6 degrees of freedom. Figure 7 shows the model being, temporarily, held in position by two electro-magnets suspended from the celling of the Ocean Basin. The frame on the model carries the tracking reflectors used by the Qualisys Motion Capture system.

Figure 7: Model in the Ocean Basin held in position by electro magnets.

A number of regular wave tests were undertaken for each heading as shown in Table 1; the regular waves covered a number of frequencies and also looked at varying the wave steepness to investigate any non-linearities in the response.

Table 1 also shows the irregular waves, looking at a range of increasing sea states all represented by the Bretschneider spectral formulation.

Table 1: Waves

Wave amplitude (m)	Wave length/boat length
a	0.63,0.90,1.15,1.50,1.88,2.43
1.5a	0.63,0.90,1.15,1.50,1.88,2.43
2a	1.50,1.88
2.5a	1.50,1.88

(a) Regular waves

Sea State	Significant wave height H1/3 (m)	Modal period T0 (s)
Mid sea state four	1.88	6.79
Upper sea state four	2.5	8.8
Mid sea state five	3.25	9.7
Mid sea state six	5.0	12.4

(b) Irregular

A series of computational simulations were conducted using FREDYN to repeat a number of tests undertaken in the Ocean Basin at QinetiQ Haslar. The same process of developing and verifying the FREDYN model, as described above, was repeated to replicate the particular model condition and to capture the free flood spaces particular to the model. Each simulation was set up manually, so that the same full scale equivalent input conditions could be set. Further zero speed roll decay tests were conducted to capture the viscous contributions to roll damping. Simulations were conducted for each of the experimental runs.

5.2 DATA ANALYSIS

A common method of analysing both the experimental data and the predictions from FREDYN was developed for both the regular and irregular wave tests. This section includes only a small proportion of the entire comparisons between model tests and FREDYN predictions. This is due to the constraints on the length of the paper – however, the presented data are representative of the whole comparison.

The analysis of the regular wave tests was looking to provide the first and second order responses. For the first order responses, because the model was allowed to move in six degrees of freedom, was to identify a section of the time histories where the yaw motion remain within +/- 10 degrees of the initial heading then a high pass filter was applied to remove the low frequency effects. An FFT was applied to extract the frequency then a linear least squares regression used to extract the amplitude. The resultant transfer functions are then plotted against non-dimensional wave frequency; see Figure 8, as an example. This figure shows some of the first order regular wave responses in bow quartering seas. The lines are from FREDYN predictions for a range of wave steepness and the markers show data from the model tests. The translational have been non-dimensionalised by wave amplitude and rotations by wave slope. This confirms the earlier observation that roll is over-predicted by about a factor of 2 using the method of extracting the linear and quadratic damping coefficients from roll decay tests.

The second order responses in regular waves were derived from the original motion time histories by extracting the mean roll, mean drift speed and final heading again plotted against non-dimensional wave frequency.

Figure 9 shows an example of the second order responses in irregular waves; here the comparisons are mixed, the mean drift speed is predicted quite well and the final heading can differ by around 40 degrees in some cases; the mean roll angle is predicted to be significantly higher than measured which maybe as a consequence of not getting the final heading correct. The drift speed and heading change is a second order response that is captured in FREDYN by the manoeuvring coefficients; these coefficients are empirical in nature based on data from frigate type hull forms. An alternative method is to use a cross flow drag coefficient that can be derived from CFD or model tests.

Figure 10 shows the RMS roll and pitch for each of the sea states considered. Predictions of pitch appear to agreed very well with the measurements; roll appears to be over predicted which reflects the results found in the regular wave tests.

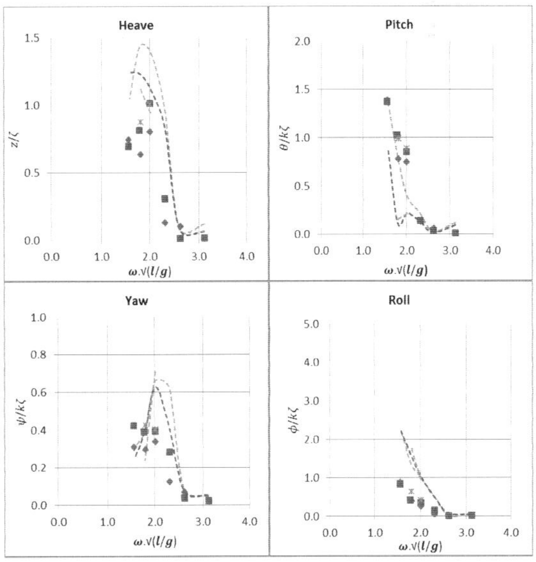

Figure 8: First order responses in regular waves for bow quartering seas

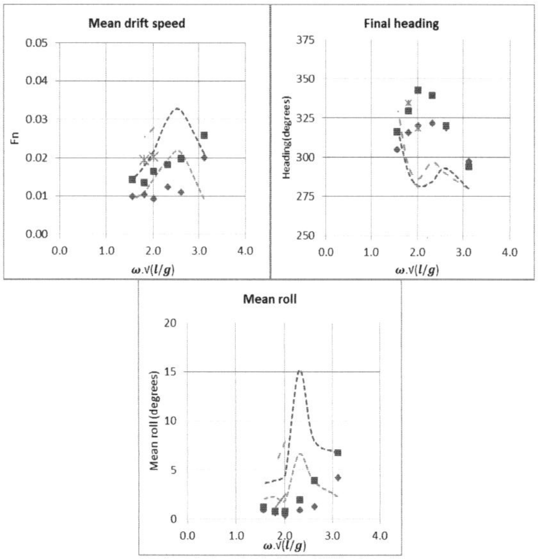

Figure 9: Second order responses in regular waves for bow quartering seas.

As described earlier, it is conjectured that the probability of failure can be derived from the assumption that the roll amplitude and durations of exceedance can be approximated by Rayleigh distributions. Further work is therefore required in the data analysis to extract these distributions and compare them with the theoretical curves.

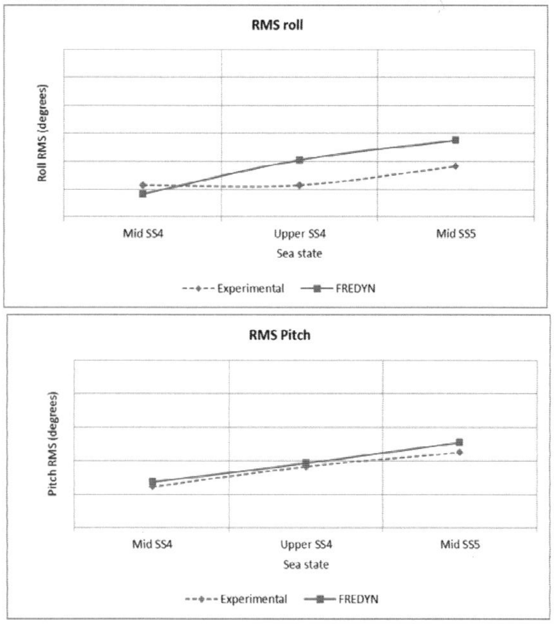

Figure 10: RMS response from irregular waves for bow quartering seas.

To conclude on the validation, differences between the measurements and FREDYN predictions were observed; some of these differences were consistent across the test conditions. However, at the same time there were similar trends in submarine behaviour observed from the experiment and simulations.

The intention of the test was to investigate the first and second order responses with the model allowed to move in the full six degrees of freedom; this makes the analysis of the data quite complex. By applying the same methodology to both the experiments and the predictions it is expected that any anomalies due to the analysis method are minimised. Clearly, roll is the dominant response when consider the failure mode and predictions appear to be out by a factor of 2. However, the second order yaw response is also important since that dictates the heading that the submarine will take and therefore directly influences the subsequent roll motion. It was seen that in the regular wave test that the predicted final heading was not entirely consistent with the experiment data. Generally, for following seas, the behaviour of the submarine in the experiment and simulations were similar. In contrast, the head sea tests showed the submarine changing heading by a greater extent in experiments when compared to FREDYN.

For regular waves, the pitch and heave motions are generally comparable across the conditions tested.

6. CONCLUSIONS

So, in summary, this paper has described a programme of work aimed at applying some of the tools and techniques developed for surface ships to quantify the feasibility of using such techniques for the dynamic stability of submarines on the surface, primarily at zero speed. The paper describes the approach to quantifying the mode of failure for a submarine on the surface in rough weather and a potential mathematical model for deriving the probabilities of such failures. To some extent the underlying ship motion prediction capability has been validated against tests with a scale model of a typical submarine. The initial results from the simulations suggest that when considering a joint probability of exceeding a given roll amplitude for a duration of time, the results appear to provide a boundary of acceptable risk that is commensurate with the boundary dictated by the existing static stability criteria

The experimental data appear to be self-consistent across the conditions tested. FREDYN first order responses in regular and irregular waves agree well with experiments except roll; roll is over predicted by approximately a factor of 2. The intention is to update these predictions using the latest version of FREDYN that allows a combination of the default Ikeda *et al* (1978) method and inputting additional linear and quadratic terms to match the experimental roll decay curves.

Second order responses in regular waves are quite good but the final heading is not predicted correctly with some consistency. There is additional analysis required to understand the statistical distribution of the roll motion and how it compares with the assumed Rayleigh distribution.

The work reported here was for a single submarine design at one displacement with 4 varying VCG positions and only at zero speed; further work is required to extend the investigation space perhaps to other designs to help understand the relationship between the existing static stability criteria and the proposed risk of failure, but also to perhaps account for a limit ability to maintain some forward speed.

7. ACKNOWLEDGEMENTS

The authors acknowledge the UK MoD for supporting this work.

8. REFERENCES

1. Bales SL, Lee WT, Voelker JM, Standardised Wave and Wind Environments for NATO Operational Areas, DTNSRDC, 1982.

2. Belenky V, de Kat JO, Umeda N, Toward Performance Based Criteria for Intact Stability. Marine Technology, Vol. 45, No. 2, April 2008, pp. 101–123.

3. Belenky V, Weems KM, Bassler CC, Dipper MJ, Campbell B, Spyrou KJ, Approaches to rare events in stochastic dynamics of ships. Probabilistic Engineering Mechanics, Vol 28, pp 30-38, 2012.

4. De Kat JO, Brouwer R, McTaggart K, Thomas WL. Intact Ship Survivability in Extreme Waves: New Criteria from a Research and Navies Perspective, STAB'94, USA, 1994.

5. De Kat JO, Paulling JR, The simulation of ship motions and capsizing in severe seas. Trans SNAME, Vol 97, 1989.

6. Graham R, Baitis AE, Meyers WG, A frequency-domain method for estimating the incidence and severity of sliding. David Taylor Research Center, DTRC/SHD-1361-01, August 1991.

7. Ikeda Y, Himeno Y, Tanaka N, A prediction method for ship roll damping. Report No. 00405, Department of Naval Architecture, University of Osaka Prefecture, 1978.

8. ITTC – Recommended procedures and guidelines for Model Test on Intact Stability, 7.5-02-07-04.1, Revision 2, 2008. http://ittc.sname.info.

9. ITTC – Recommended procedures and guidelines for Numerical estimation of roll, 7.5-02-07-04.5, 2012

10. Jong SK, Vanem R, Oyvind E Risk Evaluation Criteria, SAFEDOR-D-4.5.2-2005-10-21-DNV, October 2005.

11. Lloyd ARJM, Seakeeping: Ship behaviour in rough weather. Ellis Horwood Series in Marine Technology, ISBN 0-7458-0230-3, 1989.

12. Peters A, Tolerable capsize risk of a naval vessel, Proceedings of the 11th International Ship Stability Workshop, Jun 2010.

13. Peters W, Belenky V, Bassler C, Spyrou K, Umeda N, Bulian G, Altmayer B, The second generation of intact stability criteria. An overview of development. Trans SNAME, Vol 119, 2011.

14. Sarchin TH, Goldberg LL, Stability and Buoyancy Criteria for U. S. Naval Surface Ships, Trans SNAME, Vol 72, 1962.

9. AUTHORS BIOGRAPHY

Paul Crossland currently leads the Submarine and Propulsion Team at QinetiQ Haslar. He is responsible for the delivery of the submarine hydrodynamics programme at QinetiQ Haslar, developing experimental and computational techniques for assessing the performance of submarines, including understanding the behaviour of submarines close to the surface.

Charles Pope is currently a naval architect in the Surface Ships Team at QinetiQ Haslar. Recently he has been working onHis focus, recently, has been on experimental and computational assessment of ship stability in extreme conditions.

Stuart Machin is the current Head of Submarine Naval Architecture within the Naval Authority Group of the UK MOD Submarine Delivery Agency, where he has responsibility for naval architecture standards and design guidance and issuing certification for UK submarines. He has been a member of the Royal Corps of Naval Constructors since 1987 and amongst a wide ranging career he was responsible for writing the UK Naval Engineering Standard on submarine stability and Sea Systems Controllerate Publication on Submarine Hydromechanics.

EFFICIENT MODELLING OF THE STRUCTURAL RESPONSE OF SUBMARINE PRESSURE HULLS TO UNDERWATER EXPLOSIONS

D Graham, J Hobson, P Murphy, C Toole, S Cross and **J Farnworth,** QinetiQ Rosyth, UK

SUMMARY

The loss of equipment under UNDEX (underwater explosion) loading is usually accepted as a critical mode for ships and submarines near the surface and the effects can be mitigated by sophisticated shock mounts. Major structural damage can also be caused by whipping and, because of hydrostatic pressure the submarine pressure hull becomes more susceptible to structural damage with increasing depth.

This paper compares two approaches to modelling underwater explosions and the subsequent response of the pressure hull structure. A 'hydrocode' approach, which models the evolution of the gas bubble explicitly, and a simpler, but much quicker, approach which models the water as an acoustic medium and prescribes the pressure loading from an explosion bubble. The modelling approaches have been developed in parallel to provide appropriate verification and this has been reinforced by some experimental validation using published and legacy test data.

1. INTRODUCTION

The effects of underwater weapons on ships and submarines are an obvious concern to any navy and the subject has received much attention over many years. From an assailant's perspective, the aim is to eliminate the threat posed by enemy vessels and this can be achieved by causing sufficient structural damage to sink, or otherwise disable, a vessel, or by causing sufficient disruption to equipment and personnel on board to neutralise its capability. From the defensive point of view, it is accepted that no structure can be completely invulnerable to underwater explosion (UNDEX) and the challenge is to maintain the operability of systems, and crew, while the structure remains intact. To this end much of the research has been directed at the design of shock resistant mounts to isolate internal equipment and structures from the effects of shock loading.

Design guidance for structures, such as it is, is relatively scant and primarily directed at ships, the received wisdom being that a submarine pressure hull designed to withstand hydrostatic pressure at deep dive depth will be more than adequate to withstand shock loading. It is acknowledged that a shock load, which might not trouble the pressure hull near the surface, could potentially be more damaging to a highly stressed hull at significant depth and empirical 'hull rupture' rules are used to assess the performance of hulls at depth. However, they are based on historical data and have not been validated against more modern materials and structures; hence there is value in having a validated numerical modelling methodology.

This paper considers the response of a submarine to combined hydrostatic and UNDEX loading using two analytical approaches. This provides a degree of verification of the methods and some supporting experimental validation is also provided. The analyses incorporate some of the developments made in the modelling of static collapse of pressure hulls under hydrostatic pressure [1] and modelling combined hydrostatic pressure with low speed structural dynamics in the analysis of collision and grounding events [2].

2. UNDERSTANDING AND MODELLING AN UNDEX EVENT

2.1 THE NATURE OF AN UNDERWATER EXPLOSION

The characteristics of underwater explosions have been understood for many years and are well documented, e.g. [3]. On detonation, a solid charge is rapidly converted to gas, occupying the same volume, at high temperature and pressure and this sudden application of pressure accelerates the surrounding water outwards generating a shockwave. The shockwave is initially supersonic but soon slows (within 2 to 3 charge radii [4]) to the speed of sound of about 1500ms^{-1} in water and subsequently behaves as a linear acoustic wave. The timescale of the shock wave itself is typically of the order of milliseconds.

Behind the shockwave, the bubble of high pressure gas expands, its momentum allowing it to expand until the internal pressure is less than the surrounding hydrostatic pressure. The bubble then contracts and collapses creating a second pressure pulse. Although this is smaller in magnitude than the initial shockwave, its decay time is longer and it can carry a similar amount of energy. The timescale of the expansion and collapse of the bubble is of the order of seconds and the bubble can rebound and collapse several times before all the energy is dissipated. Reflections from the surface and / or the sea bed, cavitation, and migration of the bubble can all complicate the loading seen by the affected structure.

2.2 MODELLING AN UNDERWATER EXPLOSION

Two Finite Element (FE) analysis approaches have been taken to modelling the UNDEX event.

The 'hydrocode' LS-Dyna [5], allows an explicit coupled Euler-Lagrange analysis to be carried out. The behaviours of the gaseous explosion products, and the surrounding water, are modelled using appropriate equations of state and the progress of the bubble is modelled as part of the analysis. This level of physical realism allows effects such as bubble impingement on the structure to be modelled but can require substantial computing resources.

Using the Abaqus Finite Element (FE) analysis package [6], the surrounding fluid is modelled as an acoustic medium and the loadings from the UNDEX event applied as acoustic pressures. These are calculated in advance and account for the initial shockwave and subsequent pressure pulses as the bubble expands and collapses [7]. Although it models less physics than the hydrocode, the effects it ignores are largely confined to the earliest phase and close to the charge. It will not model any interaction between the bubble and the structure and may become less accurate if there are very large deformations in the structure however it runs much more quickly offering an efficient alternative, particularly when the response over longer timescales is required.

3. DEVELOPMENT OF THE MODELLING METHODS

The two approaches were investigated, and their capabilities developed, in parallel. The development of each method is discussed briefly in the next two sections and, as with any sophisticated analysis, there were a number of lessons to be learned in terms of both: how best to apply the analysis, and how best to model reality.

3.1 ACOUSTIC MEDIUM APPROACH

The process of modelling an UNDEX event using Abaqus with acoustic elements is, at least in principle, quite simple. The structure is modelled and meshed as normal and a region surrounding the structure is meshed with continuum acoustic elements. At the interface a tied constraint couples the fluid and structural meshes. These meshes do not have to conform to each other and either can be defined as the master or slave surfaces depending on the circumstances. The incident wave can be planar or spherical and is defined by a source point and a pressure time history at an appropriate 'standoff' point. Non-reflecting boundary conditions are applied at the fluid external surface and external reflecting surfaces, such as a free surface or the sea bed, can also be defined. This is illustrated in figure 1.

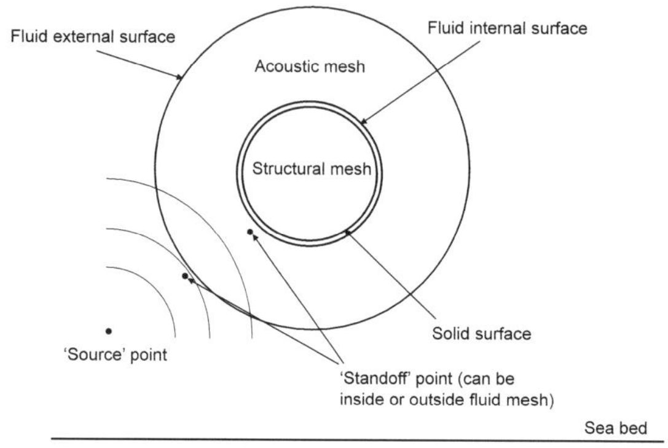

Figure 1: Schematic of Abaqus acoustic modelling of an incident pressure wave.

Abaqus provides 'total wave' and 'scattered wave' solution procedures. The former is simply the sum of the incident and scattered waves and both methods should give the same answers unless cavitation occurs, in which case the 'total wave' formulation should be used. In practice this proved not always to be the case and the solution showed sensitivities which were not explicitly mentioned in the user's manual [6]. For example, it is recommended that the 'standoff' point, where the pressure time history is defined, be placed close to the structure as this would minimise the time spent calculating the travel of the pressure wave towards the structure. However, it was found that this led to a mesh dependent error in the pressure applied to the structure when used with the total wave formulation. This was resolved by moving the 'standoff' point outside the fluid mesh.

After running a wide range of test cases, and comparing results with 'hydrocode' analyses, described below, a favoured procedure was established. The total wave formulation was used, the 'standoff' point was placed outside the fluid mesh and, where possible, the fluid was meshed with hexahedral brick elements rather than tetrahedral elements. Analysis at depth was achieved by carrying out a static analysis with hydrostatic pressure first and importing the results to the dynamic analysis model, maintaining the hydrostatic pressure as a load.

Once the basic mechanics of creating a model and obtaining a sensible solution had been addressed some validation, using published data, was attempted. Hung et. al. [8] reported on some small model tests and, usefully, they included a measured pressure time history and a sequence of high speed photographs for a test on an unstiffened aluminium cylinder.

Initially the shock wave pressure was applied analytically as an exponential decay. The peak pressure and decay constant were calculated from the charge parameters and were found to be in reasonable agreement with the measured pressure. This analysis predicted an entirely elastic response with no sign of the obvious large dent which occurred in the test. In fact the stresses were very much smaller than yield throughout. From the photographs in [8] it was apparent that the damage was occurring after the first bubble collapse so the input loading was modified to include a representation of the pressure pulse due to the bubble collapse. Although the magnitude was less than 20% of the shock pressure pulse, this analysis did induce some plastic deformation. A further refinement used the Geers-Hunter model [7] to calculate a complete time history based on the charge size. Although the predicted deformation was not quite as severe as that observed in the test, the areas of predicted peak plastic strain, shown in figure 2, coincide with the sharpest crease in the test specimen.

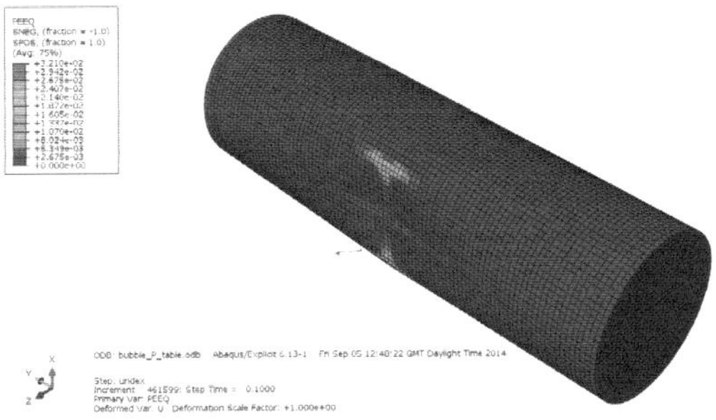

Figure 2: Deformation and plastic strain in Hung model.

This test case clearly illustrated that the bubble pulse pressure could be much more damaging to a pressure vessel than the initial shock wave.

3.2 HYDROCODE APPROACH

The basic mechanics of an Euler-Lagrange analysis are that the fluids are modelled using a spatially fixed Eulerian mesh and the structure by a deformable Lagrangian mesh. These are coupled in such a way that the Lagrangian mesh can move freely through the Eulerian mesh while interacting with the fluids. The fluid mesh may be divided into several domains, as illustrated in figure 3, in this case the air inside the submarine, the surrounding water and the gaseous explosion products. Appropriate interactions between the various components are defined.

Structural material properties are defined in the usual manner for non-linear analysis, e.g. via a stress-strain curve and the fluid behaviours are defined by appropriate equations of state (EOS). LS-Dyna provides a wide range of options for this EOS and in the current analyses the air was modelled with a Linear Polynomial EOS, the water with a Gruneisen EOS, and the gaseous explosion products using Jones-Wilkins-Lee EOS, all with appropriate parameters.

A number of analyses were carried out in parallel with the acoustic medium analysis described in the previous section. Qualitatively, the initial agreement between the two methods was very encouraging although quantitatively there were significant differences, particularly if the analyses ran beyond the first bubble expansion and collapse. Initially it was suspected that the discrepancies were because the acoustic medium approach was not modelling some feature of the interaction between the bubble and the structure. Although in most of the test cases the bubble did not impinge on the structure, at its maximum expansion it could be close to the structure and it was not impossible that this could have affected the shape, and subsequent collapse of the bubble. However, it transpired that the root cause of most of the discrepancy was a lack of refinement in the hydrocode models. Further refinement and an increase of the extent of the hydrocode model resulted in much improved agreement between the methods, even for quite a severe test case. Figure 4 compares the predicted pressures at a 5 m standoff from the charge centre.

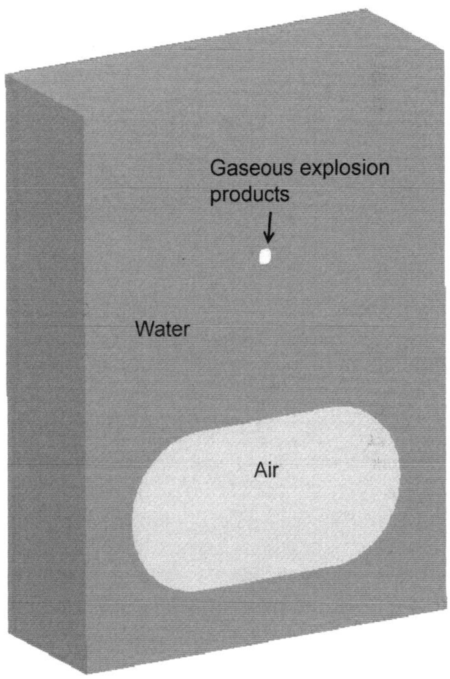

Figure 3: Fluid geometry showing division of Eulerain domains (blue water, yellow explosive and pink air).

Among the lessons learned was the fact that a non-reflecting boundary did not eliminate all reflections and a solution would be contaminated if it ran for longer than the time taken for pressure waves to travel to the boundary and back. This may not be an issue if the solution is primarily concerned with the early time response but is inconvenient if longer timescales are of interest. A potential solution was the Perfectly Matched Layer (PML) which acted as an absorbing layer and although it appeared to work well, it had its own issues. Running a hydrocode analysis over a long timescale is still a challenging problem.

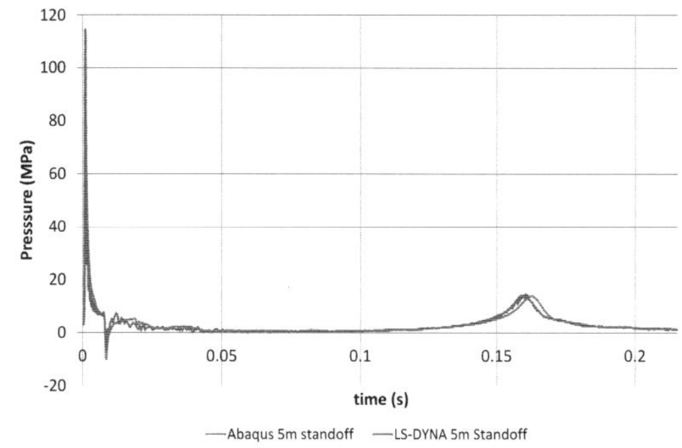

Figure 4: Comparison of LS-Dyna & Abaqus Shock & bubble pressures at 5m standoff.

4. PRACTICAL APPLICATIONS

From the investigations described above it was clear that the acoustic medium approach could be used to reliably predict the response of a pressure hull structure to a range of UNDEX loading events. The great advantage though was the time required to carry out an analysis; the difference in run times for comparable problems could be astonishing. It was not unusual for a hydrocode analysis to run for many days to replicate an analysis which took a couple of hours using the acoustic medium approach, with equivalent computing resources.

The acoustic medium method has great scope for carrying out parametric studies, and analysing large models, over extended timescales, possibly to look at behaviour such as whipping. At the same time, the limitations of the method should be remembered and there is always a place for hydrocode analysis to perform one-off analyses or providing supporting verification for alternative methods.

The following discussions expand on the developments described above to demonstrate the application of both methods to the analysis of real structural UNDEX problems.

4.1 VALIDATION AGAINST A LEGACY EXPERIMENTAL TRIAL

A report from a legacy UNDEX trial, which contained sufficient information to allow an accurate model to be built and reasonable comparisons between predictions and experimental results to be made, was sourced [9]. The test configuration is shown in figure 5. A possible limitation of the comparisons was that the test configurations involved small charges at relatively small stand-offs. The similitude equations [3] predicted a maximum bubble radius larger than the standoff which meant the bubble was almost certain to impinge on the structure. In spite of this a comparison was still considered a worthwhile exercise. Even if the acoustic medium analysis was only valid in the early stages, its performance in modelling the response to the initial shock wave could still be assessed and it would provide a rigorous test of the hydrocode capability.

Figure 5: Legacy experimental model en-route to UNDEX trial.

The trial used EDC1 explosive which was converted to an equivalent weight of TNT for the Geers-Hunter model used in the acoustic medium analysis. The hydrocode analysis modelled EDC1 explicitly with appropriate equation of state parameters but the analysis was also run with the equivalent weight of TNT to allow direct comparison with the acoustic medium analysis.

The trial results included pressures measured near the surface of the model, but offset from the closest point to the charge. Comparison with pressures calculated using the acoustic medium method showed good agreement with the initial peak shock pressure, the first bubble pulse pressure and the bubble period. The hydrocode analysis also gave good agreement with the initial shock wave pressure but there were noticeable differences with the first bubble pulse pressure and the bubble period. The most likely issue was that the small size of the physical charge was not accurately resolved, even with a relatively fine mesh, in terms of the overall size of the problem.

A photograph of the deformed model post-test is shown in figure 6 along with the calculated shape from the acoustic medium analysis (magnified by a factor of three for clarity). The magnitude of the calculated deflection is in reasonable agreement with the experiment although it appears that the frames had a greater effect than was observed in practice. In the trial, the peak deflection occurred 7.5° above the starboard centre line because of bubble migration which, at present, had not been allowed for in the analysis. Figure 7 compares the measured deflections on the centreline (90°) and at 82.5° with calculated values from the acoustic medium and hydrocode FE analyses. The hydrocode analysis gives a smoother distribution, more in line with observation, with the EDC1 analysis coming reasonably close to the maximum deflection.

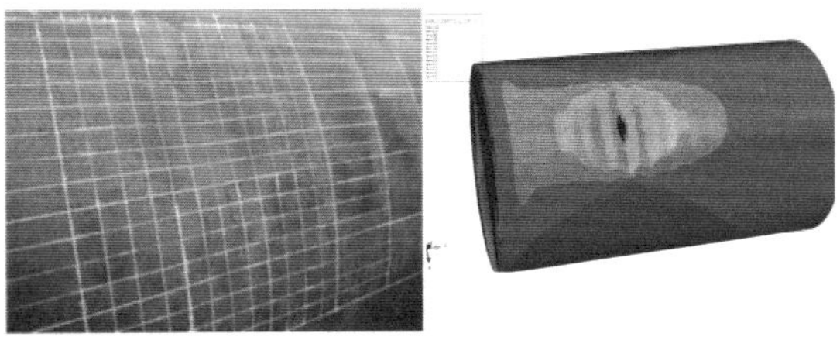

Figure 6: Photograph of deformed model from [9] and acoustic medium FE analysis (×3).

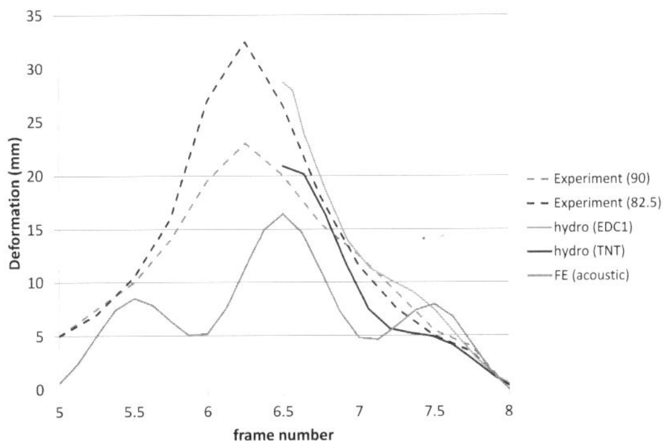

Figure 7: Measured deflection on centre line compared with acoustic medium and hydrocode FE analyses.

A comparison with strain gauge measurements showed that the acoustic medium analysis gave quite poor agreement which was attributed to the over-prediction of localised bending. Much better agreement was achieved with the hydrocode analysis.

As a test case, this was not ideal; the small charge was difficult to model accurately in the hydrocode analysis and the impingement of the bubble on the structure stretched the acoustic medium analysis beyond its limits. Despite this, the results are really quite encouraging. The acoustic medium analysis gave a reasonable description of the overall response in a very quick analysis and the hydrocode analysis provided more accuracy, albeit at a substantially increased computing effort. Improvements to the modelling can be made including: refinement of the fluid mesh around the charge in the hydrocode analysis and the inclusion of bubble migration in both analyses.

4.2 COMPARISON WITH A HULL DAMAGE LAW

The QinetiQ SURVIVE® software [10] uses a semi-empirical rule to predict hull rupture which relates critical charge size to standoff, taking account of depth. The acoustic medium approach was used to carry out a parametric study on an idealised hull compartment to effectively generate a set of collapse curves. The analysis was carried out with external pressures of 1 MPa, 2 MPa and 3 MPa respectively to represent a range of diving depths.

For a given standoff, the charge size was varied to quantify the critical charge size required to cause collapse. The hull was deemed to have survived if it showed no signs of collapse after at least two bubble pulses. Most of the collapses observed appeared to be due to the first bubble pulse although in a small number of cases the hull survived this but collapsed under the second bubble pulse. A typical collapse sequence is outlined in figure 8 which shows the deformed shape as the analysis progresses. The first frame (60 ms) is taken between the shock wave and first bubble pulse. The shock wave gives the structure an impulse and initial deformation but this does not change much before the pressure pulse from the first bubble collapse arrives. The second frame (150 ms) is taken as the pressure from the first bubble pulse reaches a peak and the third (158 ms) just after. Local deformation between the central frames is increased to such an extent that the hull can no longer resist the hydrostatic pressure and collapse ensues. Collapse could occur under the shock pulse but this was academic as it only occurred with much bigger charges than required to cause collapse as just described.

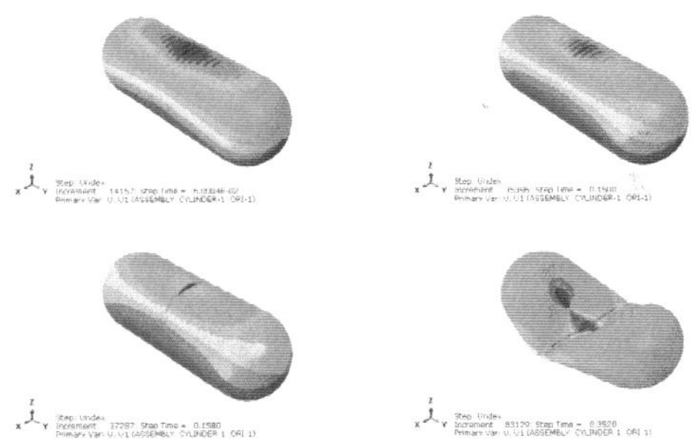

Figure 8: Sequence of deformed shapes of hull model under UNDEX and hydrostatic pressure.

The results are compared in figure 9, and appear to be in good agreement at the shallow diving depth and, although the agreement becomes poorer as the depth is increased, the correct trends are predicted. The numerical analysis appears to show a greater reduction in critical charge size with increasing depth than the empirical rule. This could be a real effect, but it could also be that the current structure is not well represented by the choice of some parameters in the empirical rule. Either way, the results are still reasonably encouraging.

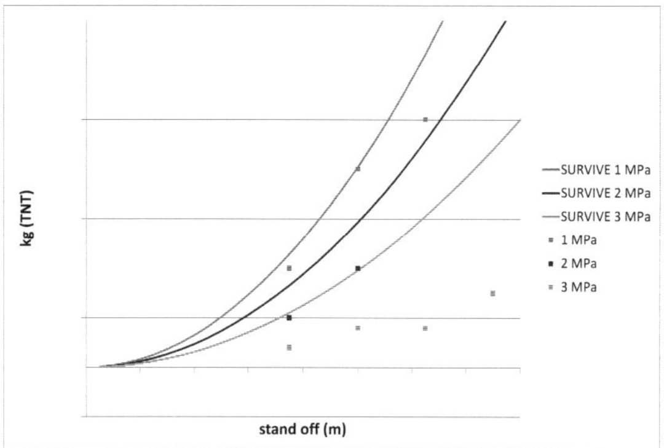

Figure 9: Comparison of FE predicted critical charge size with SURVIVE hull rupture rule.

4.3 EXTENSION TO PRACTICAL STRUCTURES

The acoustic medium method was used to investigate further cases of interest; fabrication imperfections were introduced to the idealised hull compartment model used previously, a complete hull with bulkheads was analysed, and a water backed sphere was modelled to simulate a free-flood structure such as a casing.

4.3 (a) Residual stresses and shape imperfections

Fabrication processes introduce imperfections to the pressure hull structure and two of the more significant, at least for hydrostatic collapse, are shape imperfections and residual stresses due to cold-bending of the hull plating and frames [1]. An idealised as-built shape was generated using a combination of overall and interframe shapes, as described in [2]. The shape consisted of an n = 2 overall out-of-circularity and an interframe out-of-fairness which was a combination of the axisymmetric 'hungry-horse' shape and a critical interframe buckling mode. The shape is shown (magnified) in figure 10. The residual stresses in the hull plating and a cold bent T frame were calculated using FE analysis and included in the model for UNDEX analysis as an initial condition.

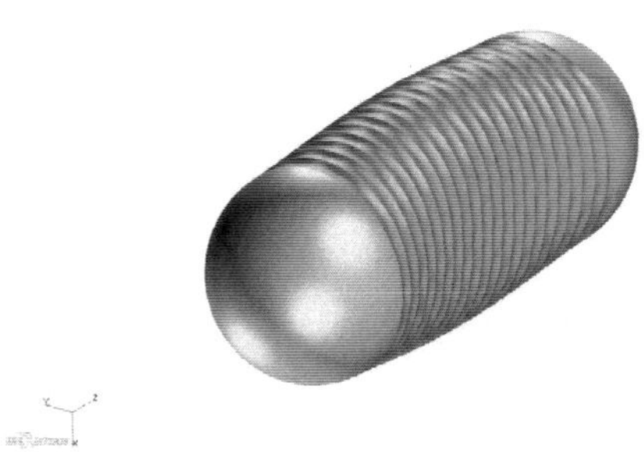

Figure 10: Magnified depiction of representative as-built shape imperfections.

Figure 11 compares the axial deflections at the dome poles (1 and 2) for a combination of charge size, standoff, and depth which the structure was predicted to survive. This shows that the shape imperfections, in the orientation shown, lead to significantly greater deflections after the bubble pulses. The residual stresses appear to moderate this slightly. There is a risk that assuming a 'perfect structure' could be non-conservative in assessing whether a structure will survive a particular UNDEX scenario, but a more extensive parametric study would be required to show whether including these imperfections makes a significant difference.

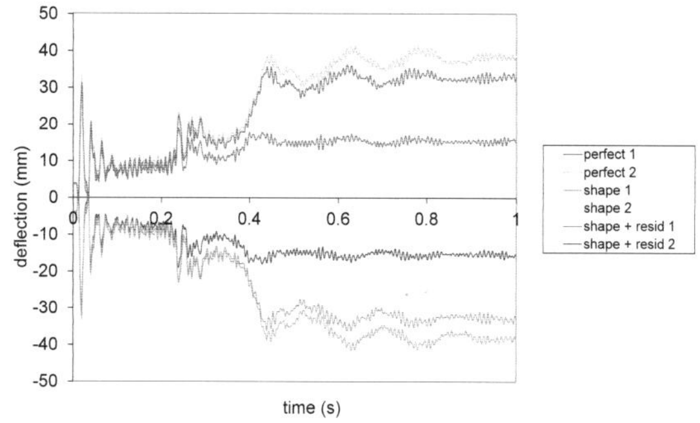

Figure 11: Axial deflections in domes without, and with as-built shape and cold bend residual stresses, under 2 MPa hydrostatic load.

4.3 (b) Whole hull model

The whole hull model, figure 12, was used to assess the effect of charges at different locations: over the centre of the longest compartment, over a central bulkhead, and over the pole of the dome.

Figure 12: Geometry of whole hull model (cross section).

This is illustrated in figure 13, which shows the incident and reflected pressure waves at an early time in the analysis, and figure 14, which shows the deformed shapes at the end of each run. This only offers a snapshot of the behaviour but it does show that the charge over a long compartment has resulted in extensive damage and collapse, over a bulkhead has caused less widespread damage but is still collapsing, and over the dome has caused significant damage but not collapse (at the depth considered). Obviously a more refined parametric study would be required to draw any more specific conclusions regarding the relative vulnerability of different parts of the hull structure, but it does illustrate that the response can be quite different at different locations, as might be expected.

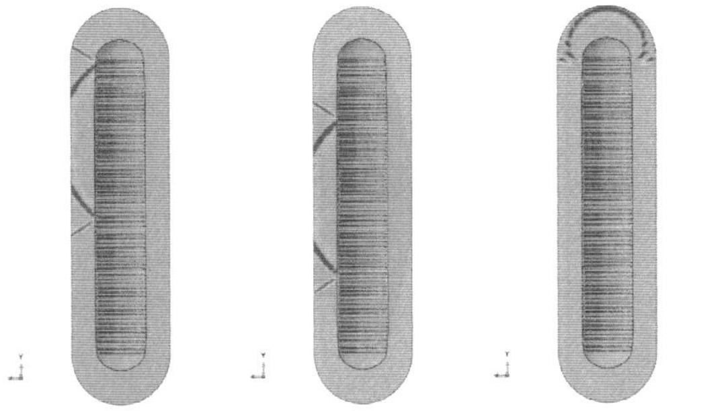

Figure 13: Early time pressures with charge over a) mid bay, b) bulkhead and c) dome.

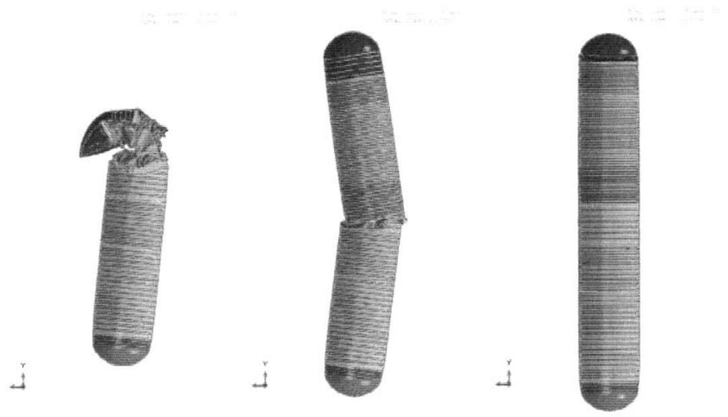

Figure 14: Deformed shapes at end of run.

4.3 (c) Free-flood structure

To calculate the complete response of a submarine it would be necessary to include the external structure, fore and aft ends, casings etc. which can include entrained water. This is relatively simple to model in principle, requiring only an internal and external fluid mesh with appropriate contacts defined between both and the structural mesh [6]. To demonstrate the capability a simple sphere was analysed and three cases considered: air filled, water filled and half air – half water filled. In the latter case the internal water and air were separated by a bulkhead; the acoustic medium does not model free surface effects such as sloshing.

Figure 15 shows the shock wave interaction with the structures at an early phase of the analysis and the behaviour appears to be much as expected. The pressure wave is reflected from the air backed sphere whereas much of the energy is transmitted through the water backed sphere. The combined case obviously shows both transmission and reflection. This exercise has demonstrated that, in principle, the acoustic medium approach could be extended to model full submarine structures including free-flooded external structure.

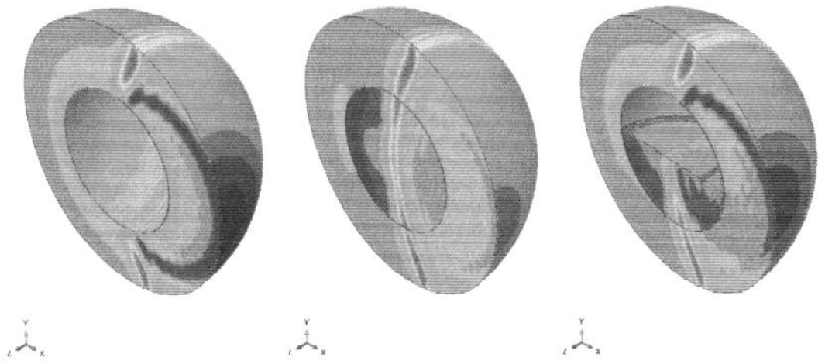

Figure 15: Shock pulse impacting air filled, water filled, and combined spheres.

5. CONCLUSIONS

This work was undertaken to address a perception that the analysis of the structural response to UNDEX loading is, possibly, underrepresented in current design guidance. The emphasis was on developing a relatively simple, but rigorously verified and validated, methodology which could be applied to practical problems. This was achieved by taking advantage of the complementary strengths of two modelling approaches i.e. a 'hydrocode' and an 'acoustic medium' feature of a FE code.

Although the hydrocode approach models the basic physics at a more fundamental level than the acoustic medium approach, it is very sensitive to modelling decisions and other user inputs. A successful model requires careful development and modelling a complex structure in a realistic scenario may be prohibitive in terms of computer resource. Understanding the limitations (and strengths) of the acoustic medium approach gives access to a practical, but still rigorous, tool which can efficiently analyse large and complex structures over sensible time periods.

Both methods performed well against a relatively severe experimental test case and the 'acoustic medium' method was then applied to a range of problems of practical interest.

6. ACKNOWLEDGEMENTS

This paper describes work carried out by QinetiQ for the MoD under the Maritime Strategic Capabilities Agreement (MSCA) contract.

7. REFERENCES

1. GRAHAM D. *Predicting the collapse of externally pressurised ring-stiffened cylinders using finite element analysis*. Marine Structures 20 (2007) 202-217. Elsevier.

2. GRAHAM D, MURPHY P, HOBSON J, CROSS S, TOOLE C AND FARNWORTH J. *Advanced numerical modelling techniques applied to submarine structures research and development, design, build, and refit and repair.* Warship 2014: Naval Submarines & UUVs. 18-19 June 2014, Bath UK.

3. MAP 01-470 Shock Design Manual.

4. COSTANZO F A. Underwater Explosion Phenomena and Shock Physics. Proceedings of the IMAC-XXVIII. February 1-4, 2010, Jacksonville, Florida USA.

5. LS-Dyna. Livermore Software technology Corporation.

6. Abaqus 2016, © Dassault Systèmes, 2015.

7. GEERS T L & HUNTER K S. An integrated wave-effects model for an underwater explosion bubble. J. Acoust. Soc. Am. 111 (4), April 2002.

8. HUNG C F, LIN B J, HWANG-FUU J J & HSU P Y. Dynamic response of cylindrical shell structures subjected to under water explosion. Ocean Engineering 36 (2009) 564-577. Elsevier 2009.

9 PENMAN J M & BUTCHER S. Unpublished report.

10 SURVIVE®, QinetiQ.

8. AUTHORS BIOGRAPHY

The authors constitute the Submarine Structures team at QinetiQ Rosyth and between them have over 175 years of experience of numerical modelling and analysis of submarine, and other, structures. Their capabilities cover most areas of structural analysis including: linear and non-linear, static and dynamic analysis; optimisation; thermal analysis; and fluid structure interaction.

ALTERNATIVES TO DOMES AS PRESSURE HULL END CLOSURES

K J Heaney, H J Hickson, P E Buckton, J M Underwood, BMT Defence Services, UK,
LCDR B Blum, P Gough, Ministry of Defence, UK

SUMMARY

Traditionally large diameter submarine pressure hulls utilise unstiffened domes as end closures. These domes are difficult to manufacture as they require precise double curvature to be pressed into thick steel plate. The domes failure mode is one of buckling collapse which is extremely sensitive to initial shape imperfections. Thus, the domes are subject to relatively high reserve factors in design, and tight manufacturing tolerances. This paper examines two alternative end closures which are constructed from flat plates; the faceted dome and flat sandwich bulkhead. The structural design and arrangement implications of both concepts are explored and compared to domed end closures. Both concepts are shown to be feasible pressure hull end closures, but result in a considerable weight penalty compared to a conventional dome. However, the reduced sensitivity to imperfections and potential arrangement benefits offsets some of the weight penalty.

NOMENCLATURE

FE	Finite Element
FEA	Finite Element Analysis
OOC	Out Of Circularity
SME	Subject Matter Expert
t	Plate thickness
TIG	Tungsten Inert Gas

1. INTRODUCTION

Traditionally, large diameter submarine pressure hulls utilise full diameter hemispherical or torispherical domes as end closures, or conical ends with reduced diameter end closures. These large domes are difficult to manufacture as they require precise double curvature to be pressed in to thick steel plate petals. Welding of the petals together to form the dome introduces further challenges. The structural failure mode of such domes when loaded with external pressure is one of buckling collapse, which is extremely sensitive to manufacturing shape imperfections; consequently they are subject to relatively high reserve factors during design and tight build tolerances in manufacture. Once the dome has been fabricated large holes are cut in to it and castings welded in for the weapons embarkation hatch, torpedo tubes and air turbine pump inserts. Welding the inserts requires a large amount of heat to be put into the dome, which increases the chance of significant distortion and imperfections being present in the final form.

This paper presents conceptual research into a number of alternative pressure hull end closure geometries that do not require double curvature, two of which are presented; faceted domes and flat bulkheads. Finite Element Analysis (FEA) has been used to examine the structural response and viability of the potential alternative end closures. This paper discusses the advantages and disadvantages of the different end closure options, including the weight and volume changes compared to a traditional dome, the potential manufacturing issues, and overall viability of the concepts.

2. PRESSURE HULL END CLOSURE OPTIONS

Faceted domes and flat bulkheads were considered as alternatives to traditional domes as they are constructed from flat plate, avoiding the manufacturing complexity of pressing double curvature into thick steel plate. Additionally, the use of flat plate may result in a failure mode other than a buckling failure, which may be less sensitive to shape imperfections. Although these end closures will not be as structurally efficient as a dome, their potential ease of manufacture and reduced sensitivity to imperfections may lead designers to consider them in future submarine designs.

Faceted or geodesic domes utilise flat triangular facets to form a dome structure. These structures have been used in many civil structures, but their use as a submarine pressure hull end closure is a novel application and will be subject to a vastly different set of criteria than civil structures. The faceted dome end closure concept originated from the UK Ministry of Defence (MoD) Naval Authority Group, who identified a need to explore alternative submarine end closures. BMT Defence Services (BMT) have carried out a number of studies for the UK MoD into the viability of faceted dome end closures. It is this work that forms the basis of the faceted dome section of this paper, in conjunction with additional work by BMT to further understanding of the concept.

Unlike faceted domes, flat bulkheads have been used previously as submarine end closures, but they are uncommon in modern large diameter submarines with significant deep diving depths. The two most notable uses of flat end closures on modern submarines are the Australian Collins Class and the Swedish Gotland Class. However, with a pressure hull diameter of approximately 8m on the Collins Class [1] and 6.2m on the Gotland Class [2] these boats are significantly smaller than nuclear submarines which typically have pressure hull diameters typically greater than 10m. Perceived wisdom suggests that flat end closures, while suitable for smaller diameter submarines, would attract such large weight penalties on large diameter submarines that they would be unfeasible.

BMT felt that this had not been definitively proven to be true. As modern large diameter nuclear submarines tend to be space driven designs rather than weight driven designs, and have a large amount of solid ballast forward, a heavier end closure may still be viable. Therefore, it was postulated that the weight penalty may not prove the design to be unfeasible if there are demonstrable advantages regarding the submarine layout and manufacture.

Previous work has been carried out on the feasibility of large diameter flat end closures by QinetiQ [3]. However, the focus of the study was primarily on the use of plate thickness and overall geometry optimisation techniques through FEA, which did not consider benefits from a more traditional grillage structure to reach a feasible comprise between weight, space, strength, stiffness and buildability. The subsequent optimised design was approximately 6 times heavier than a traditional dome and did not possess a lot of the potential arrangement benefits that can be obtained with a flat end closure. It was felt that a study investigating a traditional grillage structure as an end closure would be useful to investigate the weight increase of a grillage style flat bulkhead and explore the potential benefits regarding spatial arrangement.

3. DESIGN APPROACH

The intent of these studies was to investigate whether the faceted dome and flat bulkhead could act as feasible large diameter pressure hull end closures. However, as each end closure provides different potential benefits over traditional domes, each required a different design approach.

3.1 FACETED DOME DESIGN APPROACH

The faceted dome has the advantage of forming a dome-like structure from flat plate. As the structure is dome-like it was hypothesised that its failure was more likely to be due to a buckling collapse mode. Hence, the faceted dome was analysed using non-linear FEA to determine its collapse pressure, failure mode and sensitivity to imperfections. Subsequently consideration of suitable reserve factors was made to indicate whether suitable justification may exist to reduce the reserve factors from those applied to traditional domes.

3.2 FLAT BULKHEAD DESIGN APPROACH

The flat bulkhead is a deep grillage-like structure formed between two bulkhead plates. Therefore failure is primarily due to exceedance of allowable bending stresses or deflections. Subsequently it was considered that the flat bulkhead should be designed such that all stresses are in the elastic region at the design depth. This allowed first principal calculations to be used to develop and refine the scantlings for the flat bulkhead. FEA was then used to verify the assumptions used in the first principal calculations and identify any issues in the design. As the adopted design methodology is to ensure that all stresses remain elastic at the design depth, only linear FEA is required to verify this.

4. FACETED DOMES

4.1 CONCEPT

Torispherical or hemispherical domes are not generally stiffened, the inherent stiffness of a dome, coupled with an appropriate plate thickness adequately compensates for the lack of stiffening. Within a faceted dome the angles created by the intersections of the facets represent a type of inherent stiffener. The angles also complicate the inclusion of stiffeners. For these reasons plate-only solutions (i.e. non-stiffened) faceted domes have been considered.

4.2 GEOMETRY SELECTION

Initially 5 faceted dome geometries were considered:

- 2 frequency octahedron;
- 3 frequency octahedron;
- 4 frequency octahedron;
- 3 frequency icosahedron;
- 4 frequency icosahedron.

These domes are shown from left to right in Figure 1. The domes were all based on a 10m diameter pressure hull.

Figure 1: 5 Initial Faceted Dome Geometries

The dome variants were compared on complexity, ratio of enclosed volume to surface area, length of welding, ease of integration into the pressure hull, elastic stress within facets, and plate wastage during manufacture.
The comparison showed the four frequency icosahedron geometry to be the most promising: therefore, all further development of the faceted dome concept has been based on this geometry. The four frequency icosahedron geometry contains five different facet sizes. The maximum edge length of a facet is 1.63m and the maximum facet area is 1.14m^2.

4.3 STRUCTURAL DESIGN

During initial structural assessment the elastic stresses in the facets were considered. The largest facet in each geometry was assessed as a simply supported flat triangular plate. The elastic stresses were required to remain below 75% of the minimum material yield stress for Q1(N) steel under hydrostatic loading at the design depth. This elastic stress limit is a typical limit applied to account for fatigue in pressure hull design [4]. The faceted dome used a plate thickness of 64mm throughout this study.

The 64mm thick faceted dome was found to have a weight increase of 21% over a torispherical dome of the same thickness; however, it also had a 33% increase in enclosed volume.

4.4 MANUFACTURING

Two potential manufacturing methods were considered in this study:

- Formed/Welded Construction: A series/chain of facets is produced by performing a series of line bends in a flat plate: these chains of facets are then welded together to produce a faceted dome. This production method would significantly reduce the amount of welding required;
- Fully Welded Construction: Each facet is cut from flat plate and welded together to form a dome.

In order to assess which manufacturing method was the most feasible, Subject Matter Experts (SMEs) in heavy plate manufacture and welding were consulted.

4.4 (a) Formed/Welded

The findings from the SME consultations suggested that the formed/welded construction method was an unfeasible manufacturing method for a submarine pressure hull faceted dome due to the limitations of the plate bending process. The findings were as follows:

- Facet Chain Geometry: There is a requirement for a significant amount of excess/green material (250-300mm) on the ends of the facet chains in order to be able to exert sufficient bending moment in to the chain to produce the required bends. This would need to be removed after the bending process leading to a high degree of plate wastage;
- Accuracy of Bending Process: An accuracy of +/- 1° may be possible regarding the bend angle, and an accuracy of +/- 2mm may be possible regarding the bend location. Although these accuracy levels may seem adequate, the cumulative accuracy between adjacent facet chains becomes unacceptable. At locations in the facet dome where 4 chains of facets meet, the misalignment between facets could be over 200mm, due to the accuracy of the bend angle alone. Therefore, it would be necessary to improve the accuracy of the bending process or limit the length of the facet chains to achieve suitable tolerances in the formed/welded faceted dome;
- Plate Thinning: When a plate is bent to form facets, the plate thickness at the bend location will reduce. The extent of plate thinning the faceted dome would experience could not be determined in this study, as it isn't possible to accurately predict the extent of plate thinning without bending a plate of the required thickness and grade. To compensate for the plate thinning, the plate thickness may need to be increased; which would lead to a weight inefficient design as the majority of the plate would be thicker than necessary.

The SME consultation has shown that in order for the formed/welded production methods to be suitable for constructing a faceted dome it would require short facet chains. The shorter the facet chain the less beneficial this production method is as the amount of welding increases. Therefore, it was concluded that the formed/welded construction method was not suitable for a submarine faceted dome.

4.4 (b) Fully Welded – Traditional Welding

It is possible to weld the plate thicknesses required for a faceted dome using traditional welding techniques (e.g. TIG welding). However, it will be a time consuming process to construct a faceted dome as the required full penetration welds will require multiple passes and pre-heat. Therefore, it is highly likely that a faceted dome welded using traditional welding techniques would contain significant distortion and residual stresses which would be difficult to control or alleviate. Although weld shrinkage and the resultant distortion are issues on traditional pressure hull domes, it is much more significant on a faceted dome due to the increased amount of welding. The concept faceted dome has 354m of weld seams compared to an equivalent diameter torispherical dome, which had 61m. Therefore, fully welding a faceted dome using traditional welding was discounted as a feasible manufacturing method.

4.4 (c) Fully Welded – Advanced Welding

As a fully welded faceted dome using traditional welding was discounted, advanced welding techniques were considered. Advanced welding techniques such as electron beam welding or laser welding may prove the fully welded construction to be feasible. This study has focused only on electron beam welding due to SMEs experience.

Electron beam welding has many advantages that make it well suited for use in submarine construction; these benefits are not solely limited to faceted dome construction:

- It is possible to weld plate up to 300mm thick in a single pass;

- It is a relatively quick process compared to traditional welding. There is no need for any pre-heat and the speed of welding in 100mm thick plate is approximately 1metre/minute;
- As electron beam welding is a hydrogen free process there is no risk of hydrogen cracking;
- As there is no pre-heat requirement there is a very small heat affected zone, minimising potential distortion. Additionally the resulting thickness of an electron beam weld is 5mm regardless of plate thickness. As steel welds shrink by approximately 10% of weld thickness, the welds themselves suffer minimal shrinkage;
- There is no consumable required for electron beam welding which makes it possible to cross welds without issue. This provides greater flexibility for the welding sequence.

However, there are a number of issues which would need to be resolved before electron beam welding could be used to manufacture a faceted dome:

- It is advisable to perform electron beam welding under a vacuum. However, partial vacuum has been shown to provide a suitable level of precision to weld using a local vacuum head, and negate the need for a large vacuum chamber to house the structure;
- 30% of the power of the electron beam emerges from the back of the weld and needs to be absorbed;
- An electron beam weld leaves a rough surface finish which requires grinding back or capping to improve fatigue life;
- There is a need to eliminate residual magnetism in plate prior to welding to avoid the beam path being affected;
- The key to a high quality electron beam weld is accurate plate preparation. Electron beam welding requires a square butt joint, preferably with machined edges. The maximum allowable gap between plates is 1% of the plate thickness; however, small gaps can be filled with a suitable filler material. Due to the faceted dome geometry the required chamfer is not constant for all facets, thus the complexity of manufacturing the facets will increase.

It was concluded that the drawbacks to electron beam welding were not insurmountable and that it represents a feasible production method for the faceted dome.

4.5 ARRANGEMENT

Due to the faceted dome's similarity to a dome with regard to internal shape, it is not thought to have a significant impact on the submarine general arrangement options. However, the inclusion of penetrations into the faceted dome and the connection of the faceted dome to the pressure hull require consideration.

The major penetrations i.e. torpedo tubes, stern tube, air turbine pump insert, etc. may be able to be built in to the faceted dome during fabrication, rather than the traditional approach of completing the dome fabrication and then cutting out holes for the penetrations. Building in the penetrations eliminates the need for cutting and welding after the dome fabrication is complete, which would reduce the distortion present in the final dome. Ideally, to build in the large penetrations, the casting or forging for the penetration will be in the shape of a single facet. Where the penetrations are too large to fit in a single facet, two or more facets would be replaced by the penetration insert. However, this built in penetration insert concept relies on the penetration location aligning with the facets; this may not always be possible, particularly in the case of the torpedo tubes. The integration of small penetrations is not considered to be significantly different than for a traditional dome, other than some limitations on location.

A straight sided faceted dome cannot be directly connected to the cylindrical section of a pressure hull. Two potential solutions have been proposed. The first solution would be to project the edge of the facets adjacent to the base of the dome in the plane of the facets until they meet the cylinder, see Figure 2. This would result in a serrated edge to the cylindrical pressure hull which would likely not be conducive to an automatic welding process.

Figure 2: A method of connecting the faceted dome to the cylindrical pressure hull

The second solution is to project the facet edges adjacent to the pressure hull cylinder in the radial direction such that this bottom row of facets only have a single curvature, i.e. two straight edges and one curved edge. It should be possible to produce these facets by plate rolling rather than pressing. This was concluded to be the more viable option and was used through the remainder of the faceted dome study.

4.6 STRUCTURAL ANALYSIS

In order to compare the structural response of the faceted dome to a traditional dome a series of non-linear finite element analyses were undertaken to understand the failure mode of the faceted dome.

All analysis was undertaken using ANSYS FE software and used a 10m diameter, 64mm thick faceted dome modelled using Q1(N) high tensile steel. An elastic-perfectly plastic material model was used for the faceted dome and an elastic material model applied to the pressure hull to ensure the model failed in the faceted dome. The faceted dome FEA model also included a section of stiffened pressure hull cylinder, 4 frame bays longs, attached to the dome.

4.6(a) Faceted Dome FEA – No Imperfections or Distortions

The non-linear FEA for the faceted dome showed a ring of high stress at the base of the faceted dome, similar to a hoop stress, see Figure 3 and 4. Locally, each facet also undergoes bending typical of a flat plate with constrained edges. This hoop stress combines with the facet bending stress to cause the facet dome to collapse in the lower two rows of facets in an overall geometric collapse of the faceted dome.

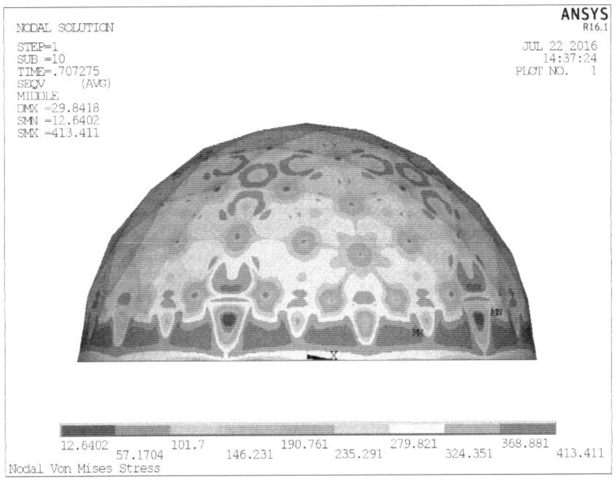

Figure 3: Von Mises stress in the faceted dome at the point of collapse

The failure pressure of the faceted dome was 7.07MPa compared to a collapse pressure of 8.09MPa for a torisphere of the same diameter and thickness which includes build imperfections [3].

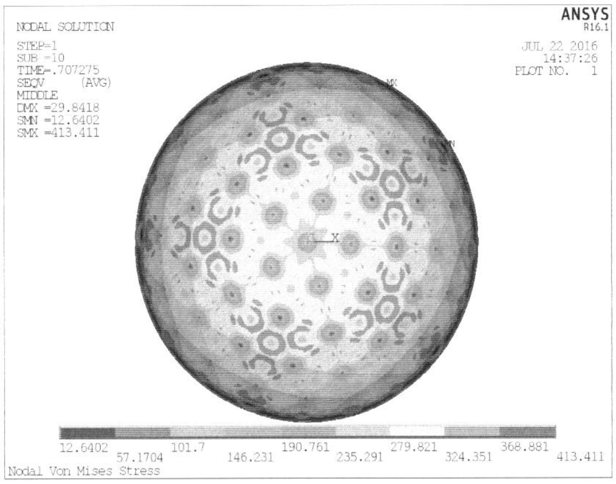

Figure 4: Von Mises stress in the faceted dome at the point of collapse

4.6(b) Faceted Dome FEA –Sensitivity Study

To ensure the area of higher stresses present at the base of the faceted dome were not an effect of the finite element modelling and would be present in reality, a series of sensitivity studies were undertaken.

A boundary condition sensitivity study was completed in which a number of different boundary conditions were analysed, which ensured that they were not influencing the failure of the faceted dome.

It was thought that the difference in thickness between the thicker faceted dome and the thinner pressure hull plate may be contributing to the area of high stress at the base of the dome. Therefore, a model was analysed using the same thickness for the faceted dome and the pressure hull. This analysis showed the same stress pattern indicating that it is not a result of differing plate thicknesses.

4.6(c) Faceted Dome FEA – Out of Circularity

As the sensitivity study showed the high stress region at the base of the faceted dome was not a result of the applied boundary conditions or the change in plate thickness, it was thought the perfectly circular base of the faceted dome was causing this region to be much stiffer than the rest of the dome. As perfect circularity will not be present in the as built faceted dome, the effect of non-circularity at the base of the dome was investigated.

A mode 5 Out Of Circularity (OOC) imperfection was applied to the pressure hull and the base of the faceted dome. The mode 5 OOC imperfection was chosen to result in a peak amplitude coinciding with the maximum displacement in each segment of the faceted dome. The maximum amplitude of the OOC was 0.5% of the pressure hull radius.

The OOC imperfection resulted in a 4.5% reduction in collapse pressure when compared to the perfect faceted dome. The base of the faceted dome still contained a region of high stiffness at its base, although it was reduced compared with the perfect faceted dome. The faceted dome still failed in a similar mechanism and region with the OOC imperfection, as can be seen in the figure below.

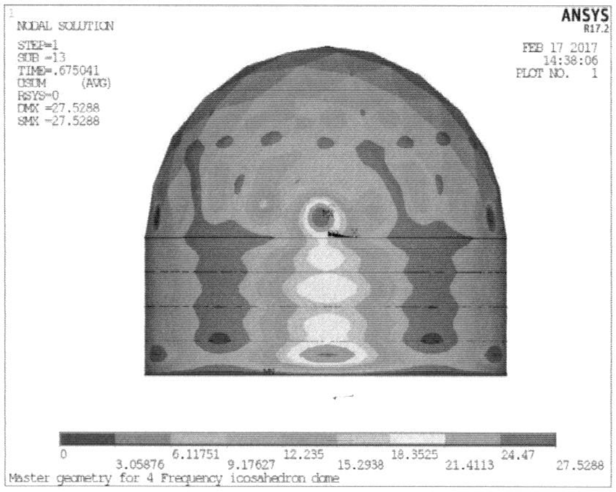

Figure 5: Faceted dome displacement with an OOC imperfection

Therefore, it has been concluded that the region of high stiffness/high stress at the base of the faceted dome is a result of the facet dome geometry and not solely an effect of the perfectly circular base.

4.7 IMPERFECTION STUDY

As traditional domes fail in a buckling collapse mode, they are extremely sensitive to initial shape imperfections. Consequently they are subject to relatively high reserve factors during design, and tight build tolerances in manufacture. If the faceted dome can be demonstrated to be more predictable in its failure mode and less sensitive to initial imperfections than traditional domes, it may be possible to justify a reduction to the reserve factors applied. A reduction in reserve factor will corresponds to a weight saving in the faceted dome. A series of FE analyses were completed to investigate the sensitivity of the faceted dome collapse to shape imperfections. The following imperfections were determined to be feasible build imperfections (where t is the faceted dome thickness):

- Case 1: One facet offset by ±0.2t, ±0.15t, ±0.1t, ±0.05t normal to the facet face at the dome crown;
- Case 2: One facet offset by ±0.2t, ±0.15t, ±0.1t, ±0.05t normal to the facet face one row in from the dome edge;
- Case 3:One facet twisted by ±0.2t, ±0.15t, ±0.1t, ±0.05t at the dome crown (i.e. one corner displaced ± and opposite side displaced ∓);
- Case 4:One facet twisted by ±0.2t, ±0.15t, ±0.1t, ±0.05t 1 row in from dome edge (i.e. one corner displaced ± and opposite side displaced ∓);
- Case 5:One node at a facet junction on the dome crown displaced inwards by 0.2t, 0.15t, 0.1t, 0.05t;
- Case 6:One node at a facet junction 1 row in from the dome edge displaced inwards by 0.2t, 0.15t, 0.1t, 0.05t;
- Case 7:One node at a facet junction on the dome crown displaced outwards by 0.2t, 0.15t, 0.1t, 0.05t, adjacent facets displaced inwards by 0.2t, 0.15t, 0.1t, 0.05t respectively (i.e. total 4 cases);
- Case 8: One node at a facet junction on the dome crown displaced inwards by 0.2t, 0.15t, 0.1t, 0.05t, adjacent facets displaced outwards by 0.2t, 0.15t, 0.1t, 0.05t respectively (i.e. total 4 cases).

The results of the imperfection study showed that imperfections at the crown of the dome had no effect on the collapse pressure, location or mechanism. Only the local stress and displacement patterns were altered by the presence of the imperfections.

The imperfections applied to the facets one row in from the dome edge did result in reduction in collapse pressure for most imperfection cases, with the largest reduction in collapse pressure being 5%. A summary of the imperfection results is below:

- Case 1: No effect on collapse pressure or location;
- Case 2: A reduction in collapse pressure of up to 5% for offsetting the facet radially outwards. Imperfections which offset the facet inwards had no effect on collapse pressure as the dome collapsed in a different segment to where the imperfection was applied: the imperfection marginally increased the strength of the dome locally;
- Case 3: No effect on collapse pressure or location;
- Case 4: A reduction in collapse pressure of up to approximately 1.5%, no change in collapse location;
- Case 5: No effect on collapse pressure or location;
- Case 6: A reduction in collapse pressure of up to approximately 1%, no change in collapse location;
- Case 7: No effect on collapse pressure or location;
- Case 8: No effect on collapse pressure or location.

The imperfection study has shown that the faceted dome is only sensitive to imperfections in the region where the perfect faceted dome fails, for the imperfection magnitudes investigated in this study. Misalignment of facets at the dome crown only alters the local stress and displacement without having any global impact. As the maximum collapse pressure reduction is only 5% for an imperfection displacement of 20% of the plate thickness, it appears that the faceted dome is much less sensitive to shape imperfections than traditional domes. Further analysis of combinations of imperfections would be required to be able to formally conclude this.

4.8 FACETED DOME CONCLUSION

The study has shown that the faceted dome is a viable pressure hull end closure constructed from flat plate and may be an alternative to traditional domes. The faceted dome would suffer a weight penalty to provide the same diving depth as a traditional dome. The FEA has shown the collapse pressure of the faceted dome is relatively insensitive to initial shape imperfection compared with traditional domes. Thus, it may be possible to reduce the weight penalty of a faceted dome by reducing the reserve factors it is subject to in design. Additionally, the study has shown that the faceted dome is only sensitive to imperfections at the base of the dome: therefore, it may be possible to optimise the plate thicknesses of the facets to reduce the weight penalty compared with traditional domes.

The viability of the faceted dome concept may ultimately hinge upon its manufacturability. The study has shown that advanced welding techniques may hold the key to the faceted domes fabrication, though the technologies must become much more widely adopted in industry before they are seriously considered for use in submarine pressure hull construction.

5. FLAT BULKHEADS

5.1 CONCEPT

As previously mentioned, flat sandwich bulkhead end closures have previously been used on conventional submarines with pressure hull diameters under 8 meters. Flat end closures have some advantages over domes such as:

- Construction from flat plate and reduced sensitivity to build imperfections compared with cold pressed double curvature domes;
- A flat bulkhead may require lower reserve factors than a traditional dome due to the different failure modes;
- The general arrangement will result in a more useable internal and external space;
- Potential for easier inclusion of penetrations into the end closures.

However, they will be heavier than traditional domes and it has been thought that on larger diameter submarines (e.g. nuclear submarines with diameters greater than 10 meters) the weight penalty will result in the flat end closure becoming unfeasible. This has been investigated in this study.

5.2 GEOMETRY SELECTION

The pressure hull diameter selected for the flat bulkhead study was 11 meters. The arrangement study considered the forward most compartment and forward under-casing arrangement on the Astute Class submarine as this is a typical SSN arrangement. This allows the study to comment on the arrangement benefits of a flat end closure and quantify the weight penalty that a large diameter flat end closure would have over a traditional dome.
A sandwich bulkhead comprising deep webs between supporting flat plate was selected for this study, as it is more structurally efficient than a single flat plate arrangement and it allows the internals of the sandwich bulkhead to be used for tank or storage space, improving the space utilisation in the submarine arrangement.

5.3 STRUCTURAL DESIGN

There is minimal published structural design guidance for flat end closures. The UK commercial code on the design of pressure vessels [5] contains guidance on the design of flat ends for internally pressurised vessels. However, the guidance is mostly focused on the design of unsupported flat ends. The UK MoD [4] guidance suggests designing flat end closures according to internal safety bulkhead design guidance, but doesn't provide specific guidance on the design of a grillage style flat end closure. Therefore, this study has based the structural design of the flat bulkhead on first principle calculations and ensured the structure remains elastic at the design depth.

The basis for the flat bulkhead structural design was a sandwich bulkhead constructed from Q1(N) steel with vertical and horizontal webs inside the sandwich structure. The structural design of the flat bulkhead aimed to minimise the structural weight while providing a buildable and maintainable end closure.

An initial study into the longitudinal extent of the sandwich bulkhead suggested that a bulkhead length of 2 meters provided a suitable weight/structural capability solution. Additionally, a bulkhead length of 2 meters provides a large enough internal space to be usable, buildable and maintainable.

To minimise the weight of the flat bulkhead, a small spacing for the vertical and horizontal webs would be desired as this will provide stiffness and reduce the required thickness for the bulkhead plates. However, in submarine design, the flat bulkhead web spacing will be driven by arrangement constraints, namely the penetrations associated with the weapons handling and discharge system (weapons embarkation hatch, torpedo tubes, air turbine pump inserts etc.) and deck locations. There is usually minimal flexibility in these systems and the end closure structure will have to be optimised around the fixed size and locations of these penetrations and the fixed locations of the decks. It is envisaged that the major penetrations will all be located in the "cells" between the vertical and horizontal webs rather than the penetrations impinging on the web structure. Additionally, the web spacing must allow sufficient space for the end closure to be constructed, surveyed and maintained. These constraints drive the required web spacing's in the order of 1.5 – 2 meters. The web spacings may be uniform to increase ease of manufacture, or differing web spacing's can be used throughout the end closure to minimise weight while satisfying the arrangement constraints.

The design methodology for the flat bulkhead is that the vertical and horizontal webs in the bulkhead along with the attached bulkhead plate act as I beams and resist the lateral bending force applied to the end closure. The plate between the webs is then assessed as a flat plate subject to a uniform pressure with suitable edge constraints. There will also be in-plane loads applied to the bulkhead structure from the hydrostatic pressure acting on the pressure hull around the end closure. However, design guidance [4] suggests that a structure designed to withstand the lateral loads will be capable of resisting these in-plane loads. Therefore, the lateral pressure load is considered to be the driving load case. Although the bulkhead webs are designed as I beams to resist bending, their large size means shear buckling of the webs is a potential failure mode and must be considered.

To further reduce the structural weight of the flat bulkhead, secondary stiffeners may be added to the bulkhead plates and/or the internal webs. When selecting the level of secondary stiffening and the stiffener geometry, the constraints they place on the arrangement and end closure fabrication must be considered. Although adding secondary stiffeners to the bulkhead plating and all internal webs will reduce the total weight of the end closure it increases the manufacturing cost and difficulty. It also significantly increases the length of welding in the end closure, which increases the through life survey and maintenance costs. The secondary stiffeners also limit the useful storage/tank space in the flat bulkhead, reducing some of the arrangement benefits. This study elected to use secondary stiffening on the bulkhead plates only.

The structural arrangement of the flat bulkhead can be seen in Figure 6 and the scantlings for the flat bulkhead can be seen in Table 1.

Figure 6: Structural arrangement of the flat bulkhead. Top left – Entire structure. Top right - aft bulkhead plate and secondary stiffeners removed for clarity. Bottom – Forward bulkhead and primary stiffening only

Table 1: Flat bulkhead scantlings

Item	Dimension (mm)
Forward bulkhead plate thickness	50
Aft bulkhead plate thickness	50
Vertical web thickness	50
Horizontal web thickness	50
Secondary stiffener web thickness	45
Secondary stiffener overall depth	450
Secondary stiffener flange thickness	60
Secondary stiffener flange width	225

5.4 FINITE ELEMENT ANALYSIS

A linear finite element analysis was performed on the flat bulkhead arrangement to determine the accuracy of the first principal calculations and highlight any areas of the structural design that may require additional consideration. The FE model included the flat bulkhead and three and a half frame bays of cylindrical pressure hull to capture the interaction between the end closure and the rest of the pressure hull. The pressure hull, bulkhead plating and bulkhead webs were modelled as shell elements, while the secondary stiffeners and ring stiffeners were modelled as beam elements. A selection of the FEA results can be seen in Figures 7 and 8.

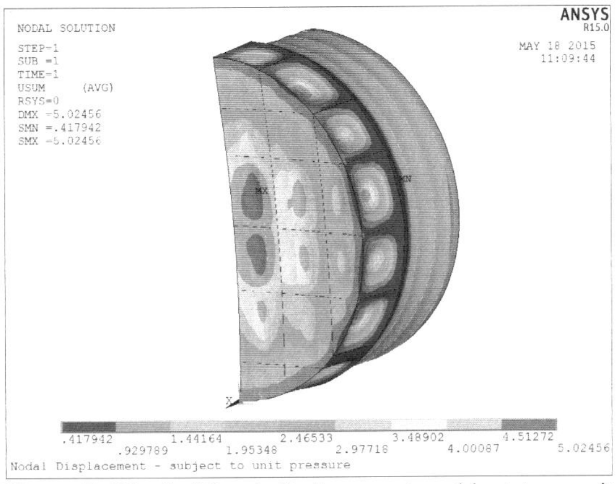

Figure 7: Flat bulkhead displacements subject to a unit load

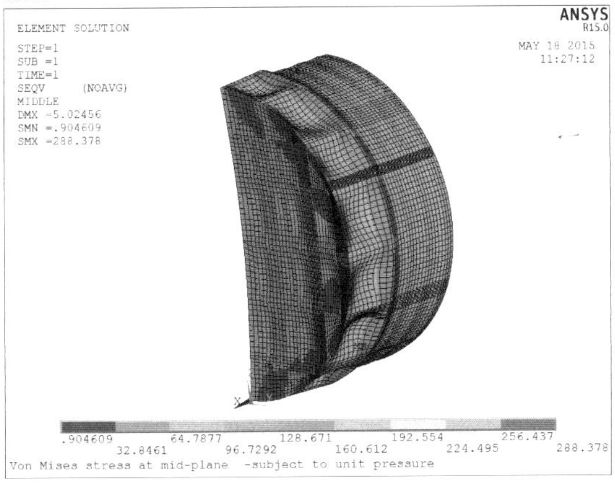

Figure 8: Flat bulkhead Von Mises stresses subject to a unit load

The FEA showed the following:

- The first principal calculations used to initially select scantlings did not accurately account for the deflection in the bulkhead plate. The calculations only considered the local plate deflection and did not account for the global deformation of the end closure. Therefore, first principal calculations under estimated the deflection in the bulkhead plate;
- The FEA showed significant deflection and stresses in the pressure hull plate between the two bulkhead plates. It was therefore determined that a ring stiffener, although not required to prevent shell buckling, was required within the sandwich bulkhead to limit stress and deflection;
- High stresses were seen at the connection of the secondary stiffeners to the bulkhead web plates, these high stresses might lead to fatigue issues through the life of the end closure as it undergoes the cyclic diving loads. Attention must be given to these connection points during the bracketing design. Bracketing design was not considered within this concept study;
- The FEA showed that the aft bulkhead plate was lightly loaded and only the plate that acts as a flange to the bulkhead webs contributed significantly to the strength of the end closure. Therefore, the secondary stiffeners and significant amounts of the aft bulkhead plate can be removed or thinned to reduce the weight of the flat bulkhead.

The FEA has shown that the first principal calculations are suitable for use in initial concept design for obtaining initial scantlings and weight. FEA must then be used to further develop the structural design and identify the areas of the structure to focus on in detailed design.

5.5 MANUFACTURING

The manufacture of the flat bulkhead is relatively straight forward compared to fabricating traditional domes. The flat bulkhead is constructed of flat steel plates welded together with secondary stiffening attached. The fabrication sequence will need to be well planned to maximise the use of automated welding and ensure as much of the fabrication as possible is complete before the end closure is attached to the rest of the pressure hull, once attached access to the flat bulkhead for welding and integration of penetrations becomes more difficult.

This design has ensured that all major penetrations fall in the bulkhead plate between the internal webs. This allows for easier integration compared to traditional domes as a single section of plate has the penetration insets welded into it, this minimises the risk and extent of potential distortion when integrating penetrations into the end closure.

5.6 ARRANGEMENT

A flat bulkhead end closure allows for a more efficient use of internal pressure hull space as deck height is not lost due to the curvature of a dome. Additionally, it leads to more regularly shaped tanks in the forward under casing space. To quantify these arrangement benefits, a typical SSN forward end arrangement, based on the Astute cut away model, see Figure 9, was adapted to include a flat bulkhead.

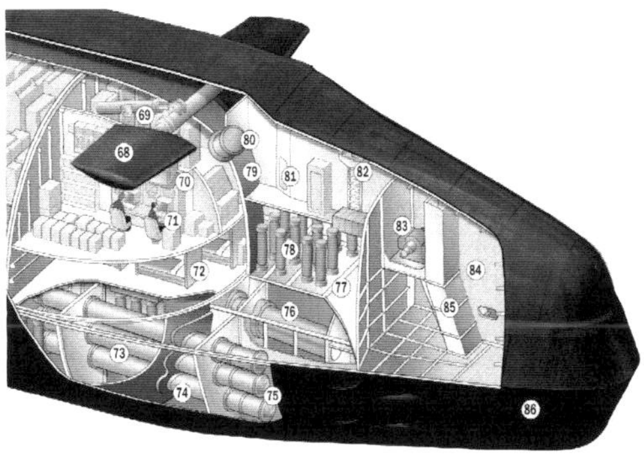

Figure 9: Astute cut-away [6]

It was found that, due to the more efficient use of deck space and use of the bulkhead internals as storage and tank space, the pressure hull length could be reduced by approximately 0.6m. The different shape of the end closure resulted in an increase of pressure hull volume of approximately 60m^3. The flat bulkhead arrangement allowed the external tanks in the fore end structure to be more regularly shaped, which are easier to build, inspect and maintain. It also allowed for a more efficient use of the space under the casing, for the same casing design, leading to the weapons embarkation space doubling in size and providing extra space that could be used for storage or additional tankage.

As the flat bulkhead leads to a change in pressure hull length, it was not deemed appropriate to solely compare the weight of the flat bulkhead structure directly to the weight of a dome structure as this disguises the arrangement benefits of the flat bulkhead. The most suitable comparison was deemed to be the total structural weight and enclosed volume of the forward pressure hull compartment and end closure. The flat bulkhead arrangement was found to be 60% heavier than the dome arrangement. However, the 60m^3 increase in the pressure hull volume on the flat bulkhead arrangement offsets the structural weight increase.

5.7 FLAT BULKHEAD CONCLUSION

The study has shown that a flat bulkhead can form a feasible end closure for a large diameter submarine. The flat plate construction is more straightforward than dome fabrication and the bending failure mode results in reduced sensitivity to build imperfections compared with traditional domes, allowing lower reserve factors to be employed in design and build.

It does incur a weight penalty over a traditional dome, but the increased weight is not prohibitive. The flat bulkhead arrangement also leads to a reduced pressure hull length and increased pressure hull volume, which largely offsets the weight increase. While not as structurally efficient a design as a domed end closure, the flat bulkhead is a feasible option that should be at least considered in future concept designs.

6. CONCLUSION

This study has shown that both faceted domes and flat bulkheads are feasible alternatives to domed end closures. As both are constructed from flat plate they avoid the manufacturing difficulties of double curvature domes, but they encounter manufacturing difficulties of their own, particularly the need for advanced welding techniques in the faceted dome. Both concepts were shown to be less sensitive to shape imperfections than traditional domes as they do not fail in buckling modes, which could lead to lower reserve factors being used in their design and build. Both concepts suffer a weight penalty over traditional domes, but neither renders the structural design unfeasible, and it is felt that both designs could have their weight decreased with further optimisation.

Ultimately, even though both designs are feasible, they are not as structurally efficient as traditional domed end closures. As domed end closures have been used successfully for decades, it will take more than their manufacturing difficulty and sensitivity to imperfections to cause designers to utilise faceted domes or flat bulkheads instead. However, the alternative concepts should be continued to be developed, particularly as manufacturing and material technology advances may lead to the concepts becoming a logical alternative to domes in the future.

7. ACKNOWLEDGEMENTS

The authors would like to thank the SMEs who assisted the faceted dome investigations, particularly TWI who provided the knowledge on electron beam welding.

8. REFERENCES

1. Naval Technology, SSK Collins Class (Type 471) Submarine, *http://www.naval-technology.com/projects/collins*, accessed 3rd May 2017.

2. SAAB, Kockums Gotland Class, *http://saab.com/naval/Submarines-and-Warships/submarines/Kockums-Gotland-Class/*, accessed 3rd May 2017.

3. W Christie and J Hobson, Designing A Flat End For A Round Hole: Innovative Solutions For Pressure Hull Ends Using Optimisation Methods, RINA Warship2008: Naval submarines, 2008.

4. Sea Systems Publication No. 74 (SSP74), Design of Submarine Structures, *Sea Technology Group, Defence Procurement Agency*, June 2001.

5. PD 5500 2012, Specification For Unfired Fusion Welded Pressure Vessels, *British Standards Institute*, 2012

6. Telegraph, Astute cutaway, *http://i.telegraph.co.uk/multimedia/archive/01786/astute_cutaway_1786751a.jpg*, accessed 2nd May 2017

9. AUTHORS BIOGRAPHY

Kevin Heaney is currently a Naval Architect at BMT Defence Services. He has predominately worked in both the submarine department and structures department carrying out a range of consultancy projects on structural design, linear and non-linear FEA and submarine in-service engineering. In his time at BMT he has been heavily involved in the development of a number of alternative submarine end closures. He holds a Master's Degree in Naval Architecture (MEng).

Henry Hickson holds the current position of Naval Architect at BMT Defence Services. He works in the structures department working predominately on consultancy projects utilising FEA. He holds a Master's Degree in Naval Architecture (MEng).

Paul Buckton holds the current position of Chief Engineer at BMT Defence Services. He is responsible for submarine and structural design and analysis projects. He spent the first decade of his career at the UK Ministry of Defence. Since joining BMT he has held a number of roles including Ship Design Manager, Head of Naval Architecture and Chief Naval Architect for various ship and submarine design and support projects.

James Underwood holds the current position of Principal Naval Architect and Structures Group Manager at BMT Defence Services. He is responsible for structural design and analysis projects with a particular focus on the implementation of non-linear FEA. He holds a Master's (MEng) degree in Naval Architecture and Engineering Doctorate (EngD) for research in to the strength of damaged steel ships. He has over 10 years' experience in ship and submarine structural design and analysis.

LCDR (RCN) Brandice Blum holds the current position of Submarine Naval Architect –Structures in the Naval Authority Group, part of the recently formed Submarine Delivery Agency. She is primarily responsible for Submarine Structural Safety. She has 18 years in the Royal Canadian Navy as a Marine Systems Engineering Officer, conducting her Head of Department tour on HMCS Vancouver, a Halifax Class Frigate. Prior to being posted to the UK she was the lead Naval Architect for the Canadian Victoria Class Submarine fleet. She holds a Master's Degree in Naval Architecture (MSc).

Patrick Gough holds the current position of Naval Architect in the Naval Authority Group, part of the recently formed Submarine Delivery Agency. He is predominantly responsible for submarine structures. He holds a Master's Degree in Mechanical engineering (MEng) and Naval Architecture (MSc).

PERFORMANCE VERIFICATION OF THE AIR-CONDITIONING SYSTEM IN A SUBMARINE

Jan D. Wilgenhof, MSc. , MecDes, The Netherlands
Jesús Molina Toledo, MSc. , Navantia, Spain

SUMMARY

One of the large energy consumers in a conventional submarine is the air-conditioning system. The system removes a large part of equipment waste heat as well as metabolic heat and water vapor produced by the crew. It also reduces the humidity of intake air during and after snorting in tropical waters. Air temperatures and humidity in the compartments are important in view of crew endurance and equipment limits (e.g. humid air in electronics).

In the first design stage the air conditioning cooling power is estimated. In the detail design stage it is desired to verify the performance capability with the then available data on structure details, insulation, deck covering, accommodation separation walls, etc., as well as ventilation duct arrangement and air flow rates. Besides cooling power verification it is also desired to know resulting temperatures in the compartments in different sailing conditions. A conventional heat balance cannot provide this because it assumes fixed (maximum allowable) space temperatures and it is performed only for steady state.

Therefore a dynamic air energy balance has been developed. It is a software tool that simulates the heat transfer with the environment and the heat transfer between the spaces in the submarine as well as the thermal contribution of the ventilation flow, based on variable space temperature. Equipment and crew are included as heat sources. Humidity is simulated as water vapor concentration.

Several types of air cooling equipment were developed: a dry air cooler model and a cooling coil model in an air conditioning unit which provides the temperature and humidity ratio difference as well as sensible and latent cooling power. A large number of tests of air conditioning units were used to develop the model. Also attention was paid to efficiency. In this way the tool enables calculation of the resulting temperature and humidity in each space in the submarine as well as power consumption.

The model has been applied on the S80 submarine now under construction. It was requested to determine the cooling power margin, power consumption and future growth margin. The installed cooling power will result in lower compartment temperatures than the maximum allowable in a certain sailing condition. By adjusting the cooling power downward the available cooling power margin can be found. Then the average power consumption for this sailing condition can be found.

The new software tool enables faster and more efficient analysis, in particular if a lot of submarine operation conditions have to be analyzed; demonstration of compliance with building specification requirements; optimization of the cooling power distribution over the spaces; finally it provides more accurate energy consumption data for the submarine endurance and range calculations.

1. INTRODUCTION

The waste heat of equipment in the submarine is removed by a number of cooling systems: the fluid cooling systems, such as the fresh water, sea water and the air cooling system. The air cooling also controls humidity, therefore it is called air-conditioning system. For this system different designs exist: a chilled water plant producing low temperature water which is distributed to the several units with a cooling coil located in the compartments; or it can exist of a number of individual air-conditioning units, where the rejected heat is removed by a fresh water circuit [1]. The chain of fluid cooling systems exists for a number of reasons: safety, efficiency, corrosion and fouling, but heat accumulating in the air cannot be avoided and air cooling is almost always required. In this paper we will focus on the air cooling system with separately installed Air Conditioning Units (ACU's), which remove a large part of the heat generated by equipment and all of the heat generated by the crew as well as humidity.

In the design stage the maximum allowable temperatures in all the spaces are known, as well as the operating conditions and maximum sea water temperature, mostly defined in a contract specification or a similar requirements defining document. A conventional heat balance is carried out to determine the required cooling capacity for each space and thus equipment sizes. The heat balance is the equality between the total amount of heat from the heat loads which is accumulated in the air and the heat removal capacity of the air conditioning system. Other definitions for a heat balance are possible e.g. if the fluid cooling systems are included, but the discussion in this paper will be limited to the energy in the air in the form of heat.

In the following stages when more details become available, such as ventilation ducting and flow rates, bulkheads and deck structure, separation walls in accommodation, coverings, isolation, etc. it is of interest to verify the installed cooling power and to know the actually occurring space temperatures in varying operating conditions. It may be required to demonstrate compliance with requirements before the installation on board is completed.

Also it is desired to have a prognosis of the electric consumption in operating conditions or in general to know the behaviour of the air conditioning system for instance to know how the temperature in a space varies in dependency of equipment power or number of crew present, or ventilation flow rate.

In this paper the aspects involved for the calculation of the occurring temperatures, humidity and the cooling power will be discussed and some example results from the new software tool that is under development.

Waste heat from equipment accumulates in the internal air; is removed by air cooling and is rejected into the environment via a fluid cooling system, see figure 1. A number of air coolers is installed in several spaces. For each space a heat balance is prepared, which in principle contains the terms equipment heat load, human heat load, ventilation heat load, heat transmission through the boundary (walls including pressure hull) and cooling power of the ACU's. Most of the terms in this thermal energy equation are temperature dependent and the software calculates the equilibrium temperature and humidity.

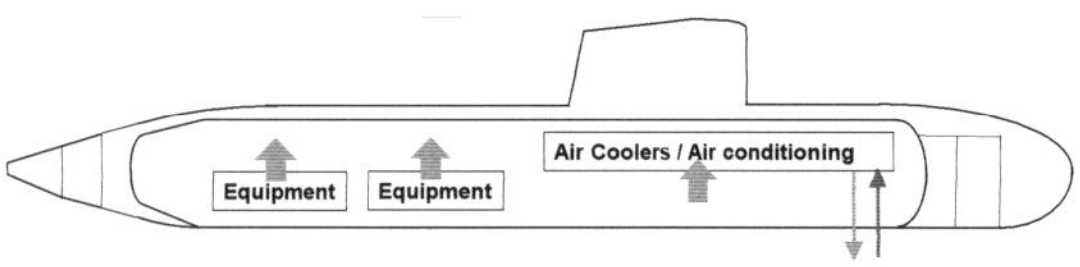

Fig. 1. Waste heat from equipment accumulates in the internal air; is removed by air cooling and rejected into the outboard seawater via fluid cooling.

2. HEAT SOURCES AND HEAT FLOWS

2.1 OVERVIEW

The dynamic air energy balance contains terms describing the heat sources, heat flows and the cooling power. In the figure 2 the situation for one particular space is schematically illustrated. The space contains heat generating equipment, crew members generating metabolic heat (sensible and latent heat) and air cooling equipment.

Furthermore the ventilation flow contributes to the internal energy of the air in the space. The model assumes a uniform temperature in the space (perfect mixing). The density of the air can be assumed constant without large error, as long as temperatures remain within a certain range. Apart from air temperature it is of course possible to take into account other variations and influences or vary other parameters such as outboard seawater temperature, number of persons in a space, the equipment operation (on/off switching and power variation).

The problem is translated into a set non-linear coupled differential equations describing the temperature changes in all the spaces, which are solved simultaneously by a numerical integration method.

The situation in the figure is somewhat simplified because the air conditioning unit may be located outside the space. In this case air is drawn from the space, mixed with ventilation and cooled down in the airco-unit and then returned into the space. In other cases one airco-unit may distribute cool air to a number of spaces.

In the following the elements of the heat transmission model of one space are concisely discussed, except for the cooling equipment, which will be discussed in the next chapter.

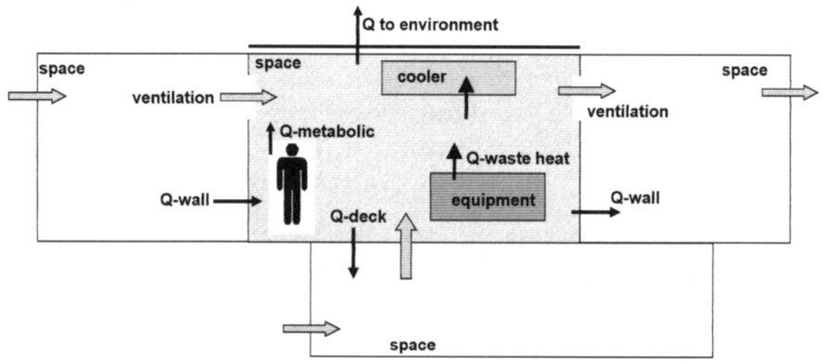

Fig. 2. Heat sources in one space and heat flows to/from adjacent spaces and environment.

2.2 HEAT SOURCES

Several types of heat sources can be defined. The first, most simple one, has the property of constant power, independent of space air temperature. This type can be used for most equipment as a first approximation. For instance cabinets with electronic equipment will generally emit a constant amount of heat, so that the difference between internal electronics temperature and space temperature will remain approximately the same. Electric motors of permanently running equipment will also emit constant heat for the same reason.

The second type has constant surface temperature, so that the heat release is dependent on the surrounding air temperature and can for instance be used to estimate the heat release of a diesel engine (apart from radiation).

The third type is metabolic heat release of human bodies existing of sensible heat and latent heat. As known, both are dependent on the surrounding air temperature and also the amount of work effort. Mathematical descriptions can be found in literature [2], [3]. The latent heat does not contribute to air temperature increase, so that it may be omitted in the thermal balance however it is of interest for the air enthalpy and thus total cooling power.

2.3 HEAT TRANSFER THROUGH WALLS

The heat transmission over boundaries of spaces involves walls, bulkheads, decks, pressure hulls and other boundaries. The basic equation is probably well-known: $q = k.A.\Delta T$ [Watt]

With k = heat transfer coefficient [W/m^2.C], A = area of heat transfer [m^2], ΔT = temperature difference over the wall [C], and q the heat flow rate [Watt]. The direction of the heat flow is dependent on the temperature of the adjacent space, which is unknown on beforehand. In the ΔT – term both temperatures are variable, which causes mutual interaction between the spaces which has to be solved by the simulation model. The determination of the heat transfer coefficient k may require relatively large effort, see [2], [3] and [4], especially when the wall is built up of several layers. If knowledge from measurements is available this will increase the final accuracy. The model is capable to take into account spaces with different density e.g. fluid filled tanks.

2.4 HEAT TRANSFER TO ENVIRONMENT

The same applies more or less to the heat transfer to the environment, it can be estimated by basic physics as first approximation. The submarine can be seen as one object with a well-defined boundary with the environment, the surrounding sea water. The average heat transfer coefficient at the boundary sea water – submarine hull is not difficult to determine, although it may be of interest to include the dependency of the ships speed.

The heat transmission through the hull occurs over different paths (in the following a single hull submarine is assumed): a direct path from sea water through pressure hull, with or without sound insulation layer at the outer side, and with or without heat insulation layer at the inside; a second path may be heat transmission via the superstructure outer hull, the (almost still) sea water in the superstructure and the pressure hull; a third path may be heat transmission via an internal tank and through the pressure hull.

In this way the submarine hull can be divided into a number of different heat transmitting areas with each their own heat transmission coefficient.

2.5 VENTILATION

A submarine is actually a confined space, divided in a large number of individual spaces, some of which are connected to each other by deck openings, a door, a hatch or passageway, while others are connected by air ducts, which together from a network of ducting and spaces. In the submerged condition the air is circulated throughout this network. The ventilation system circulates the air to maintain a uniform atmosphere, i.e. to equalize carbon dioxide and oxygen concentration; to ventilate the battery spaces (to reduce the hydrogen concentration and to contribute to cooling); and to distribute oxygen supply evenly. In snorting condition the air circulation path is different due to the air intake by the diesel engines. In chapter 4 a schematic example of a ventilation ducting network is shown.

The ventilation air flow contributes to the energy content of the air, in other words the increase or decrease of the temperature. The air flow mixes with the air in the space and in this way the ventilation system works as heat transporter (and may cool spaces without air cooler). The final mixing temperature in one space will vary in a complex way dependent on the air pathway through the spaces. The spaces interact with each other via the ventilation flow dependent on the varying temperatures. This causes mutual interaction between the spaces which has to be solved in the simulation model.

The flow rates in all the ventilated spaces need to be known. This can be determined by means of the design flow rates or by calculation of the actual flows from the fan performance curve and the pressure losses in the duct network.

3. AIR-CONDITIONING UNIT MODEL

3.1 COOLING POWER

Several cooler types may be defined. The simplest model is an air cooler which absorbs constant power (or heat). A better approximation of the air cooler behaviour is offered by the NTU-epsilon method, which is described in [2]. This method provides an exchanger model in which the heat flow depends on the air inlet temperature and temperature of some cooling fluid. Other well-known parameters are heat transmission coefficient and area, and cooling fluid properties and flow rates.

The more important cooling equipment is the air conditioning unit (ACU) model. The air cooler in the ACU is modelled as a heat exchanger with heat transfer dependent on temperature and humidity, while cooling power is based on an expression for the air enthalpy decrease. In figure 3 the orange line represents the ACU cooling line (from warm and high humidity air to cooler air with lower humidity). The relation with the total enthalpy decrease is also shown in the temperature – humidity diagram. The functions for total enthalpy, sensible and latent enthalpy in the calculation model have been derived from cooling power and air flow rate measurements on the ACU's that will be installed in the system.

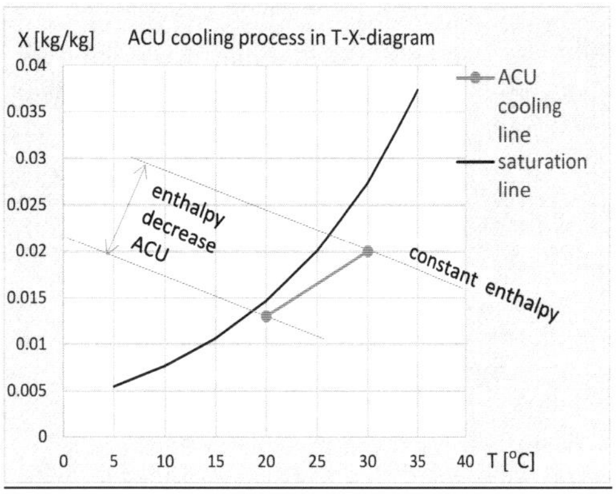

Fig. 3. Temperature – humidity ratio diagram

Figure 4 shows examples of the ACU performance: total cooling power dependent on inlet enthalpy and the total enthalpy decrease in the unit dependent on inlet temperature and humidity ratio. The sensible enthalpy is also dependent on inlet temperature. The models of the different ACU's has been developed using a large number of test data of the ACU's that will be installed on board.

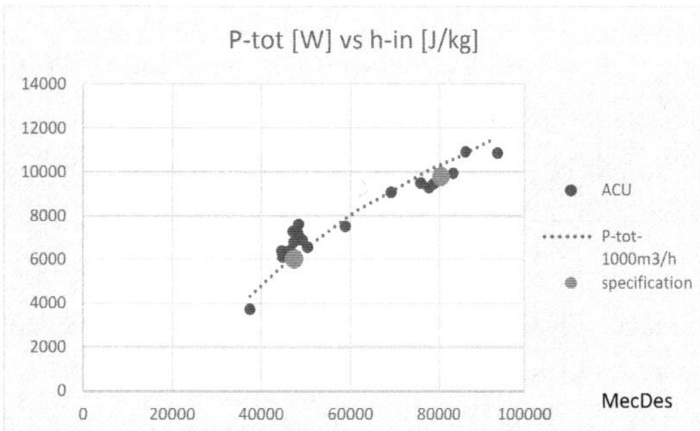

Fig. 4a. Example of cooling power (Watt) versus inlet enthalpy (J/kg), according test data. Dotted line is the model approximation. Blue dots are values acc. specification.

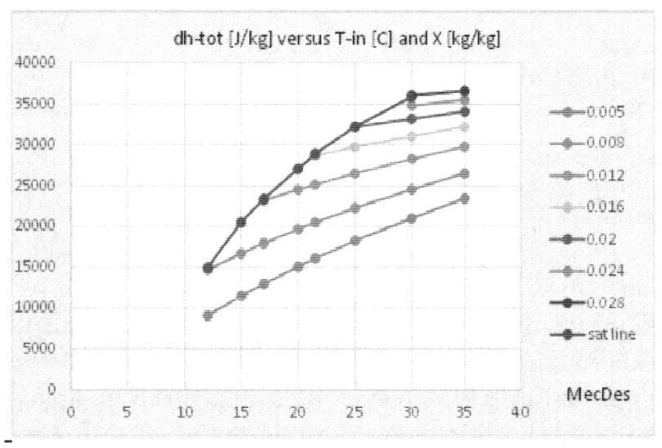

Fig. 4b. Example of an ACU model represented by total enthalpy decrease versus inlet temperature at constant humidity. The limiting curve on the left side is the saturation line.

3.2 EFFICIENCY AND ELECTRIC POWER CONSUMPTION

Equipment considered in this paper are Air Conditioning Units (ACUs) used in S80 Submarine series. Those ACUs are direct expansion equipment through a cooling fluid, consisting in one air-liquid heat exchanger (evaporator), one condenser liquid (water)-liquid (cooling fluid), one compressor and one fan. Condenser phase is done through a fresh water circuit, which is cooled by sea water available on board; therefore the whole heat dissipated through the water means an energetic consumption for the submarine.

The electric consumption of ACUs means an important percentage of the whole electrical consumption of the submarine. This has a direct impact on the discretion during the mission, therefore is of critical importance to reduce this consumption by means of high efficiency ACUs designed for the real working condition. A first step to achieve the maximum efficiency of equipment is to know all the performances in the different operation conditions on board. For that purpose a test program of the ACUs has been done. Different measurements have been taken during the tests for different running conditions. All of these tests have focused to obtain efficiency of the ACUs used on board.

This will enable us to adjust the ACUs behaviour during the submarine operation, with accuracy higher than usual for this kind of design. To know the performance of the ACUs during the first stage of submarine design allows us to obtain efficiency measures of the submarine which have been designed.

The several ACU Performances, which have been considered are:

- Total Cooling Power available (TCP)
- Sensible Cooling Power available (SCP)
- Coefficient Of Performance (COP), which is a measure of the efficiency.
- Electrical consumption of compressor (CC)
  Note: Electrical consumption for the internal fan of the ACU is not the matter of the paper, because it depends on others subject different of thermodynamic criteria.
  Variables considered to obtain the above mentioned performances are:
- Air temperature at inlet of ACU
- Relative Humidity (r.h.) of the air at inlet of ACU
- Cooling fresh water temperature (condenser phase)
  This paper includes the result for one ACU as example, since those results are representative of the whole tests conclusions. Efficiency values obtained through the COP measured in the test allow the designer to obtain electrical consumption for the different operative conditions on board.

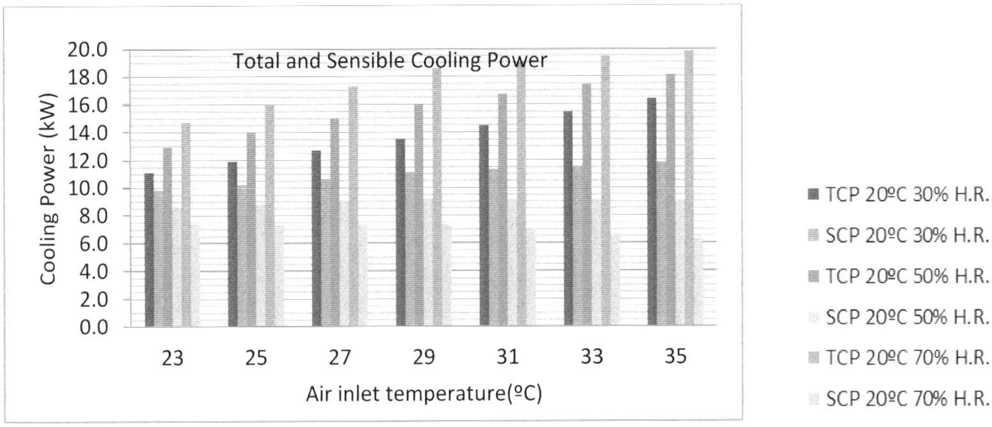

Figure 5 Relationship between TCP & SCP, for fresh cooling water of 30°C.

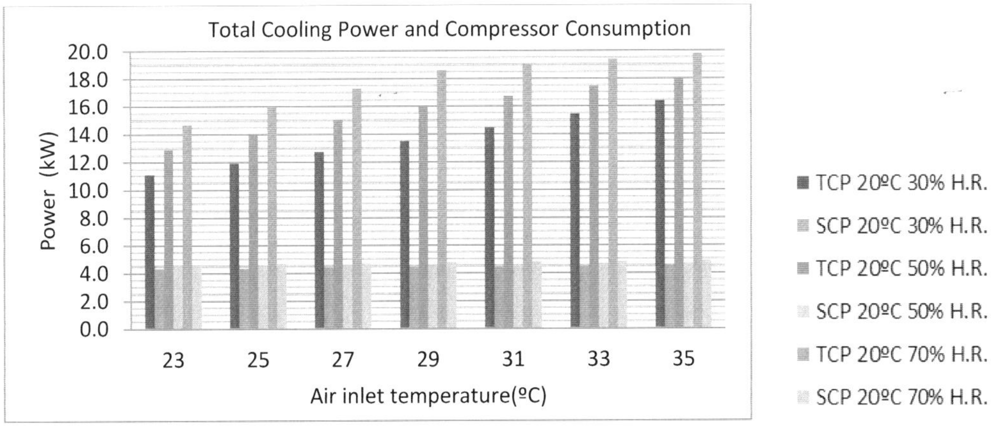

Figure 6 Relationship between TCP & CC, for fresh cooling water of 30°C.

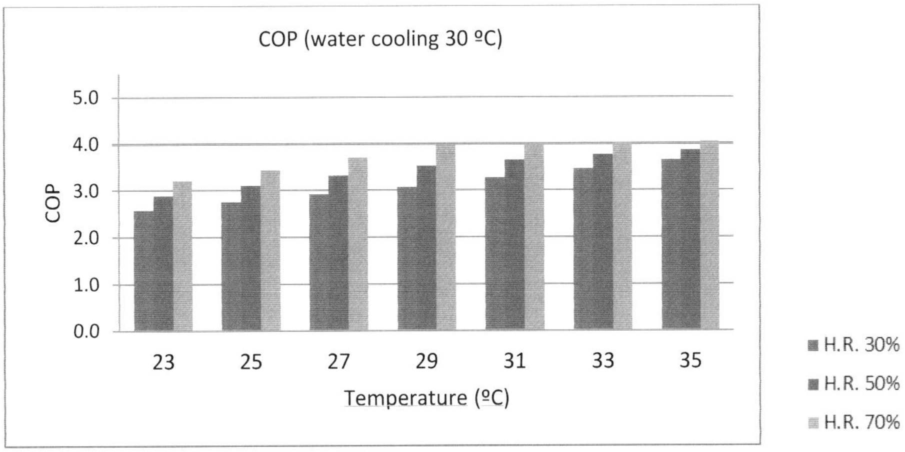

Figure 7 COP values for different temperatures and RH

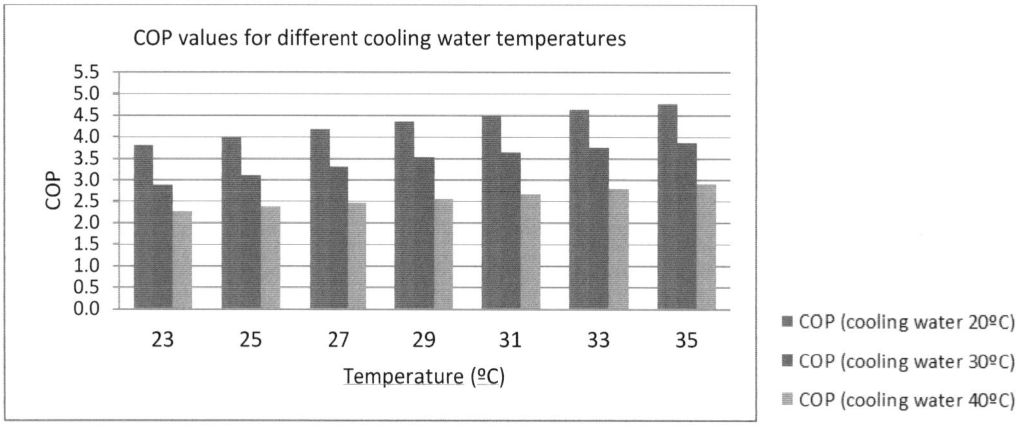

Figure 8 COP values for different cooling water temperatures

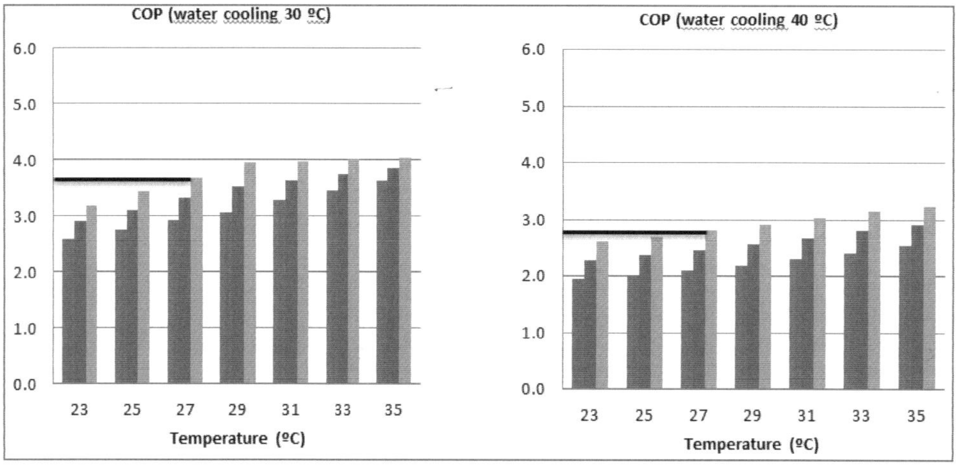

Figure 9 Example. Practical application

Figure 5 shows relationship between TCP and SCP. Both of those values depend on temperature and r.h at the air inlet. The graphic has been obtained with a fresh cooling water of 30°C; it became evident that differences between TCP and SCP are very similar for different temperatures between 20°C and 40°C. The graphic shows how the difference between TCP and SCP is as higher as the increasing of both, temperature and r. h. at the air inlet. In other words, as higher is enthalpy of air inlet as higher is the difference between TCP and SCP.

Figure 6 shows relationship between the TCP and compressor consumption CC for different air inlet temperature and r.h. The graphic has been obtained with a fresh cooling water of 30°C; it became evident that as higher fresh cooling temperature as higher are the differences between TCP and CC. If fresh water cooling temperature is getting lower, also differences between TCP and CC are lower. It is observed that differences observed between TCP and CC are proportional to the fresh water cooling temperature Figure 7shows COP values for different air inlet temperatures in the ACU. Each one of the group of three columns shows COP values for each temperature value and 3 different r.h. values. COP is increasing with both, temperature and r.h. increase, in other words, an increasing of COP (efficiency) is observed with larger air inlet enthalpy.

COP values showed in Figure 7 are for fresh water cooling of 30°C.

A fresh water cooling temperature increase means a reduction of COP values, as showed in Figure 8, where the COP values are for r.h. of 50% for the whole temperature range. Differences observed in the test results for other r.h. values are analogous.

All this information allows the designer to obtain information of the electrical consumption in the earlier stage of submarine design. As shown in the next example.

The example shows a concrete sailing condition of which the designer knows the environmental condition (inside and outside the submarine). It is observed that increasing

10°C the fresh water cooling temperature (in the condenser) the compressor consumption is increased by 1 kW (around 25% increased of each ACU).

Despite the electrical consumption increase observed the evaporator efficiency, measured as delta T, does not change (13°C±0,2). Therefore the TCP of ACU has not significant variation (about 16 kW).

It is also observed that a high r.h. value means that the most of the cooling capacity of ACU is involved in condensing of the water in the evaporator. As far as the contractual requirement allows, this phenomenon should be reduced to the minimum since it is not contributing to the maximum advantage of the available energy.

4. ACU PERFORMANCE RESULTS IN EXAMPLE SUBMARINE

4.1 EXAMPLE SUBMARINE

Almost all input data for the simulation model comes from the S80 Submarine project. The S80 submarine is currently under construction in the Navantia Shipyard in Cartagena, Spain. For reasons of confidentiality some details in the results are withheld or slightly modified. The performance verification was carried out to investigate the margins in capability of the projected ACU's in the submarine and to provide averaged operation time data.

The performance verification includes the determination of actually occurring cooling power and space temperatures in several operating conditions of the submarine, such as submerged transit and snorting, at the maximum sea water temperature.

Further it includes the determination of averaged operation time (duty factors) of individual ACU's. These duty factors depend on the number of ACU's in operation and the heat load.

Finally it includes verification of the capability of the system for the future growth margin on the heat loads.

4.2 VENTILATION NETWORK

The ventilation flow rates for each space are required as input data for the simulation model and were provided from a separate air flow rate analysis. Figure 10 shows an example ventilation network for submerged operation consisting of air ducts and spaces; the blocs indicate spaces and ACU's; the lines indicate the air flows.

4.3 ACU PERFORMANCE - SPACE TEMPERATURES

The analysis was carried out with ACU's performing 100% cooling power continuously in order to verify that they are able to maintain the space temperatures below the maximum allowable. Figure 11 shows the

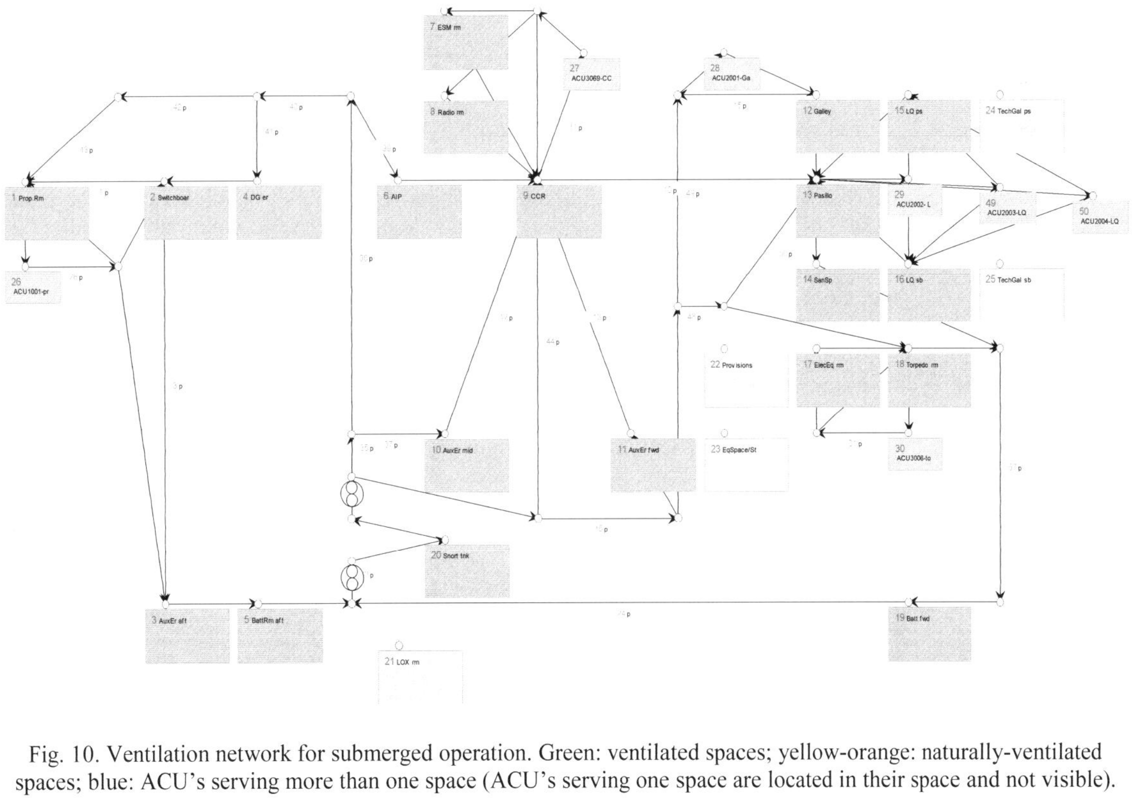

Fig. 10. Ventilation network for submerged operation. Green: ventilated spaces; yellow-orange: naturally-ventilated spaces; blue: ACU's serving more than one space (ACU's serving one space are located in their space and not visible).

equilibrium temperatures in submerged operation condition at 25 C sea water temperature. As expected in only a few spaces the temperature is close to the maximum, in other words a cooling power margin exists. In practice the ACU's do not operate continuously due to the on/off switching by the temperature control. For instance as indicated in the figure the ACU in the galley is capable to reach 17 C but normally this will not occur.

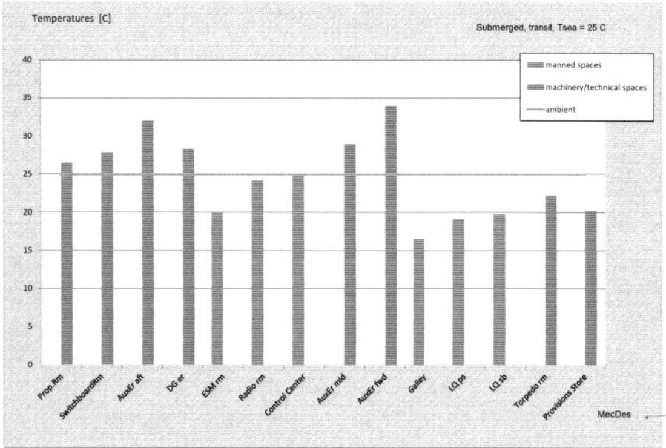

Fig 11. Calculated temperatures in manned spaces and machinery/technical spaces.

4.4 ACU PERFORMANCE - DUTY FACTORS

In the previous paragraph the results are presented of the temperature levels that are reached when the air conditioning equipment is in operation continuously, without temperature control. We have also seen that the sensible cooling power of the selected ACU's is sufficient. So, if less than full power is applied the temperature may rise up to the allowable level and this reduced power is a measure for the time that the ACU is in operation. The duty factor is defined as the fraction of time of operation of an ACU (based on on-off control of the ACU). The reduced sensible cooling power has to be found by trial and error because it is dependent on the temperature.

The table with an example of resulting duty factors of a selected number of spaces confirms that some ACU's have a significant cooling power margin, however this is valid for only one operating condition; other operating conditions have to be investigated as well of course. At lower seawater temperatures the duty factors will also be lower. In this example only temperatures were considered, thus sensible cooling power is relevant. If the calculation is extended to include the requirement of the humidity window then the ACU total cooling power has to be taken into account.

Finally, when the duty factor is known the electric power consumption of the ACU compressor and fan can be found for each ACU. It may serve as one of the consumption inputs for the calculation of underwater time.

Table Duty Factor

Room	available sensible power Watt	calculated sensible power Watt	duty factor %
Auxiliary Engine Room mid	13000	7000	54
Central Control Room	20600	19000	92
Galley	3900	1850	47
Accommodation	7900	4400	56
Torpedo Room	13900	12000	86
	59300	44250	

*) Calculated sensible power is valid for the maximum allowable room temperature

On the average about 75% of available sensible cooling power is required.

4.5 GROWTH MARGIN ON FUTURE HEAT LOAD

Often the building specification contains a requirement for the air-conditioning system in the form of a future growth margin on heat load in a number of compartments, for instance where electronic equipment is located.

In the study on performance verification it was found that if the heat load is increased, the associated increase in cooling power varies for each ACU and is always lower in terms of percentage of heat load.

A 10% growth of the complete heat load resulted in a cooling power increase of 2 to 6 % for the different ACU's. The growth in cooling power is lower than the heat load increase, because this increase is divided over the wall heat transmission, ventilation air discharge and the cooling power (therefore a margin on the heat load is not the same as a margin on the cooling power).

5. CONCLUDING REMARKS

In the detail design stage it is desired to verify the performance capability of the airconditioning with the available data on structure details, insulation, deck covering, accommodation separation walls, etc., as well as ventilation duct arrangement and air flow rates. Also the actual temperatures and cooling power that will occur in normal conditions are of interest as well as the energy consumption.

The equilibrium temperatures in the spaces depend on each other due to the mutual interaction of wall heat transmission and ventilation contribution. Moreover the ACU cooling power is temperature dependent, so that a simulation model is required to solve for the resulting temperatures in the spaces. The software tool calculates all the terms in the heat balance that are dependent on temperature and determines the temperature at which all terms are in equilibrium as well as sensible and latent cooling power of the ACU's. The software tool also includes a water vapour gas concentration model for the calculation of the relative humidity in every space.

The performance verification of the airconditioning system comprised several subjects:

- The determination of actually occurring cooling power of each ACU in several operating conditions of the submarine, such as submerged transit and snorting.
- The determination of duty factors of individual ACU's. When the duty factor is known the electric power consumption of the ACU compressor and fan can be found for each ACU. It may serve as one of the consumption inputs for the calculation of underwater time.
- The verification that the cooling power is sufficient to cope with a possible larger heat load in the future.

The software tool enables faster and more efficient analysis, in particular if a lot of submarine operation conditions have to be analysed; demonstration of compliance with building specification requirements; optimization of the cooling power distribution over the spaces; finally it provides more accurate energy consumption data for the submarine endurance and range calculations.

6. REFERENCES

1. Optimising the submarine cooling system, J. McLaughlin, 2013, ASC, Australia.

2. 2001 ASHRAE Fundamentals Handbook, chapter 8, Thermal comfort.

3. Mechanical Engineering Handbook, Chapter 9 airconditioning and refrigeration, SKWang And Z LAvan, 1999.

4. VDI-Wärmeatlas, achte auflage, 1997.

5. Heat balance S80 submarine [internal project document], Navantia, Spain.

6. Dynamic heat balance of a submarine, paper presented at MAST Asia, Yokohama, May 2015 and UDT Europe, Rotterdam, The Netherlands, June 2015.

7. Evaluation of ACU tests and determination ACU parameters, apr. 2016, [internal project document], MecDes.

7. ACKNOWLEDGEMENTS

The authors are grateful to the shipyard Navantia s.a. (Cartagena, Spain), who has given the opportunity and consent to prepare this paper as well as contributions and support during the carrying out of the work.

8. AUTHOR'S BIOGRAPHIES

Jan D. Wilgenhof is senior mechanical engineer and technical consultant in MecDes, The Netherlands. He has been working on the S-80 project for several years, on a large number of mechanical systems.

Jesús Molina Toledo is HVAC senior engineer in the mechanical engineering department in Navantia, Spain.

ASSESSMENT OF FLOW GENERATED NOISE ON NAVAL PLATFORMS

C Carter and **G Hawkes**, Frazer-Nash Consultancy Ltd, UK
C Smethurst, BAE Systems Maritime Ltd, UK

SUMMARY

Noise generated by fluid systems can play a significant role in the on-board and underwater radiated noise of a naval platform; however, until recently, attempts to manage this were challenging due to a lack of prediction capabilities and mitigation strategies.

This paper describes a programme of work led by BAE Systems with the support of Frazer-Nash Consultancy Ltd to quantify vibro-acoustic performance during the design process and develop new tools to support decision making. This new capability enables flow generated noise performance to be progressively assured against requirements and the performance to be balanced against programme, cost and risk considerations.

1. PAPER OUTLINE

Designing for low signatures and through life signature management is key to maintaining the capability of naval platforms. There are multiple facets to a platforms signature which include:

- Underwater Radiated Noise (URN)
- Thermal
- Target Echo Strength
- Visual
- Electro-magnetic
- Infra-red
- Radar cross-section

The objective of this paper is to present an approach which has been developed to assess URN from naval platforms, specifically the contribution originating from internal fluid systems (termed here *Flow Noise*). A comprehensive review of all aspects of URN signature assessment and management is given in [1].

2. INTRODUCTION TO FLOW NOISE

2.1 WHY THE FOCUS ON FLOW NOISE

To help understand risks to meeting URN targets and assigning appropriate design effort, it is convention to assign signature budgets to components, systems and platform areas. Given the complexity of such platforms and other conflicting requirements for capability, space, weight and cost, inevitably these budgets will need to be traded and actively managed.

The Noise and Vibration Engineering Department (NAVED) at BAE Systems provides guidance on the design of quiet naval platforms and their internal systems. Traditionally this has focused on machinery noise, e.g. rotating parts, which includes providing design guidance to suppliers and determining resilient mounting arrangements. The success of this guidance, and improved ability of threat sonar systems to detect acoustic signature, means that historically less significant sources of noise, such as flow through fluid systems, now warrant more detailed attention.

2.2 WHAT IS FLOW NOISE?

The majority of flow fields encountered in engineering are turbulent in nature, meaning that even under nominally steady operating conditions, flow variables such as velocity, density, temperature and pressure exhibit local, time varying fluctuations.

The pressure fluctuations within a turbulent flow field are generally short in their range of operation. They are responsible for the local accelerations of the fluid in turbulent flow. Due to viscous action, a proportion of the flow's kinetic energy of the fluid is continuously dissipated into heat. This cascade of energy gives rise to component pressure drops and the dissipation of mechanical power.

Although these pressure fluctuations can be large, they do not propagate within fluid systems and as such do not present a significant contributor to URN except in cases where these fluctuations are able to mechanically excite pipework components at the flow boundaries (Flow Induced Vibration).

However, due to the finite inertia and compressibility of the fluid a small fraction of these pressure fluctuations are propagated within the fluid as acoustic waves. Although the magnitude of these acoustic waves can be small, pipework systems act as effective conduits for this energy to propagate around platforms with relatively little loss. Acoustic waves are also particularly effective at coupling to the surrounding pipework structure at discontinuities (e.g. bends) and, hence, can become a contributor to hull vibration. This component of flow noise is often termed the *Pressure Ripple*.

The proportion of the flow's mechanical energy which is transduced to acoustic energy is small (typically <<1%). This proportion (termed the *Acoustic Efficiency*) is not fixed and is very sensitive to changes in flow conditions (i.e. flow velocity magnitude in relation to the speed of sound) and geometry (particularly effective at sharp geometric discontinuities).

Furthermore, the acoustic efficiency can also be augmented by particular flow conditions for different working fluids. For example, for gas systems there is a significant increase in noise when pressure differentials give rise to local supersonic flow and the formation of shock waves. For liquids, noise can be greater when low output pressures induce localised cavitation and bubble collapse. In this case, the impact of cavitation on flow noise occurs well before the cavitation becomes sufficiently severe to cause surface damage to components.

Although acoustic efficiencies are typically small, on a naval platform the flows of mechanical energy can be very large (of the order of MWs for the propulsion system). Hence, even small changes in acoustic efficiency can have a significant impact on URN. This provides a great opportunity to develop a low noise design, but also a significant risk.

2.3 SUSCEPTIBLE PLATFORM SYSTEMS

There are a variety of fluid systems that can affected by flow noise. These include:

- Power plant
- Fresh, seawater and chilled water cooling systems
- Hydraulic and lubrication systems
- Compressed air supplies
- HVAC systems
- Other components (e.g. Reverse Osmosis plant)

Many of these systems are intrinsically cross-connected due to their function (e.g. heat transfer) or are configurable to be cross-connected for backup in case of battle damage. Hence, these systems can be both sources of flow noise or conduits for the propagation of acoustic energy noise to the hull.

Note that although HVAC systems contribute to URN, much of the focus on these systems is concerned with ensuring compartment habitability.

2.4 FLOW NOISE CHALLENGES

In simple terms flow noise presents a particular challenge to naval platforms because it is a newer challenge than the predominantly mechanical vibration issues which have been considered historically. Hence, the industry's experience in design for low flow noise and the assessment, control and diagnosis of flow noise are still developing.

Mitigations already stipulated by NAVED for fluid systems, including isolating systems using flexible hoses and pipe clips, and limiting the flow rate in pipework, have been successful in reducing noise generation and transmission.

However, flow noise remains difficult to design against, control and diagnose because:

- Signatures is a transversal discipline that cuts across conventional platform design teams covering major equipment, functions or area.
- The acoustic efficiency is highly sensitive to the precise flow conditions and geometry.
- Flow noise sources are generally broadband in nature and may extend up to high frequencies (tens of kHz). This can make flow noise difficult to compute numerically (see Section 2.6) and provides potential opportunities for threat sonar to detect and identify platforms.
- The types of signature mitigation required to control fluid-borne noise are different from those commonly employed to control structure-borne vibrations originating from machinery.
- Fluid systems act as effective conduits for acoustic energy to propagate around platforms. Hence, the source of the noise and the point at which the energy breaks-out to the seawater may not be collocated nor within the same system.

2.5 REQUIREMENTS FOR FLOW NOISE ASSESSMENT

As flow noise is strongly influenced by flow conditions and geometry, the most beneficial time to influence design for low noise is at the Concept stage where the system architecture is being defined and many solutions are possible. Changes to reduce noise are possible in later stages of projects and indeed are necessary as risks are identified; however, fluid systems are by their very nature distributed and congested. Hence, changes become more difficult to enact due to potential conflicts with the system's core performance specification space, weight, cost, time and commissioning.

Therefore, the assessment strategy and toolset needs to be able to advise the platform programme during all stages of its life cycle from Concept through to Acceptance.

2.6 OPTIONS FOR FLOW NOISE ASSESSMENT

There are three main routes through which platforms and their associated sub-systems may be assessed for URN and flow noise risk.

2.6 (a) Trials

Platform trials are the most established and recognised means of understanding URN. This can comprise commissioning trials on major systems and/or acoustic ranging prior to the platform entering service and after major overhauls and upgrades.

Such trials provide valuable evidence as to the signature characteristics (magnitude, frequency content and directionality) for both narrow and broadband components. Using advanced signal processing techniques it is also possible for acoustic ranges to provide some localisation of noise sources across the hull which can be extremely beneficial in terms of diagnosing potential issues.

However, depending on trials alone to assess and control URN is not without challenge and limitations. Trials occur at the end of a lengthy design and procurement process spanning perhaps a decade and there may be limits to what can be achieved in terms of diagnosing and fixing URN issues after initial ranging. The diagnosis of noise from fluid systems is particularly complicated as their continual operation is often critical to maintain the health of platforms (e.g. cooling systems). This limits the ability during trials to independently switch systems on/off, to understand their impact in URN, as a controlled experiment.

2.6 (b) Experiments and Component Tests

An alternative to full platform or major systems trials is to develop bespoke test facilities (e.g. flow loops) to assess flow noise. This can be used to provide absolute inputs to platform level signature calculations, perform design trade-off studies on candidate components (e.g. valves) or develop flow noise correlations for more general application in system assessment. In turn the data acquired could be used to validate the capability of numerical simulation tools if used.

The significant advantages of experiments over trials are that they may be conducted at the system or component level, are readily controlled (i.e. independent of the whole platform's wider activities) and remain a tool throughout the design process.

The challenges of experimentation are the associated time and cost of extensive experimental programmes, the ability to practically represent a wide flow network and often simple flow loops are not capable of operating at the pressures and temperatures found in service.

2.6 (c) Numerical Simulation

The use of numerical simulation tools to assess flow-related URN and support low noise design is a relatively new approach. This is in part due to the historic focus of URN and also due to the tools required to simulate flow noise not being available for previous generations of naval platforms.

However, with the increasing focus on flow-related URN and with modern advances in computer power and multi-physics analysis capability, the prediction of flow noise is becoming practical in an industrial sense. But flow noise prediction outright remains challenging because:

- The physics involved are complicated and couple together the engineering sciences of flow-acoustics, acoustics and structural dynamics. This requires a cascade of inputs and software tools each which introduce uncertainty.
- The acoustic efficiency is very sensitive to the geometry of the system/component. Often this is not readily available early on in the design process, where the greatest potential for a low flow noise design exists.
- The library of validation evidence available to instil confidence in design decisions made based on complex numerical simulations of flow noise is currently limited.

Despite these challenges, the potential opportunities and benefits of numerical simulation are considerable:

- It does not require a physical manifestation of the system, nor the flow conditions to appraise flow noise. This affords for cost and time efficient, "what-if" studies to quantify the characteristics of system flow noise and understand how this can be controlled.
- Front-loading the assessment of flow-noise in the design process reduces the risk of additional costs and delays associated with diagnosing and fixing signature sore-thumbs during commissioning.
- The cascade of physical processes which contribute to URN, namely source generation, fluid-borne and structure-borne propagation to be assessed in isolation or unison as is appropriate. Therefore, the efficacy of noise control measures to the source or transmission path can readily be tested. This should enable space and pipe layouts to be optimised in conjunction with signatures during the design phase.
- Periodic assessments of system flow noise can be conducted at staged design gates to provide an audit trail of progressive assurance to the customer throughout the programme.

3. PROGRAMME OVERVIEW

The need for continual improvement in radiated acoustic noise has meant that the physical phenomena in fluid systems require a greater understanding. Furthermore, when a design change is requested it is vital that the resulting radiated acoustic noise impact is quantifiable in order to provide evidence for or against that change.

Given the options presented above, what is clear is that the greatest opportunity to improve URN assessment is to be able to incorporate numerical simulation tools into the more traditional trials and experimental approaches. There are two key steps to develop such an approach:

- Addressing the uncertainty and limitations associated with numerical tools to make them into a practical tool.
- Developing a strategy for how such a suite of tools could be deployed to make flow noise assessments within the design and assurance process.

To this end a programme of work was conducted led by BAE Systems Maritime's NAVED and supported by Frazer-Nash Consultancy Ltd (Frazer-Nash) to improve the understanding of flow noise, its prediction and control.

NAVED developed a flow noise test rig on which various pipe fittings are being tested and compared for noise generation or attenuation, including valves, bends, junctions, silencers, etc. Frazer-Nash focussed on improving the understanding of the underlying flow noise mechanisms, and helped in extending the range of test rig findings to real systems by developing appropriate modelling tools.

The outputs from the tools developed by Frazer-Nash Consultancy form part of the low noise design evidence, and as well as reinforcing mitigation strategies already in place they are helping to identify new palliatives. These tools are a valuable element in driving continual improvement for radiated acoustic noise.

The route to developing this toolset is described in the following subsections and the detail of the toolset and its validation are discussed in Sections 4 and 5.

3.1 TOOLSET IDENTIFICATION

This stage identified the tools that could be used to support the low noise design and to provide progressive acceptance through the design process. An initial literature review into the state-of-the-art of flow noise assessment was conducted and identified three candidate tiers of methods:

- **Basic:** Scaling laws, rules of thumb, codes and standards that could be used to quickly appraise systems for flow noise risks.
- **Hybrid:** Bespoke tools which look to extend the applicability of the basic methods (and to enable system-wide assessments) using small amounts of targeted numerical analysis.
- **High Fidelity:** Numerical simulation tools which would be used for detailed design trade-offs, appraising high risk items and final assurance activities (i.e. Computational Fluid Dynamics, Vibro-acoustic Finite Element Analysis and Statistical Energy Analysis).

Further details of the toolset are given in Section 4.

3.2 TOOLSET APPRAISAL

Once the potential tools had been identified the high fidelity tools were subject to extensive testing to understand the limits of what could be achieved within forthcoming programmes using numerical methods. To begin with this considered a suite of fundamental test cases to demonstrate and quickly benchmark the predictive capabilities of the tools. After some initially promising results this progressed quickly to consider the application of the tools to real systems. The systems were selected based on trials evidence that suggested they significantly contributed to URN, to see whether or not the tools could represent the signature features.

In particular, the latter half of this stage helped to develop the strategy for deploying the toolset to ensure that the assessments were efficient, fit-for-purpose and gave a level of insight and fidelity that was commensurate with the design maturity of a given system.

3.3 DEPLOYMENT

Once the approach and tools had been identified and appraised they were deployed in two major ways to support the platform development projects. Firstly to provide flow noise predictions (absolute levels, spectra and risk summaries) for systems at the appropriate design gates.

Secondly, the tools were used in between each design gate to help the system design teams address emerging issues and explore component and configuration trade-offs.

4. THE FLOW NOISE TOOLSET

4.1 DEPLOYMENT STRATEGY

The nature of URN assessment means that three key elements need to be assessed and combined to appraise the impact:

- The noise sources generated by the turbulent flow behaviour.
- The fluid-borne propagation through pipework systems.
- The structure-borne propagation of energy through the platform and its ultimate radiation to the surrounding seawater.

For any given system the predictive uncertainty for these elements in isolation is a function of the capabilities of the tools to represent the configuration and the uncertainty in the input parameters, which is a function of the design maturity. It is highly likely that the art-of-the-possible and quality of input data will be different for each element. Given the large number of fluid systems to potentially consider on a given platform, it is necessary to develop a strategy to ensure that efforts to appraise URN are appropriately balanced in accordance with each system's perceived signature risk (which may change through time).

The approach adopted by Frazer-Nash during the course of this work is articulated by the Onion Skin analogy shown in Figure 1. Here the onion is split into angular segments representing the three components which comprise the URN calculation. The various layers or skins of the onion represent levels of calculation fidelity which reflect the increasing precision (confidence) required at each design gate.

In order to get to the core it is necessary to remove each skin in turn, thus progressively balancing, focussing and improving the assessment of URN risk. Whilst it may be possible to reach the core more quickly, for example by conducting a high fidelity simulation in a single segment, the results will be dominated by the residual uncertainty in the other segments; hence the output will be unsatisfactory.

In practice, this means to get a balance of uncertainties of the three segments, the uncertainty in each segment needs to be understood and subsequently refined in a balanced and progressive way.

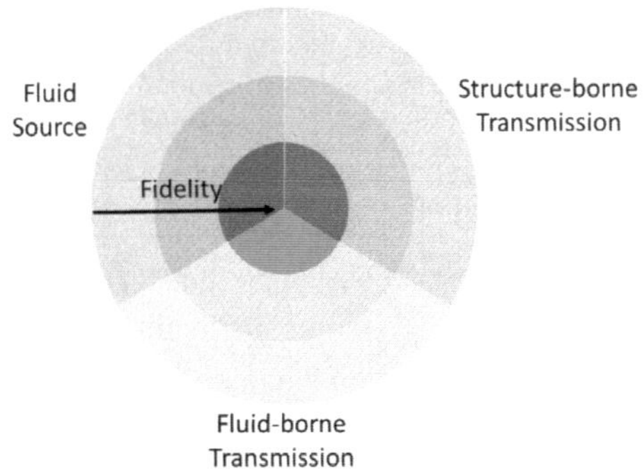

Figure 1: Onion skin analogy for the progressive use of flow noise assessments throughout the design process.

4.2 TOOLSET DETAIL

The following subsections provide an overview of the tools and methods within the toolset. They comprise:

- Basic
  - System Screening Tools
  - Source Estimation Methods
- Hybrid
  - Coupled Flow Modelling and Source Estimation
  - Stochastic Noise Generation and Radiation
- High Fidelity
  - Transient Computational Fluid Dynamics (CFD)
  - Vibro-acoustic Finite Element Analysis (FEA)
  - Statistical Energy Analysis (SEA)

4.2 (a) System Screening Tools

One of the most beneficial outcomes from this programme of work was the development of a set of simple and generic metrics to provide a first pass appraisal for flow noise risk. These metrics assessed:

- Component duties (i.e. pressure drop and flow rate).
- Estimates of acoustic efficiency – i.e. proportion of the mechanical energy that could be converted to acoustic energy.
- Threat of increased noise due to cavitation as indicated by the dimensionless cavitation number.

These metrics were readily calculated from the outputs of the hydraulic simulation tools used to design the fluid systems, which yielded the following project benefits:

- Providing an earliest possible indication to the designers of potential flow noise risks resulting from changes to the design. This would help promote a "stop and appraise" approach where a designer could consider whether the particular change was really necessary or provide the means to flag the issue to the URN assessment team for more detailed scrutiny. In effect this shifts the multiple, albeit quick, basic assessments from the URN assessment team to the system designers, which is a more efficient use of resources in the transversal team.
- Tracking the flow noise risk throughout the various design gates to assess the impact of design modifications and to see whether or not the risk is being reduced.
- Comparing the relative risk between different fluid systems under different design, functional or platform area ownership to ensure that efforts made to mitigate risks target the dominant risks at a whole platform level.

4.2 (b) Source Estimation Methods

There are a variety of methods and empirical datasets available to estimate source levels and spectra for typical noise producing components in liquid and gas systems. These cover:

- Throttling devices in liquid systems [2]
- Throttling devices in gas systems [3]

- Pumps [4]
- Pipework fittings (e.g. bends, [5])

In each case the calculation process starts with some simple definition of the hydraulic duty of the device (i.e. pressure change and flow rate). This is multiplied by an acoustic efficiency term which describes the transduction of mechanical energy into propagating acoustic energy.

Once the overall source levels are known it is possible to use maximum flow velocities and device length scales (e.g. diameters) to estimate the one-third octave band in which the sound power spectrum will peak. Typical roll-off rates per decade from this band can be used to construct the full spectrum.

4.2 (c) Coupled Flow Modelling and Source Estimation

The codes and standards described in Section 4.2 (b) are limited to simple geometries to which the underlying correlations are tuned. When the geometries deviate from these standards the approaches begin to breakdown.

Although high fidelity transient simulations can be used in this case to understand the flow noise sources (see Section 4.2(e)), it is often possible and often more efficient to use relatively simple flow modelling to assess the difference between bespoke components and those from which the correlations are derived. This analysis may be conducted assuming steady flow and the outputs, plus some physical arguments with regards to flow noise scaling laws, to extend the applicability of the standards.

4.2 (d) Stochastic Noise Generation and Radiation (SNGR)

The generation of noise by flow requires unsteady flow behaviour and turbulence. Steady flow calculations using Computational Fluid Dynamics (CFD) do not predict the temporal turbulence variation of the flow directly but do provide statistical information to describe the nature of the turbulence field. The SNGR approach uses this statistical description of the turbulence field to synthesize the time varying flow field for use in noise calculations, which is used to derive the magnitude and frequency content of the noise sources. SNGR is particularly useful to appraise the source characteristics for novel components which fall outside the scope of the standards, so long as their geometry is available.

4.2 (e) Transient CFD Simulation

In theory transient compressible CFD simulation alone should be sufficient to compute both the hydrodynamic and acoustic field components. This requires the use of a class of methods termed *Large Eddy Simulations* (and its derivatives) in which the large scale turbulent fluctuations in the flow are calculated explicitly (Figure 2).

However, in practice the numerical schemes employed to conduct the calculations are optimised for the hydrodynamic component. Hence, propagating acoustic waves tend to be rapidly and non-physically dissipated and false monopole and dipole source distributions can be predicted due to the finite levels of convergence of the underlying equations.

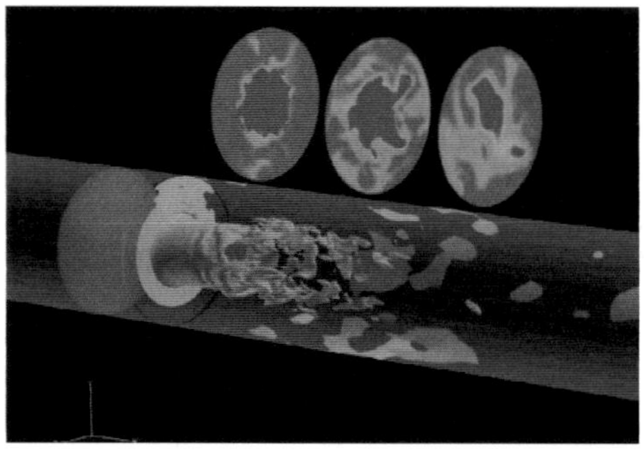

Figure 2: Large Eddy Simulation of flow through a sharp edged orifice plate, showing the instantaneous and unsteady flow structures.

Hence, it is necessary to avoid the false source terms and to extract an equivalent set of acoustic sources that can be used as inputs to a separate acoustic propagation solver. This is commonly done by the use of the Lighthill analogy [6], the consequence of which is that the acoustic sources in free turbulence flow can be represented by an arrangement of quadrupoles which are dependent on the flow's density and velocity fields.

These methods are at least an order of magnitude more computationally demanding that other methods and require significant computing hardware and expert analysts. However, the results can be extremely impressive.

Figure 3 compares the acoustic source spectrum calculated for a standard orifice plate geometry [7] using such high fidelity CFD methods with that obtained using the valve liquid noise standard [3]. Both results have been normalised to experimental data obtained from NAVED's flow-acoustic test loop.

The results indicate that the standard methods over-predict the one-third octave band levels by typically 15-25dB, whereas the error from CFD at low frequency is substantially less than 5dB. At higher frequencies some deviations are observed; however, this is attributed to resonances within the flow path due to discontinuities in the flow loop which were not represented in the CFD model.

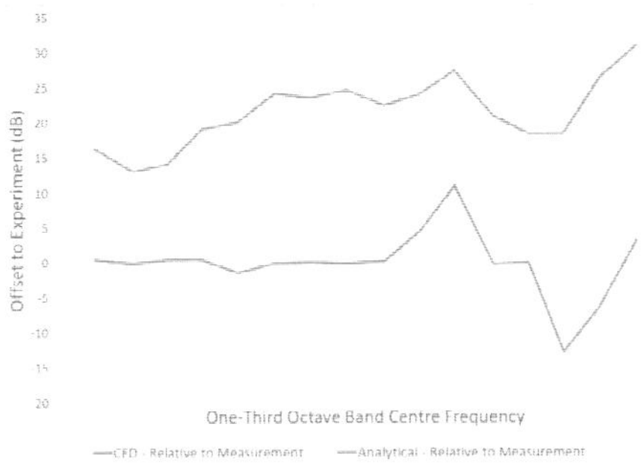

Figure 3: Comparison source spectra for an orifice plate calculated by high fidelity CFD methods and using noise standards to direct measurements from NAVED's flow test loop.

Furthermore, when the test data and CFD calculations are compared for a range of flow rates, the noise source spectra are found to exhibit the same scaling behaviour. Specifically there is a non-uniform scaling across the frequency range with respect to flow rate with higher frequencies being more sensitive. By comparison the standard approach uses a single scaling law for all frequencies.

4.2 (f) Vibro-acoustic Finite Element Analysis

Finite Element Analysis (FEA) has long been an option for the acoustic analysis. However, to date the vast majority of applications of FEA are associated with the calculation of stress, strain and the resulting fatigue or fracture of components. As a result the use of codes tailored for mechanical integrity applications to model vibro-acoustics has been somewhat cumbersome.

However, with growing interest in vibro-acoustics, particularly driven by the requirements of the automotive industry, general purpose FEA codes have become more vibro-acoustic-friendly and a number of specific Vibro-acoustic FEA codes have emerged.

Specific features of these codes which make them more readily applicable to vibro-acoustic analysis include:

- A frequency-domain analysis approach.
- The ability to define the excitation and boundary conditions of the flow path using fundamental acoustic source descriptions (e.g. volume-velocity sources), higher order modal forcing and anechoic boundary conditions.
- Ability to couple to the outputs from transient CFD to define direct pressure excitation at component walls and/or flow-noise source terms in the flow path.

The significant benefit of these methods is that it is possible to represent faithfully the component geometry and calculate both fluid and structure-borne energy propagation and coupling within a single tool. Outputs from such tools are naturally narrowband in nature and can be used to investigate the behaviour of tones; however, these are extremely sensitive to model set up and results must be interpreted with caution.

With modern computer power these models can incorporate millions of degrees of freedom. Hence, provided that the computational domain is not too large this resolution can be used to investigate transmission path performance up to high frequencies (tens of kHz). Beyond this the size of the vibro-acoustic stiffness matrix becomes difficult to solve on a practical timescale, although due to the frequency-domain decomposition, each frequency can be solved independently to benefit from modern parallel computing architectures.

4.2 (g) Statistical Energy Analysis

As mentioned in Section 4.2(f), the computational requirement for vibro-acoustic FEA of large systems up to high frequencies can be prohibitive. In this regime an alternative approach is that of Statistical Energy Analysis (SEA).

Within the SEA approach a complex system is modelled as a collection of coupled sub-systems. The transmission of energy between sub-systems is described by coupling loss factors. The technique requires the systems to span a number of wavelengths of interest (such that each frequency band is modally-dense and representative ensemble-averaged quantities can be calculated).

Although the SEA approach is specifically configured to cover the high frequency performance of vibro-acoustic systems, it is possible to combine the technique with vibro-acoustic FEA to help extend the lower and upper limits of each code respectively to cover the medium frequency range.

5. VALIDATION AND VERIFICATION

An important aspect of the work conducted to date is the need for confidence in the tools to provide the evidence necessary to make design decisions. For this work programme confidence has been built in three ways:

- Cross-verification of tools between the different levels of fidelities.
- Application of the tools to assess known URN features on legacy systems (See Section 3.2).
- Direct validation of the tools using NAVED's flow test loop (e.g. Figure 3). This facility has also been used to test the efficacy of signature palliatives to address both sources and transmission aspects of URN and to develop acoustic test data for COTS items.

A particular challenge throughout this project was how to develop a simple process with which to collect evidence for, measure and communicate the confidence in, and value added by the toolset. The approach adopted is depicted in Figure 4 and used a matrix to cover the types of systems and components that the toolset would need to assess through different stages of the design process (from concept through to final assurance).

At periodic points throughout the project, validation and verification evidence was reviewed and the matrix was populated using a 6-point scoring system based on a modified Technology Readiness Level (TRL) approach. Here the readiness of the technology referred to the confidence that could be drawn from assessment with a score of 1 reflecting a proposed but untested approach and a score of 6 reflecting a fully tested and deployed tool. Each cell within the matrix was populated both with a score and hyperlinks to documents and reports presenting the latest validation evidence. The matrix was also used to plan and prioritise method development and validation and verification work throughout the project.

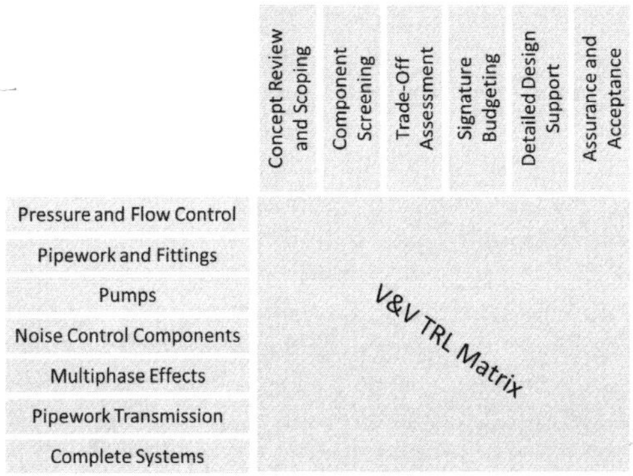

Figure 4: Matrix used to track to validation and verification status of the flow noise toolset.

6. ASSESSING FLOW NOISE PALLIATIVES

The discussion to date has focussed on the requirement to assess and mitigate flow noise risks through the design process. However, the power of the toolset has also been used to appraise the potential effectiveness and design considerations for a variety of treatments that could be utilised to control flow noise in the case that issues are identified during the platform build or commissioning. These treatments comprise a range of physical modifications and retrofits to attenuate flow noise at its source or via its transmission paths (both in the fluid and structure).

The design space for these treatments has been first explored using the numerical simulation toolset and the findings have been substantiated by experimental testing of prototypes on a bespoke flow test loop.

7. FUTURE APPLICATIONS

As noted earlier, the numerical approach has been devised to fill a gap in assessment capability that cannot readily be filled by experiments or platform trials. During the design phase, experiments and data from trials on legacy platforms can be used together to provide the necessary assurance.

The logical consequence of this is that the assessment approach is equally well suited to complement the information that can be derived from trials. Hence, it can be deployed to support platform commissioning or to diagnose issues with in-service platforms, the latter of which may make use of the palliative technologies which themselves have been developed and de-risked through numerical simulation.

8. CONCLUSIONS

This paper has explored the role that flow generated noise and fluid systems play in URN and has discussed the opportunities and challenges for their assessment. The identification, testing and successful deployment of the assessment toolset to support the design and assurance activities for naval platforms has also been discussed and its role in complementing and enhancing current practices.

The key conclusions from this work are:

- Flow noise is becoming an increasingly significant contributor to URN.
- Numerical modelling can play a role alongside traditional trials and experimental methods to help understand flow noise risks. These methods are not without their challenges but with careful and pragmatic use can significantly help to appraise and reduce flow noise risks and are particularly applicable in the early stages of the design process.
- Given the complexity of fluid systems it is important to have a method to progressively deploy the hierarchy of available methods so that the assessments are commensurate with the levels of confidence required and system maturity at successive design gates.
- The advancement in the application of both high and hybrid fidelity methods had perhaps the most visible impact within the project. However, valuable gains were made by distilling the understanding into simple screening methods which could be used to aid designers.
- Validation evidence obtained from bespoke flow test loops showed that the predictive capability of high fidelity techniques was extremely promising.
- Understanding the level of confidence that could be drawn from outputs from numerical modelling was important. The modified TRL matrix approach proved to be a simple but efficient means of tracking and visualising the status of the toolset and planning method development and validation activities.

9. ACKNOWLEDGEMENTS

The authors acknowledge the significant contribution of Kenny Aitken from the UK Ministry of Defence to this programme of work.

10. REFERENCES

1. HAMSON, K. G., 'A New Approach to Managing the Acoustic Signature Risk in Warship Design', Warship 2016: Advanced Technologies in Naval Design, Construction, & Operation, 15-16 June 2016.

2. BS EN 60534-8-3:2011, 'Industrial-process Control Valves Part 8-3 Noise Considerations - Control Valve Aerodynamic Noise Prediction method', 2011.

3. BS EN 60534-8-4:2014-02, 'Industrial-process control valves Part 8-4: Noise Considerations - Prediction of noise Generated by Hydrodynamic flow', 2014.

4. FRANCE, D. 'Pump Noise and Vibration Control', Noise and Vibration in Marine Applications, 28 September 2005.

5. BULL, M. K., and NORTON, M. P., 'On the Hydrodynamic and Acoustic Wall Pressure Fluctuations in Turbulent Pipe Flow Due to a 90° Mitred Bend', Journal of Sound Vibration (1981) 76(4), 561-586.

6. LIGHTHILL, M. J., 'Sound Generated Aerodynamically. I. General Theory', in Proceedings of the Royal Society London Series A 211:564-87, 1952.

7. EN ISO 5167-2, 'Measurement of fluid flow by means of pressure differential devices inserted in circular cross-section conduits running full - Part 2: Orifice plates', 2003.

11. AUTHORS BIOGRAPHY

Chris Carter is the Underwater Technology Business Manager at Frazer-Nash and is responsible for a portfolio of projects covering research and technology, future platforms, UUVs and propulsion technology. Chris has ten years of experience in surface ship and submarine projects and a technical background in fluid dynamics, noise and vibration and structural integrity. He has been the technical lead on a number of programmes looking at novel technology for propulsion systems. This has included rim drive systems, lightweight materials for pumpjets and the application of computational methods for the integrated design and analysis of turbomachinery.

Graham Hawkes is Flow and Thermal Performance Group Leader at Frazer-Nash Consultancy Ltd. Graham joined Frazer-Nash in 2005 following his doctorate at Oxford University in unsteady flow measurement and since has focussed on the application of numerical techniques to industrial problems involving unsteady flows, flow noise and flow induced vibration. Graham is a Chartered Engineer (CEng), European Engineer (Eur Ing) and is a Fellow of the Institution of Mechanical Engineers.

Chris Smethurst joined BAE Systems Maritime in 2008 after completing an MSc and a PhD in Computational Science at the University of Manchester, having previously completed an MSci in Theoretical Physics at the University of Durham. Based in the Noise and Vibration Engineering Department, he is part of a team that analyses systems and structures using a range of modelling techniques. These include computational fluid dynamics, finite element analysis and statistical energy analysis.

ON THE ADDED RESISTANCE OF UNDERWATER VEHICLES IN CLOSE PROXIMITY TO REGULAR WAVES

S Daum and **M Greve,** thyssenkrupp Marine Systems, Kiel, Germany
R Skejic, SINTEF Ocean & NTNU, Faculty of Engineering, Trondheim, Norway

SUMMARY

The present study is focused on performance issues of underwater vehicles near the free surface and gives insight into the analysis of added resistance and consequently the speed loss in regular deep water waves. Predictions of the drop loss are based on calculations of the total resistance and effective power in calm water and pre-selected regular wave fields. It has been assumed that the water is sufficiently deep and that the vehicle is operating in a range of small to moderate Froude numbers by moving forward on a straight-line course at constant depth. Two bare hull submarine geometries have been chosen for a comparison from an operational viewpoint of submarines cruising at periscope depth.

NOMENCLATURE

ν	Kinematic viscosity (N s m^{-2})
ρ	Density of water (kg m^{-3})
ϕ	Velocity potential
α	Angle of attack (deg)
ω	Angular wave frequency (s^{-1})
L	Maximum length of UV (m)
D	Maximum diameter (beam) of UV (m)
H	Maximum height of UV (m)
S	Hull surface area of UV (m^2)
V	Form displacement of UV (m^3)
U	Mean forward speed of UV (m s^{-1})
L_W, λ	Wave length (m)
H_W	Wave height (m)
ζ_a	Wave amplitude (m)
c	Bottom clearance (m)
s	Free surface clearance (m)
d	Submergence depth (m)
g	Ground distance (m)
Fn	Froude number
Rn	Reynolds number
BVP	Boundary Value Problem
BEM	Boundary Element Method
FVM	Finite Volume Method
VOF	Volume of Fluid
CFD	Computational Fluid Dynamics
EFD	Experimental Fluid Dynamics
SSK	Conventional Submarine
UV	Underwater Vehicle
UUV	Unmanned Underwater Vehicle

1. INTRODUCTION

Nowadays, the manoeuvring of Underwater Vehicles (UVs) in confined waters is of increasing interest for their design to accomplish operational tasks. Novel concepts of Unmanned Underwater Vehicles (UUVs) allow complex missions with- out any operator input. The fulfilment of operational requirements by ensuring safety aspects at the same time is a challenging task for such types of submersibles. Similar aspects apply to

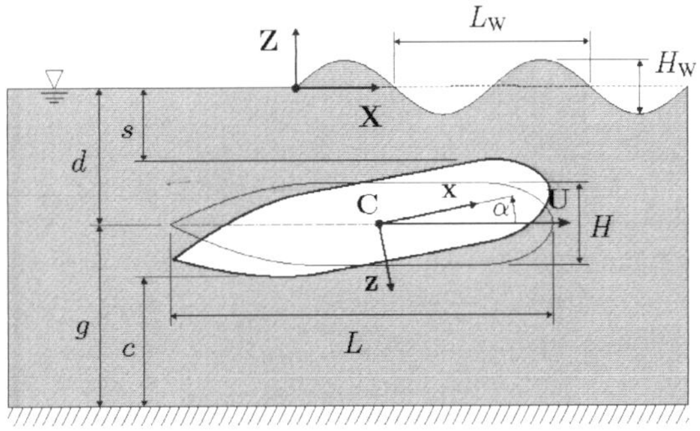

Figure 1: Co-ordinate system and characteristic parameters.

conventional submarine designs, which have to meet increasing off-design conditions for motions in proximity to waves or ice as well as for near-shore manoeuvres in (very) shallow water.

The hydrodynamics of axisymmetric bodies have been studied for decades in hydrodynamic research. Published results provide a fundamental basis to set one's sights on more complex geometries. Extreme cases, such as small submergence or low clearance to the sea bottom will be consequently addressed.
Unfortunately, there is a lack of data, when it comes to more realistic type of geometries and that is for several reasons. To close this lack there is clearly a need of profound validation by EFD and CFD.

The main objective of this paper is to demonstrate the utilization of numerical methods to determine the added resistance of UVs in regular deep water waves by determining the effective power and speed loss, respectively. The combined boundary/initial value problem is solved for two submarine hull forms of different length to diameter ratio. Therefore, classical BEM approaches such as first order panel methods are extended to account for varying distances to the free surface. A brief comparison to FVM results for the purpose of validation will complete this study by discussing some numerical aspects.

2. HYDRODYNAMIC PROBLEM

2.1 COORDINATE SYSTEM

This paper studies submersibles moving in proximity to the free surface. The underlying coordinate system is presented in figure 1. With respect to an arbitrary body fixed point C the vertical distance d to the mean free surface is called submergence depth. The origin C is chosen midship in the centreline (main axis) of a right-handed, Cartesian co-ordinate system (x, y, z) fixed with the body. The longitudinal axis x is pointing forward, y-axis is directed to starboard and z-axis vertically downwards. Main dimensions of the body are defined by the overall length L, the beam that equals the maximum hull diameter D, and the total height H that is covering all superstructure or appendages, if present.

2.2 WAVE-MAKING RESISTANCE

Estimation of the calm water resistance of a submerged under- water vehicle in a close proximity to the free surface is formulated on a simplified Doctors and Day [3] approach where the total resistance R_T is expressed as:

$$R_T = R_F + R_W \qquad (1)$$

with the frictional resistance R_F and the wave-making part R_W. Expression (1) is based on Froude's hypothesis where the frictional resistance is functionally dependent on the Reynolds number Rn = UL/ν, while the wave making resistance is dependent on the Froude number Fn = U/Lg. Here, U is the vehicle mean forward speed, L is the waterline length, g is the acceleration due to gravity (g = 9.81 m/s^2) and ν is the kinematic viscosity coefficient ($\nu = 1.35{\cdot}10^{-6}$ m^2/s of seawater at 10 °C).
The frictional (viscous) resistance R_F in (1) is calculated as:

$$R_F = C_F\, 0.5\rho SU^2 \quad (2)$$

with the frictional coefficient C_F = 0.075/(lg Rn^2) according to the ITTC'57 correlation line and water density ρ = 1026.9 kg/m^3 for sea water at 10 °C and the total wetted surface area S.

The estimation of wave resistance R_W is predicted by means of Michell's thin-ship theory, a modified version with inclusion of viscous effects according to Tuck [32] and Lazauskas [17], as well as a 3D steady BEM(RANKINE panel method, e.g. [28]), an unsteady first order 3D BEM and a fully viscous FVM.

The power needed to tow an UV without its own propulsion, which experiences the total resistance R_T given by (1) at the desired mean forward speed U is called the effective power

$$P_E = |R_T|U \qquad (3).$$

In this paper a speed loss ΔU presents the difference between initial mean forward speed and that obtained at constant effective power P_E for given non-dimensional wavelength ratios λ/L in regular deep water waves.

In the following section brief descriptions of the above stated approaches are outlined dealing with the hydrodynamic problem of moving underwater vehicles close to the free-surface.

2.3 ADDED RESISTANCE

The effects of the wave field upon the advancing UV such as a submerged submarine at selected Froude numbers are predicted by mean 2nd order wave loads in x direction (surge force). In particular, the present article uses two different approaches for the estimation of these wave loads known as the added resistance in waves.

One theory was developed by [18] while the other one is in-line with [27]. The former method uses a conservation of energy approach to express the added resistance due to presence of a regular wave field. They extended a method originally valid for head sea regular waves, see [7]. It can be recognized that the estimation of the added resistance strongly depends on the relative velocities between the ship cross sections and wave field. This fact to a certain extent limits the applicability of this theory to modern underwater vehicles since in most cases the added resistance is over-predicted compared to experimental data.

The second method according to [27] is based on the conservation of fluid momentum. It gives the mean 2nd order regular wave surge force (added resistance) in terms of the first order velocity potentials for the cross sectional two-dimensional 2D problem (strip theory) of arbitrary bodies oscillating below or in the free surface. Key steps of the theoretical formulation are omitted here, but the interested reader can find them for instance in [28]. In general, it should be kept in mind that the applicability of the two above shortly described methods depends on the regular wave characteristics, the Froude number and the slenderness of the UV.

3. APPLIED APPROACHES

3.1 MODIFIED MICHELL WAVE THEORY

A submerged UV is assumed to advance on a steady straight line course at constant speed and diving depth. Based on potential flow theory and in the reference frame moving with the vehicle, it can be shown (see for instance [29]) that the wave-making resistance R_W can be expressed as:

$$R_W = -\frac{4}{\pi}\rho U^2 \nu^2 \int_1^\infty \frac{\lambda^2}{\sqrt{\lambda^2-1}} \left|A(\lambda)\right|^2 d\lambda$$

$$A(\lambda) = -i\nu\lambda \iint_{cp} \zeta(x,z) e^{\nu z\lambda^2 + i\nu x\lambda} dz dx - \int_{-T}^{0} \zeta(x_S,z) e^{\nu z\lambda^2 + i\nu x_S\lambda} dz$$

(4), (5)

where $A(\lambda)$ represents the complex wave amplitude. In the following, if the wave resistance is estimated according to (4), the authors refer to the Michell model [21].

The expression (4) can be modified with the inclusion of viscosity effects. As shown in [29] the modification can be performed by using the model [32] and [17]. In the former one, the wave resistance R_W is given as:

$$R_W = -\frac{4}{\pi}\rho U^2 v^2 \int_1^{\infty} \frac{\lambda^2}{\sqrt{\lambda^2-1}} \left[\frac{\alpha}{\pi} \int_0^{\infty} \frac{\lambda^5 q^5}{(q-1)^2 + \alpha^2 \lambda^{10} q^6} \left| A(\lambda q) \right|^2 dq \right] d\lambda$$

$$A(\lambda q) = -i v \lambda q \iint_{cp} \zeta(x,z) e^{vz\lambda^2 q + ivx\lambda q} dz dx - \int_{-T}^{0} \zeta(x_S, z) e^{vz\lambda^2 q + ivx_S \lambda q} dz$$

(6), (7)

where α is the non-dimensional viscosity factor according to Tuck [32], which is dependent on the mean forward speed U. Similarly, the latter model according to [17], the wave resistance is formulated as:

$$R_W = -\frac{4}{\pi}\rho U^2 v^2 \int_1^{\infty} \frac{\lambda^2}{\sqrt{\lambda^2-1}} \left[\frac{\beta}{\pi} \int_0^{\infty} \frac{\lambda^3 q^4}{(q-1)^2 + \beta^2 \lambda^6 q^4} \left| A(\lambda q) \right|^2 dq \right] d\lambda$$

$$A(\lambda q) = -i v \lambda q \iint_{cp} \zeta(x,z) e^{vz\lambda^2 q + ivx\lambda q} dz dx - \int_{-T}^{0} \zeta(x_S, z) e^{vz\lambda^2 q + ivx_S \lambda q} dz$$

(8)

with $A(\lambda q)$ identical to that one of (7) where β is the non-dimensional viscosity factor according to Lazauskas [17], which is also dependent on the mean forward speed U. It should be noted that in the following section the authors refer to the Tuck viscosity model or Lazauskas viscosity model, if the wave-making resistance R_W is predicted with (6) or (8), respectively. In both expressions v is defined as $v = g/U^2$, index s indicates the transom stern section (if present) and the viscosity factors are defined as $\alpha = t/\mathrm{Fn}^5$ and $\beta = t_1/\mathrm{Fn}^3$, respectively. Here, the non-dimensional factors, t and t_1, independent of the vehicle mean forward speed U, are expressed as $t = 2(m/r)sg^{-0.5}L^{-2.5}$ and $t_1 = 4v_t g^{-0.5}L^{-1.5}$.

All values of these parameters are given in SI units. In the limits $\alpha, \beta \rightarrow 0$ it can be shown that the viscosity models expressed by (6) or (8) retrieve the ordinary Michell model (without the viscosity corrections).

The outlined expression above include a simplified formulation for the source strengths (see for instance [25] and [17]) in comparison to the Havelock sources evaluated in [8], see also [2], [12], [11] [23]. Therefore, it can be expected that evaluation of the wave resistance according to Michell wave theory is less quantitatively accurate then this more advanced method.

The l and q integrals above are numerically treated by adopting the procedure described in [30]. Accordingly, the singularity of the generalized integrand $\mu^2|A(\mu)|^2/(\mu^2-1)^{0.5}$ can be separated, so that:

$$\int_1^{\infty} \frac{\mu^2 |A(\mu)|^2}{\sqrt{\mu^2-1}} d\mu = \left[|A(1)|^2 \ln\left(2+\sqrt{3}\right) + \int_1^{2} \frac{\mu^2 |A(\mu)|^2 - |A(1)|^2}{\sqrt{\mu^2-1}} d\mu + \sum_{n=1}^{\infty} \int_{2^n}^{2^{n+1}} \frac{\mu^2 |A(\mu)|^2}{\sqrt{\mu^2-1}} d\mu \right]$$

(9)

where μ represents the generalized integration variable and $A(\mu)$ is the generalized complex amplitude. Furthermore, concerning the numerical integration of the surface integrals (double cp integrals over the UV wetted surface S) and single integrals for bodies with transom stern with a highly oscillatory integrand, the current authors also adopted the numerical technique given in [30] (Additional details of the numerical integration procedure can be found there).

3.2 BOUNDARY ELEMENT METHOD (BEM)

Submerged UVs in close proximity to the free surface are in this paper assumed to be advancing on a steady straight line course with constant mean forward speed U and diving depth, see figure 1. The fluid velocity is described by the velocity potential ϕ, which satisfies Laplace equation in the fluid domain, meaning that the fluid is assumed to be non-viscous, homogeneous and incompressible with non-rotational flow [4].

The boundary value problem (BVP) is solved through determination of the source distribution presented by the surface (panel) integral of the source density function times the Green function that is given in form of Rankine sources (singularities), as treated in [12]. This 3D Rankine panel method is further referred as BEM. With the solution of the source density at hand, the wave-making resistance R_W is found from the integration of the x-component of the hydrodynamic pressure acting on the body. While the free surface elevation $\zeta(x,y)$ follows from the linearized dynamic free surface boundary condition, see e.g. [29].

Calculations of the wave-making resistance R_W are carried out by using three different approaches. One of them is the direct pressure integration, while the other two utilize the Lagally [16] theorem by taking source distributions on the body and the free surface into account. This is done in order to recheck calculations carried out by the direct pressure method, as well as quality and quantity of the quadrilateral panel mesh discretisation of boundary surfaces.

As an alternative to the previously described stationary method the authors also an unsteady 3D panel method called *pan*MARE developed at Hamburg University of Technology.

In addition to Rankine sources a dipole is imposed on every panel when a lift-generating body is computed, see e.g. Hundemer [14]. The boundary condition is successively solved for the Kutta condition, see [15] which states that the fluid flow leaves the trailing edge smoothly and the fluid velocity is finite. This requires that the pressure on both sides of the trailing edge equal. Thus, a circulation leaves the lifting body at its trailing edge resulting in lift and drag forces in a potential fluid flow.

The unsteady free surface boundary conditions [5] are imposed by using a moving frame of reference together with the body. This formulation facilitates a robust implementation of effects due to long crested waves or wave spectra. In order to get a numerical solution of the flow-problem the boundaries are discretized with 3D panels and the boundary conditions are imposed in every collocation point of each panel centre. The unknown source strength is determined a priori of setting up the equation system by using a Neumann type boundary condition and in every panel collocation point an equation is set up using a Dirichlet type boundary condition in order to determine the unknown doublet strengths, e.g. see [14], [9]. Figure 2 presents the panel distribution for the Joubert BB2 bare hull including the deck that is used with the BEM.

3.3 FINITE VOLUME METHOD (FVM)

This section describes the applied Finite Volume method (FVM) with following two main properties. First, the method is based on discretizing the integral form of governing equations for two phase fluids (air/water) over each control volume. The basic quantities, such as mass and momentum, will therefore be conserved at the discrete level. Second, all equations are solved in a fixed Cartesian coordinate system on a mesh that does not change in time by treating the systems of partial differential equations in the segregated way.

In comparison to the used potential flow methods FVM discretises the whole fluid domain. This allows to include shear stress and turbulence models with an arbitrary range of complexity at the vehicles wake, but on costs of increasing computational effort.

For the resistance problem of submerged vehicles in general, where viscous forces are by far dominating pressure forces in deep diving conditions this balance is assumed to vary as a function of submergence when the motion in proximity to the free surface is calculated. To include viscous forces Reynolds- averaged Navier-Stokes (RANS) equations have proven to be suitable for simulations of submerged bodies from model- to full-scale, see e.g. [33], [31].

The applied RANS method in this paper are discretised using a polyhedral FVM, first to second order in space and time with cell-centred, collocated (or non-staggered) stored values. The equations are usually solved with a segregated pressure-based algorithm. Pressure-velocity coupling is achieved using a PISO approach, see e.g. [6]. For the wave resistance problem the primary wave field needs to be calculated. In this paper the so-called Volume of Fluid (VOF) approach [13] is used by solving the RANS equations for 2 incompressible, (isothermal, immiscible) fluids known as a free surface capturing approach. Therefore in each finite volume a scalar called phase-fraction is set to identify the water and air phase, respectively. To preserve a sharp interface the High-Resolution Interface Capturing (HRIC) scheme is applied, see [22, 6] for more details.

In both cases with Reynolds (Rn) number ranges between 10^6 to 10^8 of all used geometries turbulence is modelled using the standard k-omega-SST model [19, 20]. The near-wall treatment varies depending on the height of the first cell at the hull, expressed by the non-dimensional $y+$ value. For $y+ < 5$, the turbulence model is applied down to the wall and the boundary layer is considered fully resolved. For $y+ > 30$, a wall function is employed to model the turbulent flow adjacent to the wall. A blending function is applied for intermediate $y+$ values.

For the VOF-approach an unstructured mesh is used in all cases with sufficient local refinement of the air/water interface as well as the near-wall region of the UV. One mesh example can be seen in figure 3 for the DARPA SUBOFF bare hull.

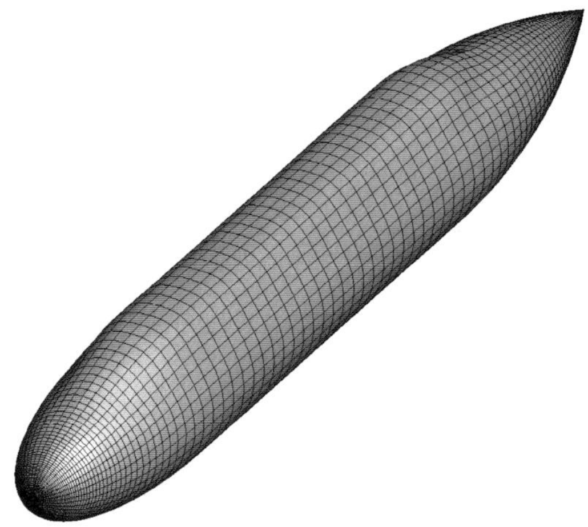

Figure 2: Discretisation of the Joubert BB2 bare hull including deck used for the steady BEM (140 x 40 panels).

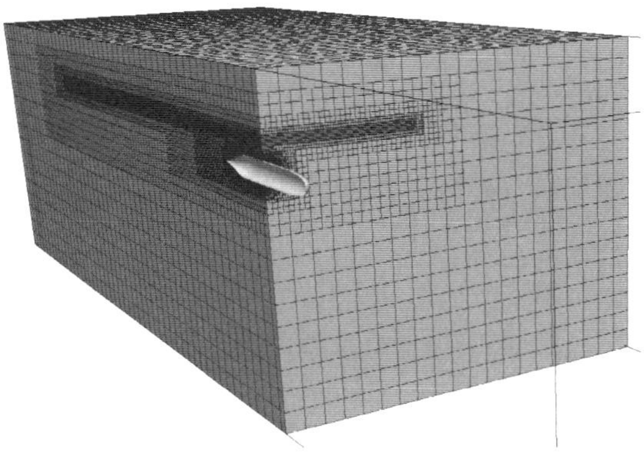

Figure 3: Longitudinal cut through the computational mesh for the FVM based VOF approach of the submerged DARPA SUBOFF.

4. CASES

In this section relevant case studies are introduced. One key element for calculating the added resistance of UVs is the de- termination of mean second order wave loads on submerged bodies. The applied potential flow method was validated for a horizontal, slender cylinder [26]. To represent arbitrary sub- marine hulls the DARPA SUBOFF and the Joubert BB2 hull form have been chosen for comparative calculations.

4.1 DARPA SUBOFF

The DARPA SUBOFF geometry was subject to a coordinated Computational Fluid Dynamics (CFD) program to assist in the development of advanced submarines for the future. For validation purposes a variety of experimental campaigns (wind tunnel and towing tank testing) were performed as a reference for numerous numerical investigations, i.e. [33], [31] to name some recent ones.

In this paper only the axisymmetric bare hull results are presented. The main advantage of the DARPA SUBOFF is that all hull lines are based on analytical expressions and can be reproduced at any time. Main particulars of the geometry are described in [10]. At model-scale the submarine has a length of L = 4.356 m, a diameter of D = 0.508 m and pocess form displacement of V = 0.708 m^3 with a total wetted surface area of S = 5.998 m^2.

4.2 JOUBERT BB2

During this investigation the need of a suitable asymmetric geometry of practical importance and for calculation purposes was raised. Conventional single hull submarines are for in- stance appended with a sail mounted on the deck superstructure. Additionally, lower casings are often part of a novel design. These structures lead to a certain degree of asymmetry. To include a realistic hull shape that is more familiar to conventional submarines the BB2 model of Prof. Joubert was recently made available by MARIN [24]. The present investigation addresses exclusively the bare hull with deck, see figure 2. At full-scale the BB2 hull has a length of L = 70.2 m, a diameter of D = 9.6 m and process form displacement of V = 4231 m^3 with a total wetted surface area of S = 1887.9 m^2.

5. RESULTS

This section discusses estimations of the wave-making and total resistance of UVs according to used theoretical models. For all introduced geometries studies have been performed to analyse the calm water resistance depending on speed and submergence as well as the added resistance in regular deep water waves.

Based on these results operative consequences in terms of effective power increase or speed loss are briefly discussed.

5.1 WAVE-MAKING RESISTANCE

The wave-making or calm water resistance of the DARPA SUBOFF was experimentally tested at Australian Maritime College (AMC) underpinned by calculations that are published in [34], [1], and [8]. In the full range of Froude numbers relevant for UVs a comparison of the Michell model with the applied steady and unsteady BEM results to published data is shown in figure 4 for a non-dimensional submergence depth of d/D = d/H = 1.1. The non-dimensional calm water resistance is defined by $C_W = R_W/0.5\rho U^2 S$.

It can be recognized that the Michell model significantly underestimates the wave resistance, but is still able to identify the local extrema. Results based on the steady 3D Rankine panel method agree with published data in [8]. Both mildly underestimate the first local extremum at Fn = 0.215 and then matches experimental data starting from the first local minimum at Fn = 0.26 until to approx. Fn = 0.4 before starting to underestimate again at higher speeds. For low Froude numbers around 0.2, where the wave making starts to develop, potential flow results are in general underestimating the resistance.
Since conventional submarines for instance are usually cruising at these speeds an improvement is needed. As already pointed out by Tuck [32], the influence of surface viscosity is highest at low Froude numbers. To account for frictional effects artificial corrections in form of viscosity coefficients added to the Michell model were introduced in [32] and [17]. Figure 5 shows family of curves for different viscosity coefficients of both models in the Froude range up to Fn = 0.3 for the Joubert BB2 bare hull including deck. The figure illustrates that presence of viscosity effects has a significant influence on the resistance curve behaviour. Humps and hollows in the Michell model are diminishing with gradually increased (constant) values of viscosity coefficients α and β. At low Froude numbers Fn < 0.2 the modified Michell curves gives more physical results in terms of C_W in

magnitude as compared to FVM simulations, but tends to dampen local extrema in the middle Froude number range 0.2 Fn < 0.5. Thus, a correction by constant viscosity coefficients cannot be applied at higher speeds.

All above described cases have also been studied for the Joubert BB2 model. Figure 6 compares the wave field for the Joubert BB2 bare hull submerged at $d/H = 1.1$ with a speed equal to Fn = 0.3. The upper result was calculated with the steady BEM at model scale (L = 2 m) while the lower result was produced with FVM at full scale (L = 70.2 m). The non- dimensional wave heights are comparable, but the characteristic wave length is slightly stretched in full-scale simulations.

5.2 ADDED RESISTANCE

The estimation of the added resistance R_{AW} for the DARPA SUBOFF and Joubert BB2 bare hull are carried out according to [18] and [27], as shortly described in section 2.3. It has been assumed that both submarines while advancing for- ward at Froude numbers Fn = 0.25 and 0.3 on the straight line course are experiencing regular head waves defined by the interval of the non-dimensional wavelength ratios $\lambda/L = 0, \ldots, 2.5$. As shown in figure 7 the estimation of the added resistance among the two applied theories [27, 18] differs significantly although the same trend of the added resistance curves can be observed. These differences can be expected since the foundation of the two above mentioned theories are diverse. An increase of the forward speed U for two wavelength ratios shows a decrease of the added resistance in waves, is followed for both theories. In comparison to the free surface piercing bodies with forward speed this behaviour of the added resistance curve is completely opposite, where an increase of the forward speed leads to an increase of the added resistance in waves, respectively.

From a comparison in figure 8 it can be seen that added resistance of the DARPA SUBOFF is significantly higher than those of the Joubert BB2. The main reason lays in the different hull design and the chosen submergence with respect to the total height H, but more importantly in the fact that the vertical motions (surge, heave and pitch) of the asymmetric Joubert BB2 configuration are different compared to the DARPA SUBOFF body of revolution, which are directly contributing to the added resistance. When the submergence of the Joubert BB2 submarine is further increased the obtained results, although not presented here, showed that the added resistance faster approaches zero and can be neglected.

5.3 EFFECTIVE POWER AND SPEED LOSS

As a consequence of added resistance the effective power P_E and the speed drop ΔU are discussed from an operational viewpoint at periscope depth. Especially, lowering the speed can be of importance for submarines' controllability or maintaining speed will lead to an unfavourable increase of propulsive power with a direct influence on mission performance. Discussing latter effects would need to include propeller characteristics. For the DARPA SUBOFF no published propeller data are known to the authors. Thus, to provide a clear comparison between both submarine hull forms the speed loss by the procedure of constant effective power is further presented.

The analysis of operative capabilities of the DARPA SUBOFF and Joubert BB2 are presented for regular head waves with preselected wave lengths $\lambda = L_W$ and heights H_W. It has been assumed that both submarines are advancing forward at constant depth on the straight line course. It should be noted that in this section results are presented for a rather big SSK variant with a length of L = 104.5 m constant for both submarines.

The effective power is proportional to the total resistance R_T given by (1) and the forward speed U. The Joubert BB2 submarine performs better in the range of chosen Froude numbers Fn = 0.25 and Fn = 0.3 since the estimated effective power is significantly lower to that of the DARPA SUBOFF. The effects of the regular wave field upon the analysed submarines are inversely proportional to the increase of their forward speed, because of the added resistance behaviour discussed in section 5.2. This means when the forward speed in-creases the added resistance in regular head waves decreases and consequently the estimation of the effective power for both submerged submarines approaches the magnitude in calm deep water. This is once again opposite to the well-known fact of free surface piercing bodies with the forward speed.

Based on a fixed effective power P_E, constant at selected Froude numbers Fn = 0.25 and Fn = 0.30 the speed loss can be calculated. Figure 9 shows the distribution of speed losses ΔU in an interval of regular wave amplitudes $\zeta_a = 1, \ldots, 6$ m. When $\zeta = 0$ m the submarine is advancing forward at initial speed. The speed drop curve are quite different. Consequently, the Joubert BB2 bare hull performs better than the DARPA SUBOFF for the same regular wave conditions exemplary defined by two different non-dimensional wavelength ratios λ/L at Fn = 0.30. Obviously by increasing the submergence, effects of regular waves diminishes and the sped drop is almost zero that could be shown in further plots.

6. CONCLUSION

This study addresses the theoretical prediction of the added resistance of underwater vehicles in close proximity to regular waves with operational consequences in terms of effective power and speed loss.

The power requirements have been derived from predictions of the total resistance. The introduced modifications are mainly related to different methods of estimating the calm water resistance according to various wave theories in which the influence of viscosity effects was studied in some detail. In particular, the modified Michell wave theory by inclusion of viscosity correction factors was introduced for the first time with applications to underwater vehicles and their effects on the wave-making resistance. It has been analysed that the influence of viscosity effects become important, especially in the lower range of Froude numbers. Insight into the free surface wave profiles for selected cases were provided by two 3D Rankine panel methods supplemented with fully viscous simulation results based on FVM. At low speeds the accuracy of predicting the wave-making resistance is challenging for all used numerical methods.

Validation and verification were performed by comparison to experimental and numerical data for axisymmetric under- water hull forms, such as the DARPA SUBOFF. Additionally, the bare hull of the Joubert BB2 submarine was used as one example of an asymmetric geometry.

In close proximity to the free surface seaway effects are significant for the motion of underwater vehicles. The added resistance in regular deep water waves is calculated for selected cases. Remarkably, longer waves at lower speeds lead to higher added resistance of a submerged body at constant forward motion. This fact is opposite to surface vessels or surface piercing bodies in general. Based on these results the effective power and speed loss in regular deep water waves are studied at selected operational speeds and submergence.

It can be concluded that the design of the Joubert BB2 bare hull provides significantly better operating performances in comparison to the DARPA SUBOFF. Attention has to be paid to the parameter that define the non-dimensional sub- mergence, total height H in this case of an asymmetric hull as the Joubert BB2 configuration with a deck. Furthermore, all findings for the bare hull forms may significantly vary if the submarine sail is included. This is subject to future investigations. While this paper is dealing with submerged vehicles in regular waves, projections for more realistic scenarios in natural seaways are drawn.

7. ACKNOWLEDGEMENTS

Current work was performed at SINTEF Ocean and thyssenkrupp Marine Systems to which the authors express their gratitude for financial support of the R&D project on theoretical investigations of underwater vehicle performances in deep water close to waves.

8. REFERENCES

1. DAWSON, E., 'An Investigation into the Effects of Submergence Depth, Speed and Hull Length-to-Diameter Ratio on the Near-surface Operation of Conventional Submarines', *University of Tasmania*, 2014.

2. DOCTORS, L. J. and BECK, R. F., 'Convergence properties of the Neumann-Kelvin problem for a submerged body', *Journal of Ship Research*, 1987.

3. DOCTORS, L. J. and DAY, A. H., 'Resistance Prediction for Transom-Stern Vessels', *Proceedings of FAST97*, 1997.

4. FALTINSEN, O. M., 'Hydrodynamics of High-Speed Marine Vehicles', *Cambridge Univ. Press*, 2007.

5. FERREIRA GONZALES D., Bechthold J., and Abdel-Maksoud, M., 'APPLICATION OF A BOUNDARY ELEMENT METHOD FOR WAVE-BODY INTERACTION PROBLEMS CONSIDERING THE NON-LINIEAR WATER SURFACE'. *ASME 2017 36st International Conference on Ocean, Offshore and Arctic Engineering*, 2017.

6. FERZIGER, J. H. and PERIC, M., 'Computational Methods for Fluid Dynamics', *Springer International Publishing*, 1999.

7. GERRITSMA, J. and BEUKELMAN, W., 'Analysis of the Resistance Increase in Waves of a Fast Cargo Ship', *International Shipbuilding Progress*, 1972.

8. GOURLAY, T. and DAWSON, E., 'A Havelock Source Panel Method for Near-Surface Submarines', *Journal of Marine Science and Application*, 2015.

9. GREVE, M., 'Non-viscous Calculation of Propeller Forces under Consideration of Free Surface Effects', *Hamburg University of Technology*, 2015.

10. GROVES, N. C., HUANG, T., and CHANG, M. S., 'Geometric characteristics of DARPA SUBOFF models (DTRC Model Nos. 5470 and 5471)', *David Taylor Research Centre*, 1989.

11. HAVELOCK, T. H., 'The Theory of Wave Resistance', *Proceedings of the Royal Society of London*, 1932.

12. HESS, J. L. and SMITH, A. M. O., 'Calculation of Non-Lifting Potential Flow about Arbitrary Three- Dimensional Bodies'. *Douglas Aircraft Company*, 1962.

13. HIRT, C. W. and NICHOLS, B. D., 'Volume of Fluid (VOF) Method for the Dynamics of Free Boundaries', *JOURNAL OF COMPUTATIONAL PHYSICS*, 1981.

14. HUNDEMER, J., 'Entwicklung eines Verfahrens zur Berechnung der instationaeren potenzialtheoretischen Propellerumstroemung', *Hamburg University of Technology*, 2013.

15. KATZ, J. and PLOTKIN, A., 'Low-Speed Aerodynamics', *Cambridge University Press*, 2001.

16. LAGALLY, M., 'Berechnung der Kraefte und Momente, die stroemende Fluessigkeiten auf ihre Begrenzung ausueben', *ZAMM-Journal of Applied Mathematics and Mechanics*, 1922.

17. LAZAUSKAS, L. V., 'Resistance, Wave-Making and Wave- Decay of Thin Ships, with Emphasis on the Effects of Viscosity', *University of Adelaide, School of Mathematical Sciences*, 2009.

18. LOUKAKIS, T A. and SCLAVOUNOS, P. D., 'Some Extensions of the Classical Approach to Strip Theory of Ship Motions, Including the Calculation of Mean Added Forces and Moments', *Journal of Ship Research*, 1978.

19. MENTER, F. R., 'Two-equation eddy-viscosity turbulence models for engineering applications', *AIAA journal*, 1994.

20. MENTER, F. R., KUNTZ, M., and LANGTRY, R., 'Ten years of industrial experience with the SST turbulence model', *Turbulence, heat and mass transfer*, 2003.

21. MICHELL, J. H., 'The Wave-Resistance of a Ship', *The London, Edinburgh, and Dublin Philosophical Magazine and Journal of Science*, 1898.

22. MUZAFERIJA, S. and PERIC, M., 'Computation of free-surface flows using interface tracking and interface-capturing methods', *Advances in Fluid Mechanics*, 1999.

23. NEWMAN, J. N., 'EVALUATION OF THE WAVE- RESISTANCE GREEN FUNCTION, PART 2: THE SINGLE INTEGRAL ON THE CENTERPLANE', *Journal of Ship Research*, 1987.

24. OVERPELT, B., NIENHUIS, B., and ANDERSON, B., 'Free running manoeuvring model tests on a modern generic SSK class submarine (BB2)', *PACIFIC 2015 International Maritime Conference*, 2015.

25. PENG, H., 'Numerical Computation of Multi-hull Ship Resistance and Motion', *Dalhousie University*, 2001.

26. PINKSTER, J. A., 'Low Frequency Second Order Wave Exciting Forces on Floating Structures', *TU Delft*, 1980.

27. SALVESEN, N., 'Second-order steady state forces and moments on surface ships in oblique regular waves', *In Proceedings of the international Symposium on dynamics of marine vehicles and structures in waves*, 1974.

28. SKEJIC, R., 'Maneuvering and Seakeeping of a Single Ship and of Two Ships in Interaction', *Norwegian University of Science and Technology*, 2008.

29. SKEJIC, R. and JULLUMSTRØ, E., 'Power Performance and Environmental Footprint of High-Speed Vessels in Calm Deep Water', *ASME 2012 31st International Conference on Ocean, Offshore and Arctic Engineering*, 2012.

30. SOEDING, H., 'Towards Fully Non-linear Floating Body Simulations by a Potential Method' , *ASME 2012 31st International Conference on Ocean, Offshore and Arctic Engineering*, 2012.

31. TOXOPEUS, S., ATSAVAPRANEE, P., WOLF, E., DAUM, S., PATTENDEN, R., WIDJAJA, R., ZHANG, J. T., and GERBER, A., 'COLLABORATIVE CFD EXERCISE FOR A SUBMARINE IN A STEADY TURN', *ASME 2012 31st International Conference on Ocean, Offshore and Arctic Engineering*, 2012.

32. TUCK, E. O., 'The Effect of a Surface Layer of Viscous Fluid on the Wave Resistance of a Thin Ship', *Journal of Ship Research*, ,1974.

33. VAZ, G., TOXOPEUS, S., and HOLMES, S., 'CALCULATION OF MANOEUVRING FORCES ON SUBMARINES USING TWO VISCOUS-FLOW SOLVERS'. *ASME 2010*

29th International Conference on Ocean, Offshore and Arctic Engineering, 2010.

34. WILSON-HAFFENDEN, S., RENILSON, M., RANMUTHUGALA, D., and DAWSON, E., 'An Investigation into the Wave Making Resistance of a Submarine Travelling Below the Free Surface', *PACIFIC 2010 International Maritime Conference*, 2010.

9. AUTHORS BIOGRAPHY

Renato Skejic holds the current position of Senior Research Engineer at SINTEF Ocean, Ocean Hydrodynamics department and Post-doctoral fellow position at NTNU, Faculty of Engineering, Department of Marine Technology in Trondheim, Norway. He is responsible for a development of new theoretical and numerical models closely related to the hydrodynamics of single or multiple floating and/or submerged non-lifting and lifting bodies operating in different environmental conditions. His previous experience includes development of the numerical seakeeping and maneuvering ocean basin, resistance analysis in open or confined calm waters, deep and finite/shallow water marine and offshore hydrodynamics, hydrodynamics of subsea installations, hydrodynamics of the high-speed vehicles.

Martin Greve holds the current position of naval architect and design engineer at thyssenkrupp Marine Systems. He is responsible for propeller design and propulsion of submarines. His previous experience includes development of numerical methods and lecturer both in the field of ship hydrodynamics.

Stefan Daum holds the current position of naval architect and design engineer at thyssenkrupp Marine Systems. He is responsible for submarine hydrodynamics at the Department of Theoretical Engineering. His experience includes applications of distinct numerical and experimental methods for hydrodynamic problems related to conventional submarines.

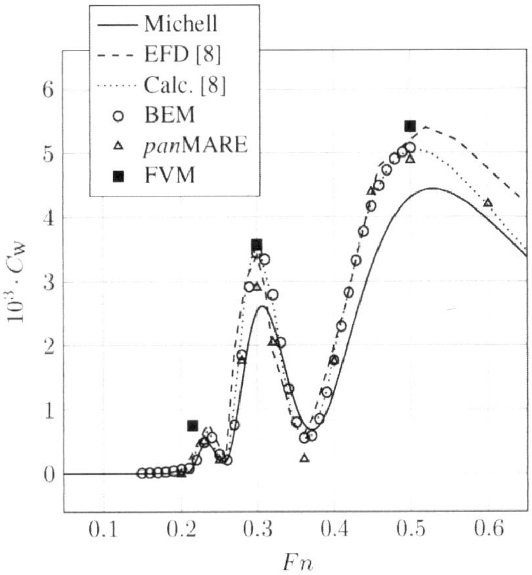

Figure 4: Non-dimensional wave-making resistance, DARPA SUBOFF at non-dimensional submergence of $d/H = 1.1$.

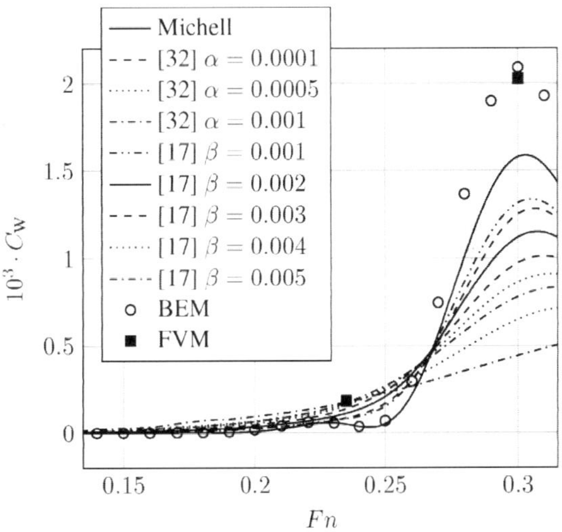

Figure 5: Non-dimensional wave-making resistance of the modified Michell model in comparison to BEM and FVM, Joubert BB2 at non-dimensional submergence of $d/H = 1.1$.

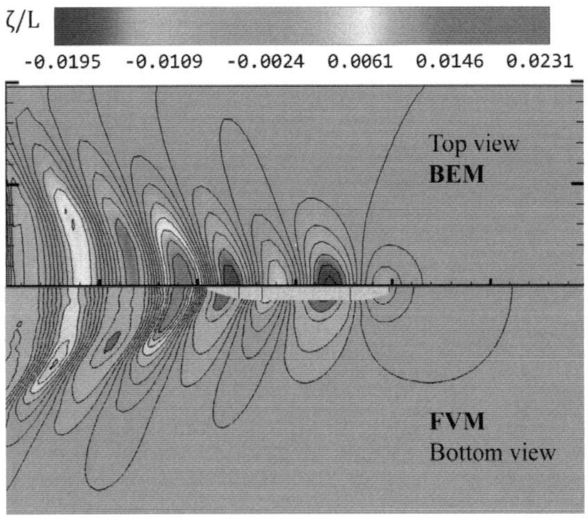

Figure 6: Comparison of the steady BEM calculated wave field (non-dimensional wave heights) with a transient FVM simulation in calm deep water for the Joubert BB2 bare hull including deck, submerged at *d*/*H* = 1.1 with a speed equal to Fn = 0.3. BEM was applied in model scale (*L* = 2 m), FVM was performed in full scale (*L* = 70.2 m).

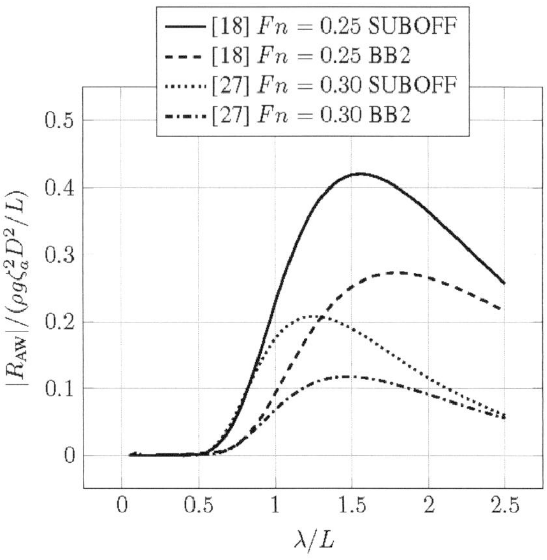

Figure 8: Comparison of DARPA SUBOFF and Joubert BB2 added resistance at *d*/*H* = 1.1 for Fn = 0.3.

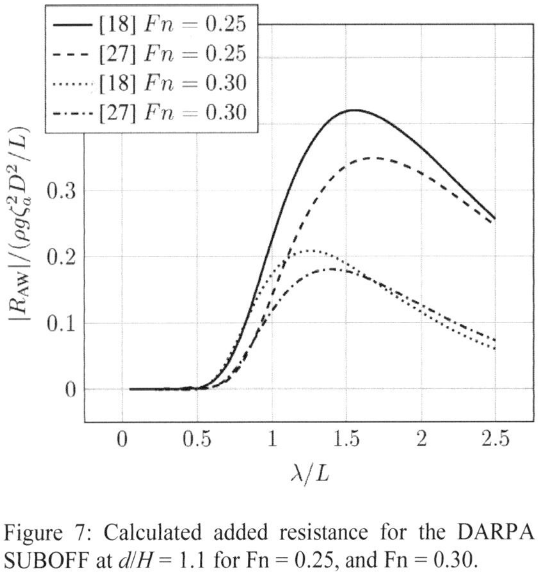

Figure 7: Calculated added resistance for the DARPA SUBOFF at *d*/*H* = 1.1 for Fn = 0.25, and Fn = 0.30.

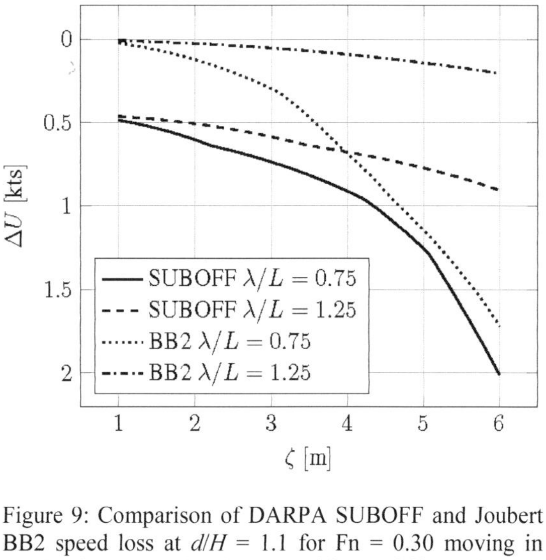

Figure 9: Comparison of DARPA SUBOFF and Joubert BB2 speed loss at *d*/*H* = 1.1 for Fn = 0.30 moving in regular deep water head waves with two different lengths λ/L = 0.75 and 1.25. Here dimensional values are presented for an equal length of *L* = 104.5 m for both hulls.

PROBABILISTIC SUBMARINE FLOOD RECOVERY ANALYSIS

N R Horn, Royal Navy / UCL, UK

SUMMARY

A review of UK MoD's existing deterministic methodology for submarine flood hazard assessment is presented, along with recommendations as to how an alternative, probabilistic approach could be implemented.

An illustrative example of a holistic flood risk estimation process is included, in which particular emphasis is placed on the handling of significant sources of uncertainty. An example set of deliverables illustrates how a distillation of the resulting data could be conveyed to key stakeholders, including Duty Holders and Operators, in order to support informed decision-making on matters such as the tolerability of major defects, or the suitability of alternative emergency procedures.

A gap analysis identifies areas of modelling software capability and technical knowledge that would need to be developed further in order to implement and support the proposed methodology.

NOMENCLATURE

CO	Commanding Officer
DEMCW	Double-Ended Main Circulating Water [system fracture]
EOP	Emergency Operating Procedure
FAFS	Flood Alarm Float Switch
FAZ	Flood Avoid Zone
FDS	Fracture Detection System
JSP	Joint Service Publication
MAP	Military Acquisition Publication
MBT	Main Ballast Tank
MCW	Main Circulating Water [system]
MED	Maximum Excursion Depth
MoD	(United Kingdom) Ministry of Defence
PDH	Platform Duty Holder
PPPY	Per Platform Per Year
RPM	(Shaft) Revolutions per Minute
SOE	Safe Operating Envelope
SSBN	Ballistic missile submarine, nuclear-powered
SSK	Conventional (i.e. non-nuclear powered) submarine
SSN	Attack submarine, nuclear-powered
UCL	University College London
VCT	Vacuum Collapse Time
$F_{i,j}$	Factor i in state j
L	Notional *loss-of-platform* event (equivalent to 100 military fatalities or 10 civilian fatalities)
O	Occurrence of a particular combination of an initiating event and other factors
$P(X)$	Absolute probability of X being true
$P_{1yr}(...)$	Probability of a discrete event per year of continuous submarine operation
$P_{oyr}(...)$	Probability per operating year (i.e. adjusted for actual/predicted time at risk)
$Q(X)$	Absolute probability of X not being true

1. INTRODUCTION

Operating a dived submarine is an inherently risky activity, which presents a broad spectrum of hazards not only to the submarine and its occupants but also to other shipping. A significant subset of these hazards relates to loss of depth control, which can be caused either by flooding or by propulsion and/or manoeuvring system failures (e.g. plane jams), and which has the potential to result in:

- Excursion beyond crush depth, leading to catastrophic failure of the pressure hull and loss of the submarine (deep water operations only);
- Collision with the sea bed, leading to a range of possible damage scenarios up to and including loss of the submarine (shallow water operations); or,
- Uncontrolled broaching, leading to a possible collision and consequential damage to, or loss of, the submarine and/or another vessel.

This paper examines UK Ministry of Defence (MoD)'s current methodology for assessing and mitigating flood-related hazards critically in relation to MoD safety policy. Weaknesses in the current analysis process will be identified in terms of its inability to:

- Capture the full complexity of the problem;
- Deal with uncertainty;

- Quantify risk, and therefore support appropriate and proportionate trade-offs between conflicting options and requirements.

An alternative, probabilistic approach to flood hazard analysis will be proposed and demonstrated. This will build upon the existing body of corporate knowledge and simulation tools, whilst re-structuring the problem to capture complexity and uncertainty, and presenting the results in a richer and more informative way than at present.

Although the investigation will centre on nuclear-powered submarines (SSNs and SSBNs) of the generic type operated by the UK MoD, many of the concepts that will be explored are equally applicable to conventionally-powered submarines (SSKs).

2. PRINCIPLES OF SUBMARINE FLOOD RECOVERY

Uncontrolled flooding of water into any submarine presents a potentially catastrophic hazard. If appropriate action is not taken, the additional weight of flood water will eventually exceed the submarine's capacity to generate extra buoyancy and/or hydrodynamic lift to compensate, and as a result, the submarine will sink uncontrollably.

In order to minimise the possibility of a catastrophic outcome, flooding incidents are generally addressed immediately using a prescribed emergency operating procedure (EOP), which aims to minimises the addition of flood mass, and then counteract its effect with additional buoyancy and/or hydrodynamic forces. EOPs may comprise some or all of the following three principal elements:

- **Isolate** the flood;
- **Deballast** (either by pumping or blowing internal tanks and/or by blowing MBTs);
- **Accelerate** (to increase hydrodynamic lift).

The following complicating factors are highlighted:

- Failure of the main circulating water (MCW) system is one of the most significant design basis floods on nuclear submarines; however, shutting the MCW hull valves compromises main condenser vacuum, and can lead to loss of propulsion.
- Blowing the MBTs commits the submarine to surfacing with very limited control over positioning, irrespective of surface traffic. The risk of collision is therefore increased.

It should be noted that both of the factors above potentially introduce risks even when the flood EOP has been initiated in response to a false alarm or a small, manageable flood (as is the true in the overwhelming majority of cases of in-service flood EOP initiation [2]).

A flood recovery can be deemed successful once the bodily weight and the trajectory of the submarine are fully under control. If the submarine has successfully surfaced, then it is generally considered that any flood will be manageable thereafter; however, it is also possible to reach an acceptable conclusion whilst remaining dived, provided that the flood has been isolated or is small enough to be compensated for by ballast pumps, and that the MBTs have not been blown.

3. HISTORICAL BACKGROUND

Two particular cases are highlighted, each of which illustrates one of two mutually-conflicting, flood-related hazards, namely *loss of depth control* and *collision on surfacing*.

3.1 USS THRESHER FLOODING INCIDENT

On 10 April 1963, during a deep dive test, USS THRESHER suffered a full-bore failure of a brazed joint in her machinery space sea water pipework [3]. Collateral damage from the flood caused loss of propulsion which, combined with the limited effectiveness of MBT blows at depth, left the THRESHER unable to recover, resulting in the loss of the submarine along with all 129 servicemen and civilians on board.

Although some of the key links in this tragic chain of events were specific to THRESHER's design, full-bore sea water pipework failures case still dominate the UK MoD's deterministic flood mitigation policy as the design basis accidents.

3.2 COLLISION BETWEEN USS GREENEVILLE AND EHIME MARU

Ehime Maru was sunk when the US nuclear submarine USS GREENEVILLE surfaced beneath her on 9 February 2001, resulting in the deaths of nine on board, including four schoolchildren [4]. GREENEVILLE, which was damaged in the incident but survived with no casualties, had been demonstrating an emergency surfacing manoeuvre to embarked VIPs.

Although the manoeuvre had not been in response a flood on this occasion, the incident provided a grim illustration of the potential dangers associated with the UK MoD's current policy of emergency surfacing immediately in response to any flood alarm.

4. CURRENT UK FLOOD MITIGATION AND ANALYSIS METHODOLOGY

A distillation of the current policy and technical guidance for the design, operation and assessment of UK submarines from a manoeuvring, hydrodynamics and hydrostatics perspective is presented in MAP 01-073 [5]. This includes information on flood mitigation, and on the modelling of flooding incidents. Standards for UK submarine manoeuvring are set out in DSTAN 08-102 [6].

4.1 REGULATORY FRAMEWORK

The safety of UK MoD ships and submarines is regulated in accordance with JSP 430 [7, 8], which charges the Platform Duty Holder (PDH) with "*Minimis[ing] the risk of exceeding the deep diving depth, broaching, grounding and collision*" due to "*All credible submarine Manoeuvring & Control scenarios, including the impact of damage or malfunction of key manoeuvring & control systems / equipment*". It also specifies that a justified risk assessment must be conducted, upon which operator guidance must be based.

Thus, it is clear that the PDH is responsible for assessing and controlling the whole spectrum of risks associated with flooding events, including both depth excursions (see THRESHER), and potential collisions as a result of emergency surfacing (see GREENEVILLE). Furthermore, all credible scenarios must be considered, including those involving equipment failures.

4.2 FLOODING EOP

UK Flooding EOPs contain minor variations between classes, but generally adhere to the following profile, colloquially known within the submarine community as *blow and go*. The EOP philosophy is summarised below:

- **Initial alerting** can take the form of:
    - *Flood alarm*: Automatic (Flood water-operated) flood alarm float switch (FAFS) or Manual (push buttons), by compartment;
    - *Verbal report*: By compartment and casualty system if identified;
    - *Unexpected Depth / Pitch perturbation*: +20m / ±5°.
- **EOP Initiation**: By default, the full EOP is initiated immediately; however, depending on the operational profile, the depth of the submarine, and/or the nature of the initial alert, a command decision may be made as to whether or not to proceed with the full EOP. (Nevertheless, this should be commenced as soon as the flood has been confirmed as *genuine*.)
- **Full EOP Response (concurrent actions)**:
    - **Blow MBTs**: Once MBTs are blown, the submarine is committed to emergency surfacing. A red grenade is fired in order to warn any surface vessels;
    - **Secure Automatic Hull Valves**: All automatic hull valves are shut within the casualty compartment. This nominally includes MCW hull valves, and thus eventually leads to loss of main propulsion if the flood is in the Main Machinery Space (MMS); however, recent work [9, 10] has led to the suspension of this policy on in-service platforms, and MCW hull valves are now only shut on the casualty side where a flood has been reported on that system.
    - **Increase Speed**: Shaft RPM is increased commensurate with the current reactor power state. A further limit is imposed if the MCW hull valves have been shut.

4.3 PRESENTATION OF THE FLOOD HAZARD

The principal aim of the current manoeuvring hazard assessment process (for both floods and plane jams) is to generate operator guidance in the form of a safe operating envelope (SOE), and thus provide implicit assurance that a submarine operating within the SOE is deterministically safe in the event of any credible manoeuvring incident (subject to all embedded assumptions within the assessment being correct).

An SOE is presented for each submarine class in terms of depth and speed. This paper will focus only on the flood avoid zone (FAZ) limitations at the slow/deep boundary of the SOE.

Figure 1 shows the various regions that make up a typical FAZ. The boundary of each region is defined by the limiting conditions at which the submarine just fails to recover from particular 'standard' flood cases (as defined in MAP 01-073), and the overall FAZ is a composite of all of these regions. The standard cases include double-ended MCW (DEMCW) floods and aft, unsecured floods. The SOE is also bounded by maximum excursion depth (MED).

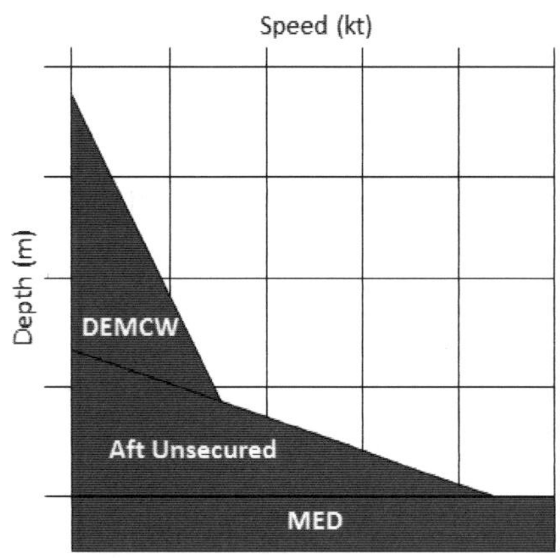

Figure 1: Illustrative FAZ

4.4 GENERATION OF THE FAZ

The SOE (including the FAZ) is generated by a computer programme known as *ARTEMIS*, which is a UK MoD-specific front-end for QinetiQ's commercial *Paramarine* software package. ARTEMIS applies the methodology framework specified in MAP 01-073 to the underlying 6 degree of freedom submarine manoeuvring model.

The manoeuvring model employs a combination of linearised derivatives and empirical corrections that are based on data from model testing and sea trials [11, 12]. Its validity is therefore confined to a limited range of validation cases, and thus generally excludes extreme manoeuvres (e.g. those involving high crossflow components and/or rapid changes in flow conditions).

4.5 EMBEDDED ASSUMPTIONS WITHIN THE DETERMINISTIC ANALYSIS

The MoD's analysis aims to produce a deterministic set of results (presented as a FAZ) that is representative of all possible cases whilst also being suitably conservative. To support this aim, several significant assumptions and simplifications have been made.

Significant conservative assumptions include:

- Exceeding a pitch limit criterion (in the order of 30°) is assumed to result in loss of platform;
- The most pessimistic main condenser vacuum collapse time (VCT) is assumed.

Significant assumptions that are potentially non-conservative include:

- The submarine is initially free from any defects;
- The EOP is initiated without delay and followed correctly, and all human reaction times are 7 seconds;
- Large flood cases with an individual probability of occurrence of less than 10^{-6} per platform per year (PPPY) are deemed *incredible* and therefore disregarded;
- Flood initiation during manoeuvring transients is not considered.

5. DISCUSSION OF SHORTFALLS OF THE CURRENT FLOOD ANALYSIS METHODOLOGY

5.1 SENSITIVITY OF THE FAZ TO SUBJECTIVE ASSUMPTIONS

A number of the assumptions identified in Section 4.5 have a significant impact on the shape and size of the FAZ. For example, the respective impacts of changing the reaction time and the pitch angle failure criterion (both within their credible ranges of variation) are illustrated by the results of a sensitivity study [1], as shown in Figure 2.

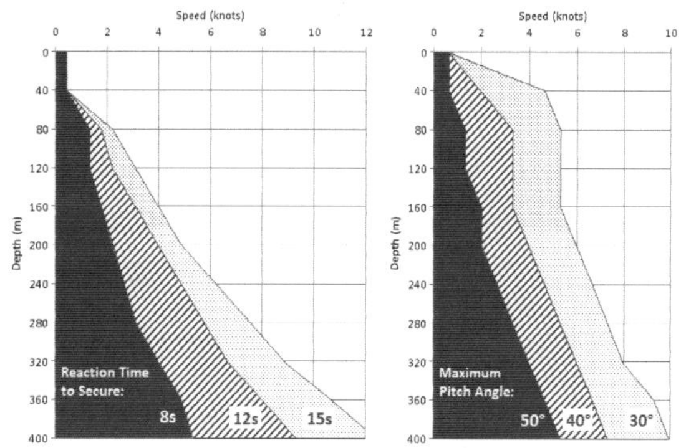

Figure 2: Sensitivity of FAZ to reaction time and pitch angle failure criterion.

Sensitivity to reaction time poses a particular problem in terms of generating a conservative FAZ, since the current 7 second assumption is potentially optimistic compared with the observed performance of real operators. Figure 3 shows an idealised distribution [13] of the measured reaction times that operators took to secure the MCW hull valves over a number of simulated flooding events during a trial [14] in the Astute Class Manoeuvring Room Trainer.

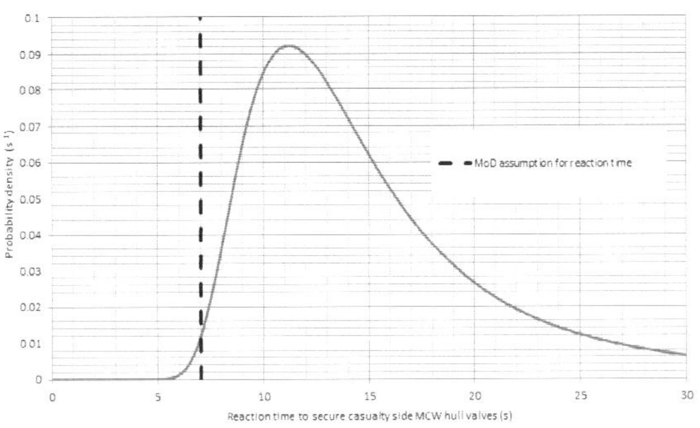

Figure 3: Idealised Distribution of Reaction Time to Secure MCW Hull Valves [1]

Given how sensitive the FAZ is to even the small handful of selected assumptions discussed above, some of which are apparently non-conservative, its validity as a boundary of a *deterministically safe* operating envelope is arguably liable to be challenged.

5.2 INABILITY TO QUANTIFY RISK

5.2 (a) Risk Trade-Offs

Whilst JSP 430 [8] requires the MoD to assess the risks posed by key hazards, the current methodology falls short of quantifying the risks that are posed both directly (e.g. sinking due to flood water) and indirectly (e.g. collisions as a result of frequent emergency surfacing) by flooding. Thus, where a conflict exists between the optimal mitigation strategies for two or more sub-hazards, it is not currently possible to make an informed trade-off between them. Risk trade-offs worthy of potential consideration include:

- The risk of delaying the blowing of MBTs for a Command decision vs. the risk of collision posed by blowing MBTs for every flood alert;
- The risk of delaying MCW isolation for MMS floods unless flooding is confirmed on the MCW system vs. the risk of isolating MCW immediately and consequently losing propulsion frequently whilst dived.

In considering risk trade-offs, it cannot be over-emphasised how frequently submarines currently conduct flood EOPs for suspected or trivial floods; i.e. in the order of 0.3 times PPPY [2], compared with the estimated frequency of genuine design-basis MCW flood events (in the order of 10^{-5} failures PPPY) – some ten-thousand-fold difference.

5.2 (b) Off-Design Operation

The SOE fails to provide any guidance on either:

- How to operate safely when the submarine is in a degraded material state (as is not an uncommon situation);

or,

- How to assess the risk of operating outside the SOE (which MAP 01-073 acknowledges is sometimes required, since the actual operating profile is sometimes constrained by operational requirements, irrespective of the published SOE guidance).

Where the SOE provides inadequate guidance, MAP 01-073 places responsibility on Commanding Officers alone to assess the risk to the submarine; however, it is not clear that they are adequately equipped to do so other than in an entirely subjective way.

5.3 SKEWED PERCEPTION OF RISK

As well as failing to quantify risk, the current SOE presentation potentially actively promotes a skewed view of risk due to its binary nature.

In some cases, risk is over-stated; for example, a point subject to a *soft* failure (e.g. relating to a transient pitch angle excursion) might fall just outside the SOE and thus be deemed unsafe, whereas in fact the risk associated with that operating point might actually be very low, especially if it also relates to an unlikely initiating event.

Conversely, in other areas, risk is under-stated. For example, firstly, some very large initiating events (e.g. flooding through main hatches or torpedo tubes) are completely neglected because they are deemed incredible. However, the difference in probability between a credible event and an incredible one might only be a single order of magnitude. This may lead to some comparably significant risks being hidden by the arbitrary credibility threshold, especially near the margins of the FAZ.

Also, as previously discussed, the risk of collision due to emergency surfacing exists uniformly within the entire operating envelope by virtue of the prevailing, conservative *blow and go* philosophy. However, because the risk is (incorrectly) seen as unavoidable by some, and also difficult to quantify, it tends to be neglected. Curiously, by comparison, one of the failure criteria for a plane jam is uncontrolled broaching, leading to a risk of collision. This appears to betray a fundamental inconsistency within the SOE philosophy itself.

5.4 INFLEXIBILITY

Investigation of possible alternative EOPs is discouraged if they involve decisions that cannot be deterministically codified [5]. Yet COs are nevertheless called upon to make judgment calls on non-standard EOP procedures and operating conditions that would simply be unacceptable from a deterministic SOE point of view. Again, this raises the question of whether more could be done to provide at least some guidance for these operationally inevitable situations.

6. PROPOSED ALTERNATIVE FLOOD ANALYSIS METHODOLOGY

6.1 OVERALL AIM

The overall aim of the proposed methodology is make a balanced, quantitative assessment of all flood-related risks posed to the submarine across the whole operating envelope, taking into account the broadest possible range of factors and uncertainties.

The resulting assessment is intended to support informed, effective decision making wherever flood-related risks are a consideration.

6.2 CHARACTERISING THE PROBLEM

Figure 4 aims to summarise the problem domain, showing the key factors that influence the overall risk, and illustrating the flow of causality from the initiating event through to a number of possible meaningful consequences.

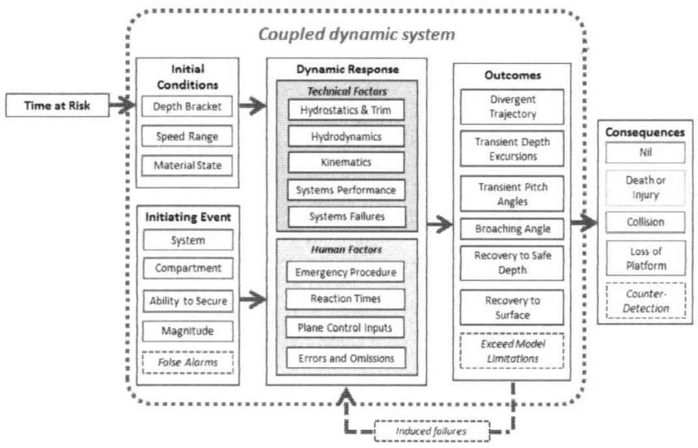

Figure 4: Structure of the flood recovery problem domain

Most of the factors that govern the model take effect within a coupled dynamic system, which means that a new simulation must be conducted for every unique combination of variables. However, some of the factors that govern the overall risk (i.e. time at risk, and any link between physical outcomes and absolute risk) fall outside the dynamic problem, and can therefore be examined without re-running the simulations.

As demonstrated in previous sections, several key factors are subject to a significant degree of uncertainty, and it is therefore deemed appropriate to treat these in a probabilistic manner. Fortunately, most of these vary independently from one another.

Following a review of various existing approaches to probabilistic safety analysis (and bearing in mind the coupled nature of the dynamic problem, along with the mutual independence of the key stochastic variables), it was decided that the most appropriate way to approach the problem was as a multifactorial experiment [15].

6.3 MULTIFACTORIAL EXPERIMENT APPROACH

In multifactorial experiments, the full set of possible combinations of independent variables can be described in terms of a *hypercuboid grid*, which has an orthogonal dimension for each independent variable. Thus, for a simple, hypothetical experiment that has f factors, each of with s possible discrete states, the number of possible combinations, C , is given by:

$$C = s^{\mathrm{f}} \qquad 1$$

From this relationship, it is clear that the number of experiments grows very rapidly as additional factors are included – a problem known as *the curse of dimensionality* [16]. However, this seemingly unmanageable quantity of individual runs can be reduced dramatically by discarding combinations of factors that result in a very low overall probability of occurrence, as illustrated in Figure 5.

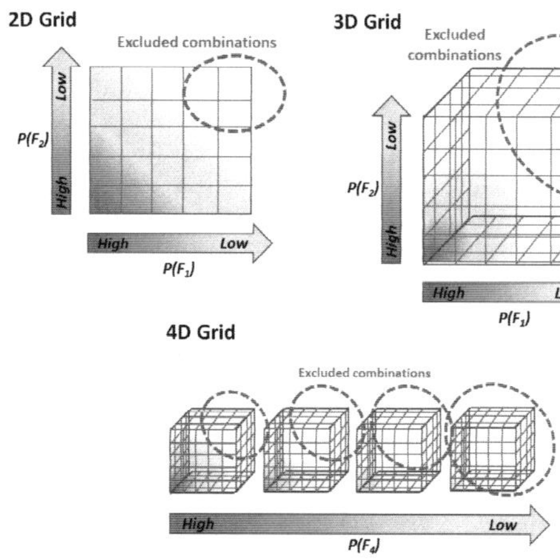

Figure 5: Increasing sparseness of hypercuboid grid with increasing dimensionality

6.4 IMPLEMENTATION

6.4 (a) Management of Input Variables

For the demonstration that is presented in this paper, a relatively small number of the factors was selected for investigation (including hole size, reaction times, vacuum collapse time and blow valve failures). Continuous variables such as hole size and reaction time were discretised coarsely into a handful of representative cases, each with an associated probability, as illustrated in Figure 5.

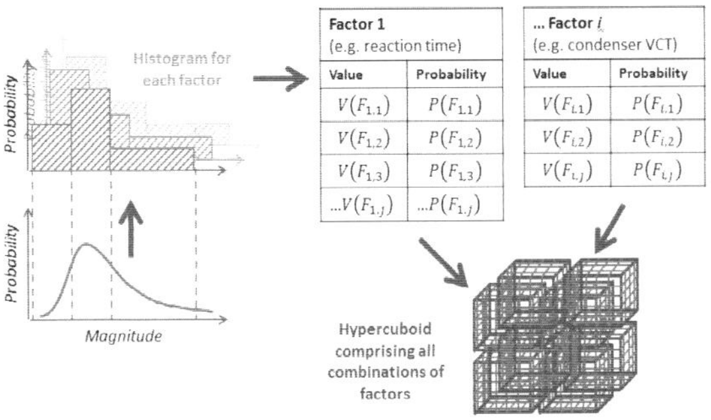

Figure 6: Discretisation of continuous variables and assembly of the hypercuboid

Crucially, for each variable, the sum the probabilities of all possible states was made to be equal to 1. (For example, for hole size, the case with the highest probability of occurrence – almost equal to 1 – is the *zero hole size* case).

A Visual Basic computer code was written to assemble every possible combination of variables and thus produce a comprehensive schedule of recovery runs to be simulated (as illustrated in Figure 6). At this point, the overall probability of occurrence, $P(O)$, for each scenario was calculated by multiplying the probability of each constituent factor. Scenarios with a vanishingly small probability of occurrence (e.g. 10^{-9}) were discarded in the process.

6.4 (b) Simulation and Analysis

The Visual Basic code utilised ARTEMIS to execute each run and compile a record of the *outcomes* (i.e. selected raw physical results such as maximum depth and maximum transient pitch angle).

A separate piece of code was then used to post-process the resulting outcomes and infer the likely *consequences*, ultimately ascribing a notional risk score to each run. The risk scores ranged from 0 for a completely benign recovery to 1 for certain loss of platform (i.e. sinking), whilst fractional risk scores were applied to represent the increased risk associated with indeterminate outcomes such as broaching at an excessive pitch angle, as illustrated in Figure 7.

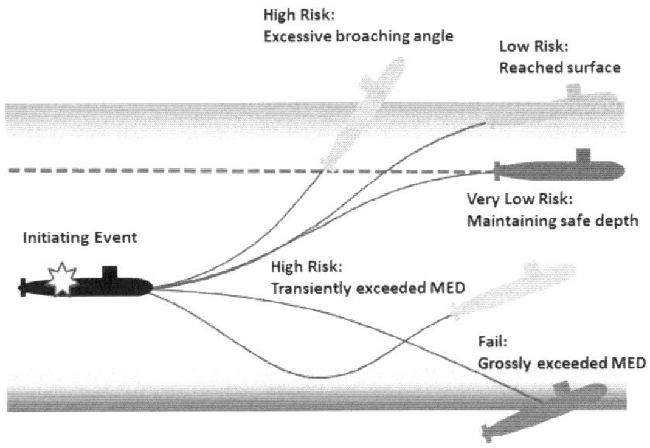

Figure 7: Relationships Between Outcome and Risk

By multiplying the probability of occurrence of each scenario with the risk score derived from the outcome of its simulated recovery run, an absolute risk score (in terms of notional probability of loss of platform) was generated for each combination of variables within the scope of the experiment.

The total risk (per operating year) of operating in each depth and speed bracket was then calculated by summing all of the absolute risk scores in each bracket.

6.4 (c) Output Visualisation

In order to make the results relatable to existing stakeholders in terms of the traditional SOE presentation, they were presented visually in the depth-speed domain, as shown in Figure 8.

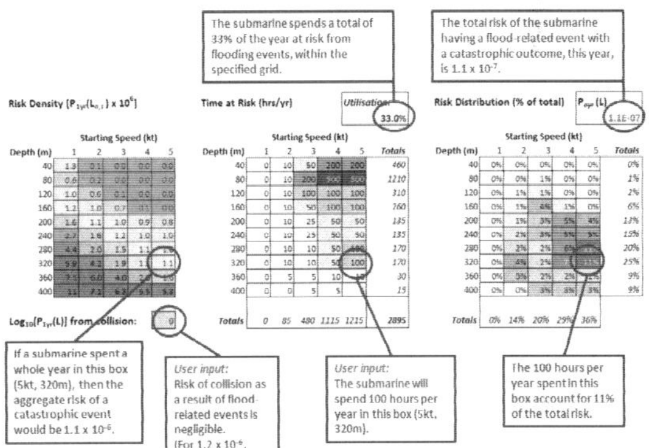

Figure 8: Example Risk Map

The risk map comprises a number of components:

- **Risk Density Plot**. Map of the relative probability (i.e. probability per full year spent at risk) of a loss-of-platform event as a result of all credible floods within each discrete depth and speed bracket, $P_{1yr}\left(L_{d,u}\right)$;
- **Time at Risk Map**. User-defined map of hours per year spent in each depth / speed bracket;
- **Risk Distribution Plot**. Map of weighted probability (i.e. probability per operating year, accounting for actual or predicted time at risk) of a loss-of-platform event as a result of all credible floods within each depth and speed bracket, $P_{oyr}\left(L_{d,u}\right)$;
- Absolute probability of a loss-of-platform event per operating year, taking into account the total time spent at risk across the entire operating envelope, $P_{oyr}\left(L\right)$.

7. EXAMPLE PROBABILISTIC FLOOD ANALYSIS

7.1 CASE STUDIES

Example risk maps were generated the following hypothetical EOP philosophies:

-
- **EOP A** involves shutting the MCW Hull Valves and blowing MBTs immediately (thereby incurring a high risk of loss of propulsion and an elevated risk of collision as a result of frequent emergency surfacing).
- **EOP B** involves shutting MCW Hull Valves only if flooding is identified on the MCW system (which incurs a delay in isolating MCW floods). MBTs are only blown if and when the submarine is responding abnormally to control inputs due to gross flooding. (This dramatically reduces the risk of collision by limiting the frequency of emergency surfacing).

7.2 EOP A RISK PLOTS

As expected, the risk map for EOP A is dominated by the DEMCW case, and shows a gradual increase in risk density towards slow and deep conditions.

The very high risk at 1kt, 40m is related to slow, shallow blow-related pitch excursions. Because this phenomenon is driven primarily by the operator action (specifically the planesman's response) rather than by the flood itself, the risk is multiplied by the higher probability associated with low-level events; this drives the high risk figure.

Risk Density [$P_{1yr}(L_{d,s}) \times 10^6$]

Depth (m) \ Starting Speed (kt)	1	2	3	4	5
40	24	0.1	0.0	0.0	0.0
80	0.7	0.1	0.0	0.0	0.0
120	0.2	0.0	0.0	0.0	0.0
160	0.5	0.0	0.0	0.0	0.0
200	0.5	0.1	0.0	0.0	0.0
240	1.3	0.5	0.1	0.0	0.0
280	1.4	0.6	0.3	0.1	0.0
320	2.2	1.2	0.5	0.1	0.0
360	2.5	1.4	0.6	0.3	0.1
400	6.2	2.1	1.3	0.6	0.3

Time at Risk (hrs/yr) — Utilisation: 33.0%

Depth (m) \ Starting Speed (kt)	1	2	3	4	5	Totals
40	0	10	50	200	200	460
80	0	10	200	500	500	1210
120	0	10	100	100	100	310
160	0	10	50	100	100	260
200	0	10	25	50	50	135
240	0	10	25	50	50	135
280	0	10	10	50	100	170
320	0	10	10	50	100	170
360	0	5	5	10	10	30
400	0	0	5	5	5	15
Totals	0	85	480	1115	1215	2895

Risk Distribution (% of total) — P_{oyr} (L) 9.4E-09

Depth (m) \ Starting Speed (kt)	1	2	3	4	5	Totals
40	0%	1%	0%	0%	0%	1%
80	0%	1%	0%	0%	0%	1%
120	0%	0%	0%	0%	0%	1%
160	0%	0%	0%	0%	0%	1%
200	0%	2%	1%	0%	0%	3%
240	0%	6%	4%	2%	0%	12%
280	0%	7%	4%	4%	2%	17%
320	0%	[illegible]	6%	9%	4%	34%
360	0%	8%	3%	4%	1%	17%
400	0%	0%	8%	3%	2%	13%
Totals	0%	40%	27%	23%	10%	

Figure 9: EOP A Risk Map (Excluding collision risk)

With benchmark Time at Risk map populated as shown, and with no risk of collision included, the absolute risk PPPY is infinitesimal (~10^{-7} PPPY). However, when a modest risk of catastrophic collision is added, at 10^{-6} PPPY, (Figure 10) the complexion of the Risk Distribution map changes considerably. Note the ~30-fold increase in total risk; the distribution of which is now aligned largely with the time at risk grid, rather than with risk density.

Risk Density [$P_{1yr}(L_{d,s}) \times 10^6$]

Depth (m) \ Starting Speed (kt)	1	2	3	4	5
40	25	1.1	1.0	1.0	1.0
80	1.7	1.1	1.0	1.0	1.0
120	1.2	1.0	1.0	1.0	1.0
160	1.3	1.0	1.0	1.0	1.0
200	1.5	1.1	1.0	1.0	1.0
240	2.3	1.5	1.1	1.0	1.0
280	2.4	1.6	1.3	1.1	1.0
320	3.2	2.2	1.5	1.1	1.0
360	3.3	2.4	1.6	1.3	1.1
400	7.2	3.1	2.3	1.6	1.3

Log10[P_{1yr}(L)] from collision: -6

Time at Risk (hrs/yr) — Utilisation: 33.0%

Depth (m) \ Starting Speed (kt)	1	2	3	4	5	Totals
40	0	10	50	200	200	460
80	0	10	200	500	500	1210
120	0	10	100	100	100	310
160	0	10	50	100	100	260
200	0	10	25	50	50	135
240	0	10	25	50	50	135
280	0	10	10	50	100	170
320	0	10	10	50	100	170
360	0	5	5	10	10	30
400	0	0	5	5	5	15
Totals	0	85	480	1115	1215	2895

Risk Distribution (% of total) — P_{oyr} (L) 3.4E-07

Depth (m) \ Starting Speed (kt)	1	2	3	4	5	Totals
40	0%	0%	2%	7%	7%	15%
80	0%	0%	7%	17%	17%	41%
120	0%	0%	3%	3%	3%	10%
160	0%	0%	2%	3%	3%	9%
200	0%	0%	1%	2%	2%	5%
240	0%	0%	1%	2%	2%	5%
280	0%	1%	0%	2%	3%	6%
320	0%	1%	0%	2%	3%	7%
360	0%	0%	0%	0%	0%	1%
400	0%	0%	0%	0%	0%	1%
Totals	0%	4%	17%	38%	41%	

Figure 10: EOP A risk map (Including collision risk)

7.2 EOP B RISK PLOTS

EOP B has been assessed both with and without the inclusion of a hypothetical system that immediately detects MCW floods larger than a finite flood rate threshold (known as Fracture Detection System (FDS)).

Risk Density [$P_{1yr}(L_{d,s}) \times 10^6$]

Depth (m) \ Starting Speed (kt)	1	2	3	4	5
40	1.8	0.3	0.0	0.0	0.0
80	3.6	1.5	0.4	0.0	0.0
120	6.1	2.2	0.7	0.2	0.0
160	9.9	7.4	5.4	1.1	0.4
200	9.9	9.5	7.4	5.6	2.0
240	11	10	9.2	9.1	7.4
280	15	11	10	9.2	9.1
320	21	13	11	10	9.2
360	21	14	12	10	10
400	25	21	14	13	12

Log10[P_{1yr}(L)] from collision: 0

Time at Risk (hrs/yr) — Utilisation: 33.0%

Depth (m) \ Starting Speed (kt)	1	2	3	4	5	Totals
40	0	10	50	200	200	460
80	0	10	200	500	500	1210
120	0	10	100	100	100	310
160	0	10	50	100	100	260
200	0	10	25	50	50	135
240	0	10	25	50	50	135
280	0	10	10	50	100	170
320	0	10	10	50	100	170
360	0	5	5	10	10	30
400	0	0	5	5	5	15
Totals	0	85	480	1115	1215	2895

Risk Distribution (% of total) — P_{oyr} (L) 7.2E-07

Depth (m) \ Starting Speed (kt)	1	2	3	4	5	Totals
40	0%	0%	0%	0%	0%	0%
80	0%	0%	1%	0%	0%	2%
120	0%	0%	1%	0%	0%	2%
160	0%	1%	4%	2%	1%	8%
200	0%	1%	3%	4%	2%	10%
240	0%	2%	4%	7%	6%	18%
280	0%	2%	2%	7%	[illegible]	25%
320	0%	2%	2%	8%	[illegible]	26%
360	0%	1%	1%	2%	2%	5%
400	0%	0%	1%	1%	1%	3%
Totals	0%	10%	19%	32%	40%	

Figure 11: EOP B risk map (FDS not fitted)

Without FDS, the DEMCW risk dominates the risk density matrix deep and slow; however, the overall risk figure is nevertheless comparable with EOP A when collision risk is taken into account.

Risk Density [$P_{1yr}(L_{d,s}) \times 10^6$]

Depth (m) \ Starting Speed (kt)	1	2	3	4	5
40	1.3	0.1	0.0	0.0	0.0
80	0.6	0.2	0.0	0.0	0.0
120	1.0	0.6	0.1	0.0	0.0
160	1.2	1.0	0.7	0.1	0.0
200	1.6	1.1	1.0	0.9	0.8
240	2.7	1.6	1.2	1.0	1.0
280	4.4	2.0	1.5	1.1	1.0
320	5.9	4.2	1.9	1.3	1.1
360	7.1	6.0	4.0	2.0	1.6
400	12	7.1	6.5	5.5	5.5

Log10[P_{1yr}(L)] from collision: 0

Time at Risk (hrs/yr) — Utilisation: 33.0%

Depth (m) \ Starting Speed (kt)	1	2	3	4	5	Totals
40	0	10	50	200	200	460
80	0	10	200	500	500	1210
120	0	10	100	100	100	310
160	0	10	50	100	100	260
200	0	10	25	50	50	135
240	0	10	25	50	50	135
280	0	10	10	50	100	170
320	0	10	10	50	100	170
360	0	5	5	10	10	30
400	0	0	5	5	5	15
Totals	0	85	480	1115	1215	2895

Risk Distribution (% of total) — P_{oyr} (L) 1.1E-07

Depth (m) \ Starting Speed (kt)	1	2	3	4	5	Totals
40	0%	0%	0%	0%	0%	0%
80	0%	0%	1%	0%	0%	1%
120	0%	1%	1%	0%	0%	2%
160	0%	1%	4%	1%	0%	6%
200	0%	1%	3%	5%	4%	13%
240	0%	2%	3%	5%	5%	15%
280	0%	2%	2%	5%	[illegible]	20%
320	0%	4%	2%	[illegible]	[illegible]	25%
360	0%	3%	2%	2%	2%	9%
400	0%	0%	3%	3%	3%	9%
Totals	0%	14%	20%	29%	36%	

Figure 12: EOP B risk map (FDS 100% reliable)

With FDS fitted (and assuming 100% reliability), the risk density plot for EOP B is broadly comparable with EOP A, although the band of low-level MCW risk is broader. However, in this scenario, the increase in MCW risk is more than compensated for by the reduction in collision risk.

7.4 DISCUSSION

Notwithstanding the significant assumptions, simplifications and inaccuracies that are embedded in the example analysis, it is clear that the overall probability of loss of platform due to genuine flooding, given a representative operating profile, is potentially very low. If this is at all representative of reality, then it suggests that the collision risk could potentially be more significant than flooding in terms of the overall balance of risk. (Even at a catastrophic probability of 10^{-6} PPPY, it dominates the overall platform risk associated with EOP A in the example presented).

The distribution of risk density for EOP B suggests that there may be sound justification for adopting a different EOP for different areas of operation, possibly involving temporary reversion to EOP A when slow and deep. It would therefore seem prudent to assess a range of valid EOP options, under a range of material states, and then give COs the latitude to choose the most appropriate one for the circumstances, armed with map of the relative risks.

8. PRACTICAL APPLICATION OF THE PROBABILISTIC METHOD

Figure 13 illustrates how the probabilistic risk assessment process might be applied to generate a versatile suite of guidance to operators and advice to Command.

A sufficiently diverse set of tailored risk maps would enable Command to select appropriately-justified procedures and operating parameters to suit a broad range of operational demands and degraded material states, and thus manage all manoeuvring-related risks effectively. Quantitative, evidence-based assurance would also be provided to relevant stakeholders.

In order to provide a usable output, it should be noted that the probabilistic method would need to be supported by a substantial corpus of background information and evidence that is lacking in the present example.

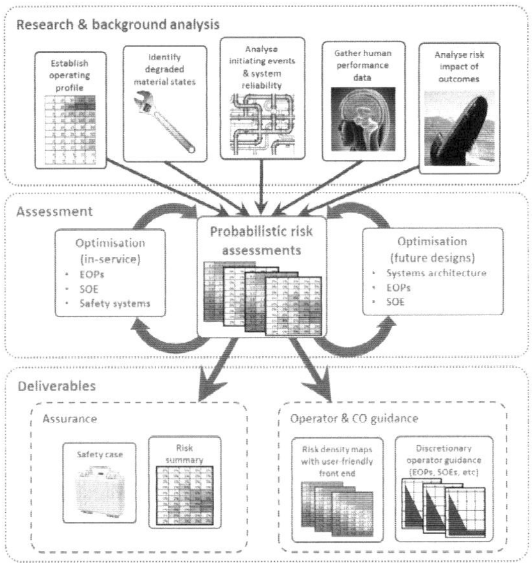

Figure 13: Application of the probabilistic flood hazard analysis in context

9. FURTHER WORK

Some of the key areas of development that would be required to bridge the gap between the illustrative example presented in this paper and a usable analytical too are presented in this section.

9.1 QUANTIFY STOCHASTIC INPUTS

This is pivotal aspect of the probabilistic approach, which has not yet been tackled in any meaningful way. The following require particular attention:

9.1 (a) Human Reaction Times

Further long-term trials are suggested, in which measurements are taken covertly during routine simulator continuation training (as opposed to during a dedicated trial);

9.1 (b) Failure rates of pipework and fittings

The assumptions used for this paper are far too simplistic, and only identify selected credible bounding cases. A comprehensive analysis of the entire sea water boundary would be needed to provide a profile of all credible initiating events. Existing methodology and data from the nuclear industry could be applied [17], in which failure rates are estimated as a function of length of pipe, material quality, environment, etc. The sharp distinction between securable and unsecured events could also be decomposed in terms of failure rates of isolation valves.

9.2 EXTEND THE VALIDITY OF THE MANOEUVRING MODEL

In order to meaningfully evaluate the risk of a catastrophic outcome, the manoeuvring model must examine the behaviour between the edges of normal operation and the complete loss of depth control. This is significantly beyond ARTEMIS's current range of validity.

Since a large number of simulations are required, it is recognised that it would not be feasible to abandon the current empirical model in favour of a CFD or physical model-based regime. The focus must therefore be on significantly extending the existing empirical approach.

It is suggested that a model that remains broadly representative throughout the typical range of marginally-survivable flood cases is preferable to one that is more finely tuned but limited in scope.

As a starting point, it is noted that the lookup table facility in ARTEMIS, which references empirical hydroplane derivatives datasets (and which crucially allows hydroplane stall to be represented) currently only applies to the bowplanes, and excludes the sternplanes and the hull form. Furthermore, if lookup tables are used, then the simulation will stop when it reaches their limits. The following modifications to ARTEMIS are therefore proposed:

- Extend the lookup tables facility to the sternplanes and hullform;
- Introduce a stall angle variable for the bowplanes and sternplanes, up to which the current small-angle derivatives would apply, and after which the force or moment would plateau. (This would quickly remove the gross optimism that is inherent in the model at very high local angles of incidence);
- Allow for a coarse computation of the outcome of bow-heavy, buoyant ascents with negative surge velocity.

9.3 INTRODUCE CONSEQUENTIAL EVENTS

For the example analysis presented, instances of excessive pitch angle have been handled by crudely increasing the risk weighting of the affected run as a function of maximum angle reached. However, the actual additional risk depends on the circumstances of the event; therefore, ideally the consequences of the high pitch angle would be modelled directly (e.g. in terms of an elevated risk of main engine trip, or some other propulsion-limiting event).

Figure 14 illustrates a scheme by which the risk of main engine trip increases at successive points (i.e. 30°, 40° and 50° pitch angle). Although no trip occurs in the parent scenario (with probability of occurrence), additional runs are then conducted, in which a trip occurs at each branching point. The probability of each case occurring (including the parent) is then calculated such that they all add to $P(O)$. The actual increase in risk from the pitch excursion then depends on whether depth control is ultimately lost in each child case.

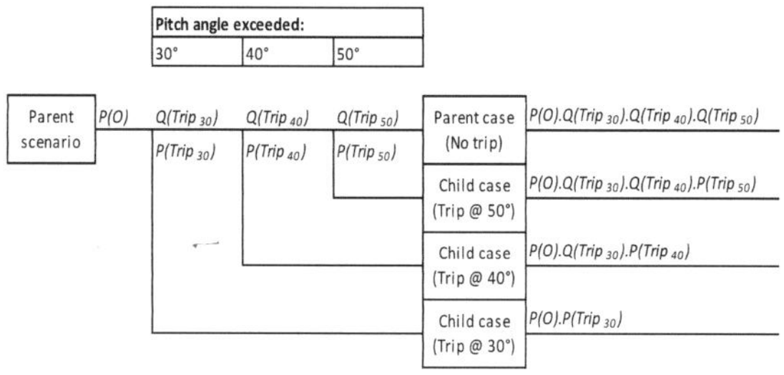

Figure 14: Additional runs created by a proposed consequential event handling routine

The impact of sub-optimal broaching conditions could be handled using a different approach, with reference to a matrix of deterministic broaching simulations at different angles and surge velocities.

9.4 INTRODUCE NON-TIME-BASED RISK FACTORS

Considering the prevalence of fatigue as a mechanism for mechanical failures of pressure boundary components, it should be considered whether it would be more appropriate to quantify risk taken in terms of depth cycles, rather than simply time at risk. This also raises the question of whether the default starting condition for some flood cases should be a descending depth transient (during which a fatigued component might be expected to reach catastrophic failure).

10. CONCLUSIONS

This study has shown that it is potentially feasible to quantify the various risks that are associated with submarine flooding, taking account of the complexity and uncertainty that is inherent in the problem, and that it is therefore feasible to make informed trade-offs between conflicting risks and operating philosophies in pursuit of a philosophy that is optimised overall. This approach relies on a move away from the use of conservative assumptions and towards the adoption of more realistic and representative assumptions to support the analytical process.

Notwithstanding the purely illustrative nature of this study, the sample results suggest that the general risk from flooding whilst operating within the safer regions of the SOE is negligible. Depending on actual operating profiles, this potentially means that some of the complexities and low-probability events that are currently ignored, such as collision on emergency surfacing, could potentially affect the overall risk profoundly. This, in turn, could affect the optimum operating philosophy.

The current SOE format may be user-friendly for operators, but it is of limited value for Duty Holders and COs, who sometimes need to assess the risk of conducting operations outside deterministically *safe* depth and speed ranges. The *risk map* presentation demonstrated in this paper provides a direct, graphical means of presenting this level of information.

This study falls far short of providing a ready-made method for producing a usable probabilistic analysis of flood risk, but it does illustrate a potentially workable approach for doing so, and identifies some of the significant areas of further work that would be needed to support it.

11. ACKNOWLEDGEMENTS

The author wishes to thank the many individuals at the UK MoD, UCL and QinetiQ who have supported and contributed to the wider pieces of work that support this paper.

The material in this paper was originally presented as part of a thesis by the author for the degree of MSc in Naval Architecture at UCL [1].

Any views expressed are those of the author and do not necessarily represent those of the Royal Navy, the UK Ministry of Defence or UCL.

12. REFERENCES

1. HORN, N., 'Probabilistic Submarine Flood Recovery Analysis - Thesis for the Degree of MSc in Naval Architecture', *University College London*, 2016.

2. Compilation of UK Naval Reactor Plant Authority (NRPA) Event Reports - 2003-2013', *UK Ministry of Defence*, 2013.

3. 'USS Thresher Disaster' [Internet], *http://www.arlingtoncemetery.net/uss-thresher.htm*, cited 2016.

4. 'Honolulu Star-Bulletin Hawaii News' [Internet]. *http://archives.starbulletin.com/2001/02/20/news/story1.html*, cited 2016

5. 'Marine Acquisition Publication 01-073 (MAP 01-073) Submarine Hydrodynamics, Issue 2.0', *UK Ministry of Defence*, 2012.

6. 'DSTAN 08-102 - Requirements and Acceptance Criteria for Submarine Hydrodynamics, Issue 12', *UK Ministry of Defence*, 2012.

7. 'Joint Service Publication 430 - Management of Ship Safety & Environmental Protection, Part 1: Regulations, Issue 5', *UK Ministry of Defence*, 2013.

8. 'Joint Service Publication 430 - Management of Ship Safety & Environmental Protection, Part 3: Naval Authority Certification, Issue 6', *UK Ministry of Defence*, 2014.

9. HORN, N. 'Technical Note - Probabilistic Analysis of Submarine Flood Recovery', *UK Ministry of Defence*, 2015.

10. RILEY, J. 'H697 Flood Hazard Risk Assessment', *UK Ministry of Defence*, 2015.

11. KIMBER, N., CROSSLAND, P., 'Advances in Experimental Techniques for Understanding the Manoeuvring Performance of Submarines', *The Royal Institute of Naval Architects*, 2008.

12. KIMBER, N., 'Submarine Manoeuvring: Correlating Simulation with Model Tests and Full Scale Trials', *The Royal Institute of Naval Architects*, 2011.

13. CARPENTER, R., WILLIAMS, M., 'Neural computation of log likelihood in control of saccadic eye movements', *Nature*, 1995.

14. RILEY, J., TREVOR-JONES, H., 'Recovery Assist Gauges FASMAT Trial 10th-14th June 2013', *UK Ministry of Defence*, 2014.

15. PLACKETT, R., BURMAN, J., 'The Design of Optimum Multifactorial Experiments', *Biometrika*, 1946.

16. KEOGH, E., MUEEN, A., 'Curse of Dimensionality' *Encyclopedia of Machine Learning. Springer US*, 2011.

17. FULWOOD, R., 'Review of Pipe Break Probability Assessment Methods and Data for Applicability to the Advanced Neutron Source Project for Oak Ridge National Laboratory', *Associated Universities, Inc.*, 1989.

13. AUTHOR'S BIOGRAPHY

Neil Horn is the PWR2 Safety Desk Officer in the Nuclear Propulsion Project Team at MoD Abbey Wood. He is currently responsible for managing the delivery of nuclear safety for Vanguard Class submarines. He first became involved in submarine flood hazard mitigation in 2013 in his previous role within the Propulsion Design Authority for In-Service Submarines, during which time he proposed a number of design and procedural changes relating to flood detection, analysis and recovery. He continued to develop some of his ideas on flood recovery analysis as part of his Naval Architecture MSc thesis at UCL. He is currently a serving Royal Navy officer.

TEMPIS FUGIT: BUT WHERE DOES IT GO

M.J. Cook, BAESYSTEMS, UK

SUMMARY

Human Reliability Assessment (HRA) has developed largely in response to safety critical tasks in industries which combine high impact hazards and an opportunity for extended diagnostic activity prior to mitigation of the safety critical event. Assessments of submarine platform design in terms of task performance, failure frequency and time needed to complete tasks are all closely coupled, because performance is influenced by the time available. The speed-accuracy trade-off in human performance reflects the inherent limits to success because of required processing time to complete the task. Increasing task complexity results in more processing and this takes more time, as well as the negative influence of time pressure on operator performance amplifying the error rates in fast responses. In submarines, there are safety critical tasks (e.g. flooding, steering and diving and plant management) which need to be accomplished in a limited time to ensure the response of the platform to safety critical events. Modelling of the operator performance in this response must become increasingly sophisticated by recognising the operator performance limitations and by understanding of the tasks they are expected to undertake; this is compounded by automated systems where transitions may occur unexpectedly. Human factors analyses suggest that characteristics such as information requirements help to estimate the likelihood of failure and the time required to complete tasks. However, some of the most contentious decisions are associated with situation awareness and expectations of what information operators may be expected to maintain over time. This paper discusses the increasing importance of human factors evaluations of system designs.

1. INTRODUCTION

Human Reliability Assessment (HRA) has developed largely in response to safety critical tasks in industries which combine high impact hazards and an opportunity for extended diagnostic activity prior to mitigation of the safety critical event [27; 16]. Analysts attempting to predict the impact of time are aware of its presence as a task-related requirement but may not be solely focussed on time as a key variable because there is a speed-accuracy trade-off inherent in human performance [39].

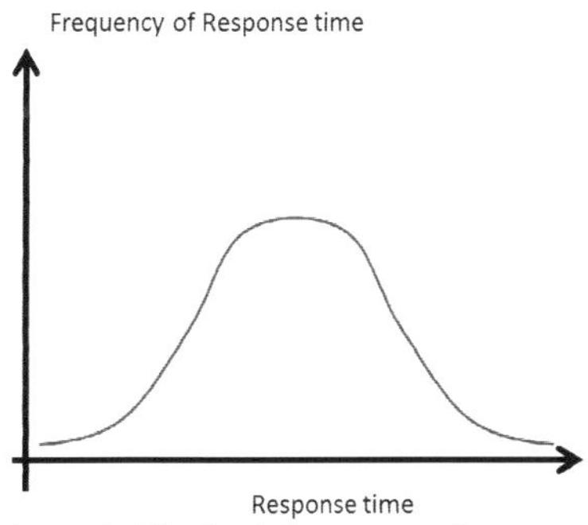

Figure 1: Idealised frequency of response times

As Figure 1 indicates any operator will normally exhibit a variable response time to a trigger stimulus, and although idealised as a normal distribution, it is more likely that the distribution will be skewed such that the longer response times create an extended tail of response times. Figure 2 indicates that attempts at faster response times (an emphasis on speed) results in more errors and emphasis on accuracy results in fewer errors but extended times. Thus, compressing human performance in time to maximise the performance envelope will naturally result in increasing error rates. This failure rate and the consequences need to be matched to the desired limits within the expressed safety case. It is suggested that the methodological components of HRA techniques, which involve identifying errors, modelling errors and quantifying them are significant in adding value to the evaluative process [34].

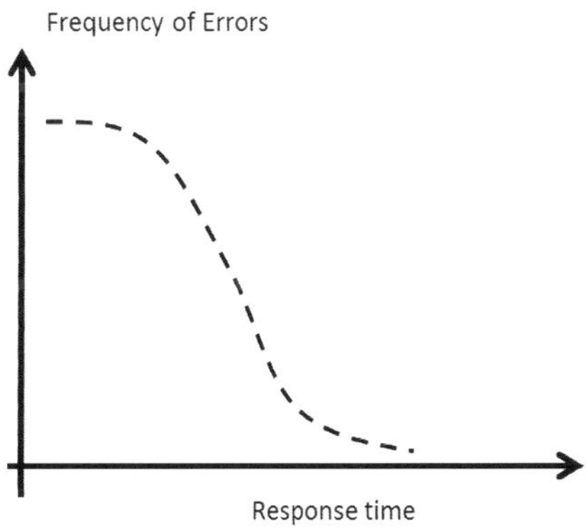

Figure 2: Frequency of errors against response times

2. ASSESSING HUMAN RELIABILITY

Human Reliability Methodologies have three critical parts identification of error, modelling of the error and quantification of the error in relation to performance. This requires that the task is analysed into task elements so that the task structure can be assessed and at the same time it may reveal the information requirements for the various elements of the task. For example, there are limits to human working memory which acts as a short term store where information is processed. Working memory can trade information complexity and complexity of the operations applied but working memory cannot accommodate large amounts of information to be processed with a complex set of operations. Thus, during the task analysis the processing choke points within the task can be assessed and this has been addressed in the GOMS modelling approach (24; 25) and combined with HRA (4). Once the task analysis is complete the error modes associated with the various elemental pieces of the task can be assessed. The elements of the task may be perceptual, cognitive (thinking, decision making) or executory action (movement or speech). Each task element may be associated with one or more independent error modes, for example individuals can commit errors by commission or omission. Individuals can forget information. Individuals can mishear or misread information. The error modes can be associated with error rates, or frequencies and these can be combined to calculate overall frequencies of errors across the task.

2.1 RECOVERY IN TIME LIMITED TASKS

The latter stages of analysis may involve an event tree which includes recovery loops. It is generally agreed that human operators make mistakes with a high frequency but often identify the errors and recover. In the case of safety critical time limited tasks opportunities for recovery by the active operator or through supervision may not be possible, driving the expected error rate higher. Thus, performance for safety critical tasks which are time limited and need to be right first time is often poor, as the normal recovery routes are not available.

3. HRA METHODS

There are a number of methods for assessing human reliability such as THERP [36], CREAM [16] and HEART [40] which essentially have identified basic core values for tasks or task elements and then are moderated through error producing conditions or performance shaping factors to identify the specific contextual factors which amplify or moderate erroneous performance. Time pressure or time stress and the awareness that limited time is available for tasks are judged to have negative consequences for performance. Thus, in addition to time limited information processing effects on performance there are additional stress-related effects of time. For this reason human performance may only achieve performance failure on demand rates of between 1 in 100 to 1in 10, or worse in some cases [2].

Individual HRA methods have different strengths and weaknesses and are arguably more or less applicable to different types of task. For example, it has been argued that CREAM is more suitable for cognitive based tasks, whilst THERP with its detailed and atomistic perception-action sequence is useful and provides a more thorough understanding of the task, to aid recommendations for improvement. THERP had been in use for fifty years in 2012 [3] and was widely used in analysis of civil nuclear power plant operation, establishing a strong pedigree for the method in non-time critical tasks. THERP and HEART have been independently validated as an effective method of human reliability assessment [26, 27, 28]. THERP and other methods can be pessimistic where sufficient detail of the task is not available but this is not the case when a detailed task analysis and event tree are developed.

3.1 LINKING HRA AND GOMS

The advantage of THERP in respect of estimating time required to complete tasks is the micro-analysis of the task which can be easily linked to a timing analysis using methods such as GOMS [8; 18; 19] or derivatives thereof. Task analyses are a core of most usability assessments because of the use of the analysis in evaluating procedural aspects of the task and information requirements of the task elements [25]. Although there have been critical reviews of these methods [20], finding that estimates may be misleading in one third of cases it is likely that initial estimates can be qualified by empirical measurement of part or whole task trials. Others have argued that the links between HRA methods and GOMS can be successfully used to explore design [6; 7].

4. HRA & SAFETY CASE

The methods identified will not form the entire evidence base for a claim-argument-evidence based safety case [22; 23; 3]. However, this approach does link to the use of safety arguments throughout design because it allows instantiation of claims, supporting an argument and based upon model evidence that may be substantiated by user trials at a later date. In respect of safety-critical tasks with limited time for task execution the claim would be that it supports safe operation of the system, this would be based on an argument related to the time to execute the task and the number of errors that would occur in relation to the task. The modelling described in this paper supports the arguments by provision of evidence estimating performance failure on demand and time required for task execution. A modular approach to safety case development [9] supports progressive refinement of the safety case as more evidence is accrued but lessens the risk at the early stages of development when the critical developmental path is chosen. Although not an exact science, the focus on more successful lines of development and exclusion of problematic lines of development using user modelling increases the likelihood of arriving at successful designs which meet the necessary safety claims.

5. TIME, MITIGATION, AUTOMATION & RECOVERY

Failure to perform a mitigating task correctly may be more significant and the probability density function for successful task completion may not simply follow task completion within a fixed time. Operators can complete tasks successfully or incorrectly (failing), or they can fail to complete a task in time. Even successful task completion can be independent of time because operators may fail to complete the task correctly by accomplishing all the necessary task elements, but guess the event type. Thus, there is always residual uncertainty in the interpretation of actual observed performance in empirical trials and in reality.

The discriminative analysis of timed performance is a long-established topic of study and assumptions about performance can be subject to very significant caveats. In experimental analysis some have argued that very long response times should be removed because they represent the execution of a task other than that assigned. One might argue that the same is true of very short reaction times. One is not comparing, or averaging like with like in some cases. However, in the real-world these erroneous performances are not as easily excluded and should be examined as a part of a robust safety-case.

Normally the time available will not limit task execution to a single pass without an opportunity for human errors and recovery paths. It has been argued that human operators frequently participate in preventing failures and human errors from resulting in catastrophic system breakdowns [37]. It is noted that rarely is it the occurrence of failure but more often the negative consequences of failure that are to be avoided [37], in time limited responses to safety critical events the occurrence of failure and the realisation of the consequences are almost indistinguishable. Recovery is a core element of many HRA assessments and analysts recognise the uncertainties of human actions which often contain errors, with later recovery. The initial assessment of safety critical tasks may consider to what degree there is sufficient time to recover from errors, for others to provide supervisory input and time to check the end result of the executed task. As noted above it is likely that safety critical tasks with limited available time may afford few opportunities for recovery in any of its guises.

6. AUTOMATION

It may be that automated tasks are inherently likely to produce time limited safety critical tasks after automation failure because of the Machines Are Better At / Men Are Better At (MABA-MABA) allocation of function doctrine (after Fitts, 1951) [14]. Some authors have criticised the MABA-MABA because it does not address the need for human-system coordination as a joint cognitive system [10]. Fitt's list which was originally developed for automation in another domain identified tasks that either humans or machines were better at and it has been argued that its use has persisted because it was intrinsically useful as an approximation [11]. The interest in allocation of function as a technique has been re-visited and critiqued by others who did not consider the intrinsic paradox in dividing tasks between human and machines without coordination [32].

The allocation of tasks on this basis focuses on the observation that automation and machine intelligence is good at fast processing of large volumes of data in integrative tasks. The capability of automation to process information quickly and to integrate the analysis of that information is frequently used to justify expansion of the performance envelope. By creeping nearer to the edge of safe performance, through the use of automation, designers may introduce a latent error pathway by making use of Fitt's List. If the automated system is unavailable the human operator may be expected to assume control in a time critical response where realisation of the negative outcomes is no longer avoidable in the time available because humans do not have the capacity that the automation has, essentially QED proof of the Fitt's List.

7. VARIABILITY OF RESPONSE

Even simple two choice reaction times exhibit variability that needs to be interpreted and it can logically justified that this process is complex [35]. The different error and recovery pathways create opportunities for temporal variability of recorded responses, with some instances of task completion occurring in shorter periods of time and others in longer periods of time. Temporal variability is a more significant issue for tasks that involve some combination of decision-making, diagnosis, problem solving or other significant cognitive process as it increases the difficulty of assessing the time required for task completion. Donders' developed the subtractive reaction time method in 1869, for estimating mental chronometry of specific processes. It provides a tool that is sufficient for examination of short-duration perceptual and cognitive processes and the response selection and motoric action that follows.

7.1 THINKING, PROBLEM SOLVING AND DIAGNOSIS

The central cognitive processes associated with elaborate decision-making, problem solving and diagnosis are less easily or accurately quantified in terms of time required to process information. However, advocates of HRA emphasise the need for modelling of the cognitive processes to accurately reflect the psychological process influencing error outcomes [4]. The identification of Naturalistic Decision Making models [17] based upon a form of pattern recognition and Daniel Kahneman's [21] work on thinking fast and thinking slow suggests that human cognition can show a significant amount of temporal variability, as well as variations in performance.

If one considers the HRA method CREAM [16] it affords three alternative levels of diagnostic behaviour, which implies different time requirements, from the thorough *diagnostic* level of processing to the very limited *evaluative* level, via the relatively superficial *identification* process. These might be equated with Rasmussen's [33] skill-rule-knowledge based processing strategies, with diagnostic processing akin to knowledge–based processing and evaluation and identification equivalent to rule or skill-based processing depending on the experience and cognitive demand on the operator. As noted safety critical operator tasks need to be assessed in terms of the ability to complete the processing required within the time available set by physical parameters of the managed process. It is necessary to estimate both the time to complete a task and the temporal variability in responses as a consequence of error processes, recovery and other cognitive parameters.

Estimation of time required to complete tasks is both accurate and feasible with simple perceptually driven response tasks but depending on the level of cognitive processing, it is significantly less effective at establishing a reasonable estimate of time required, even with trained operators. Operators experiencing events which are amenable to skill or rule-based processing may provide characteristics that limit the variability of estimates but this is clearly not the case with knowledge-based processing. It should be recalled that Rasmussen's [33] model of skilled behaviour accepts that even the most skilled operators can in some situations find themselves attempting rule-based processing or even slow effortful knowledge-based processing.

Before considering the estimation process in detail one can identify three categorical outcomes, the first is clear separation of time required and time available allowing the task to complete potentially with limited residual error rates, 1 in 100 cases of failure. The second is the situation where the time to execute the task is greater than the available time and the operator is likely to fail, either by hastily completing the task or running out of time. Finally, there are cases where the margin of error in the estimation process makes the task execution times almost equivalent to the available time. Thus, in the latter case where successful task completion is ambiguous there is a significant design problem, not least of which is the likely increase in failure rates when the task completion is combined with estimation of successful completion. The obvious conclusion is to assess the complete task within an experimental setting to determine the mean response time and more importantly the dispersion of the results. Most response times are positively skewed distributions with the median significantly smaller than the mean and the distribution has a long tail of slower response times. This makes estimation of confidence limits in the traditional sense problematic based they are upon normally distributed populations of response symmetrically distributed about a mean value, which is rarely the case with actual distributions.

When empirical measures of such safety critical tasks are proposed they are often criticised for their low levels of ecological validity. Operators given a task within a specific scenario may guess what the purpose of the task is and self-prime themselves and their speeded performance in relation to the task. Critics would then suggest that the self-primed performance provides an optimistic estimate of the time to complete the task. However, other critics may choose to reject long times where the operators fail to understand the nature of the task and spend an inordinate amount of time trying to determine what task they should prioritise or focus upon. Skilled performance is largely determined by familiarity with the critical cues and the overt selection of information based on past experience. However, design seeks to ensure that high impact hazards are relatively low frequency events, often managed with an alarms and warnings strategy like those described in EEMUA 191 [12] and even though training may occur one can argue that actual events are never as simple as those which have occurred in training. Thus, training for high impact hazards will attempt to ensure that operators are highly skilled (Suitably Qualified and Experienced Personnel) but one of the critical aspects of skilled performance is the high degree of automaticity and low levels of flexibility. It may be that instinctive rigidity that undermines performance in some critical tasks. Thematic vagabonding and cognitive lockup are symptomatic of training failures but more importantly they are indicative of time inefficient processing, which is hard to estimate in modelling estimated times.

8. HUMAN FACTORS ENGINEERING

There are clearly ecological arguments supportive of the different positions. A well-designed interface should present a clear representation of the on-going events to maintain situational awareness, so that operators are primed as to the current state, when presented with a change in state and then transition smoothly to the relevant safety critical operator task. The development of ecologically appropriate interfaces [38] is not a new idea and yet if the relevant criteria were applied to many interface designs it can be argued that they would potentially fail to achieve a satisfactory grading. Ecological interface design is essentially a method for removing or reducing the need for priming to achieve adequate operator performance by presenting information in an intuitive user-friendly mode. However, interface driven priming is not the same as that occurring in a laboratory based trial, where the situation assessment task is short-circuited. The natural occurrence of a critical event may trigger a cascade of information changes which present the operator with a difficult task of determining what the most likely trigger event is, sorting the information wheat from the chaff. In some respect this illustrates what is known about problem solving, in that a significant part in successful timely completion depends on the accurate situation assessment at the start of the task [1; 15]. Impaired situation awareness, which is linked to effective situation assessment, is part of the error processes leading to accidents [13]. One example of the application of both situation assessment and situation awareness was presented in a study of air traffic control tasks where the mental model of the controllers was linked to dimensional analysis of problems (situation assessment) and preparedness to act [30]. It is clear that as operators become more experienced the methods of information processing and the implicit structure of information storage become defined, contributing to skilled performance.

In some accident sequences leading to extreme negative consequences there is evidence of various failed memory processes from cognitive lockup, fixation on an incorrect mental model, to thematic vagabonding taking place [41], the latter is essentially a failure to locate an adequate working model. Both such processes suggest thought processes which consume the time available without necessarily identifying the true nature of the event, or developing an adequate mitigating response.

It is perhaps worth noting that those collecting reaction times have debated this issue, in an analogous way. The common practice of trimming reaction times, to discard the long tail which is a common part of the asymmetrically skewed distributions, narrows the estimate of dispersion and conveniently enables a greater likelihood of statistically significant differences between populations of responses. If one accepts the longer responses as valid responses from the same potential distribution then it is possible to accept that measures of central tendency need greater scrutiny and cautious application. The published estimates in the literature, based on trimmed distributions may generate misleading expectations about times to complete tasks.

A potential argument is that more complex tasks, both those with higher and more complex cognitive demands, are likely to produce greater variability than estimates based on individual components suggesting higher failure rates. Complex tasks will not involve simple operations in working memory and complex interleaved tasks may make demands on executive functions which exhibit the kinds of failures observed in high stress tasks. Executive management of tasks essentially makes serial sequential analysis necessary, which is slow. Prospective memory errors may occur where a place keeping mental note is forgotten and a task is left uncompleted. Perseverative errors may occur where the operators cognitive lockup on a specific thread or action to the detriment of the overall process. In simple terms complex tasks may involve metacognitive skills which have less temporal certainty and higher situational variability. It may prove impossible to reduce all tasks to the essence of a gun drill by removing all discriminative analysis.

If one accepts the argument posed above about available data the cognitive elements of tasks, which possess the greatest temporal variability, and are likely to influence responses on a trial by trial basis more substantially in terms of variability than sensory or motoric processing. The current rejection of very long response times may mean that that genuine variability is significantly reduced through sub-sampling of the response time populations and its impact is under-estimated in terms of the calculation of measures of central tendency, resulting in an optimistic calculation of response times.

9. MEASURES OF CENTRAL TENDENCY

In positively skewed distributions, the use of a median measure to place the estimated time for 50% of the population to complete the task is often at a shorter time than the mean response, because the latter is influenced by the longest response times in its calculation. Thus, median estimates give an optimistic estimate of the time for task completion relative to arithmetic means. If one combines estimates of dispersion, which are based on trimmed population samples, and median estimates of response time one is likely to calculate a distribution of response times for a population which is both optimistic and misleading.

It can be argued that the way in which the current estimates of the time required to complete tasks are calculated may be extremely optimistic, especially when estimating the 95th percentile speeded response. Thus, in assessment situations where the time available and the time to complete a task are very close then empirical corroboration of the time using trial data is required. This is particularly important with regard to mitigation responses to high impact hazardous events. Even though empirical estimates are potentially compromised by a number of the features of laboratory-based testing they represent the best method for providing definitive assurance of task completion in respect of the safety critical tasks.

There is a significant body of evidence on chronometric analysis of human performance and it seems logical to apply this to estimating the time required to complete safety critical operator tasks, especially when the available time may be very limited and afford no opportunities for error recovery. It is argued that many aspects of the chronometric analysis and the data are optimistic. Thus, where the time calculated as required by 95% of the population and that available are almost equivalent (differences of less than 20%) it is suggested that empirical data is necessary to distinguish cases where the operators completion of the safety critical task will fail because they do not have sufficient time to complete the actions required.

Finally, it is important to re-assert that time and performance failure rates are not equivalent. Operators can complete tasks but still complete a proportion of them incorrectly. For example, a large population of users have probably inadvertently deleted a file they wanted to keep but they have accomplished the task in the time available. This is perhaps particularly true with high impact hazards and high consequence operator tasks. Impulsive responding can be driven by the apparent threat and result in responding based on inconclusive or incomplete cognitive processing, sometimes correctly, sometimes incorrectly. It is equally true that even when operators are aware of the time available and the significant consequences of their actions they can delay responding, as if exploring the possibility of the high impact event and seeking confirmation that it is what is expected.

This paper seeks to question some of the basis for the cognitive modelling and estimation process, not to simply reject it but to temper the certainty with which it is applied. For example, there may be an assumption that training will both shorten mean response times and narrow the distribution of responses. It is argued that this assumption is plausible for simple perceptual-response tasks where there is little evidence of cognition. However, most of the high impact hazards are a result of more complex cognitive processes that are not easily captured within cognitive modelling. In particular, the dispersion of responses can be highly variable even with training because it is in part a function of the match between training and the event occurrence. This is the reason why recognition-primed decision making explains what is occurring when the event meets expectations.

Real events occur in open systems and the pattern of occurrence varies or is ambiguous to some degree, so that additional discriminative work is required to identify what event oriented solution is required. Thus, real responses are more variable and that variability is compounded with the local event probabilities of related and un-related events, the clarity of the cues, and even with operator characteristics. For example, training attempts to ensure that the operator's prior experience is homogenised to assure their response but in reality this aim may not be met. Then there are additional situational variations such as time of day, fatigue effects or stress, which impact cognition.

It is argued and accepted that estimation of time to complete tasks and performance failure are not exact sciences and should not be treated as such within an engineering framework. They are not nugatory though because they can help establish if a task is feasible or not. Often the task analytic process and associated cognitive task analysis can determine the cognitive chokepoints and identify poor design which does not intuitively support decision making. Thus, the need to hold significant amounts of information in memory and to operate on the information by arithmetic or comparative operators, or the need to hold time series information for a later comparison, might be remedied by simple changes in interface design, enabling more effective decision making and saving time in the process. It should be noted that stress and fatigue both influence working memory capacity so that a design which operates 24/7 is even more likely to exhibit performance variability based on working memory.

It is argued that a categorical analysis of time to complete a task provides two categories which can be used to assure system designs for safety critical operator tasks, those where it is possible to complete the tasks in the available time and those where it is not. A third category of task exists where the available time and the time required are so similar that empirical evidence to support the analysis and/or re-design of the human machine interface is required to both save time and reduce errors. It should be noted that automation is not proposed for the third category because of the inflexibility of automation, which does not solve the discriminative ambiguity of real events and it tends to result in exclusion of the operator from the situation assessment process leaving them less able to recover if the automation fails.

10. CONCLUSIONS

The estimation of performance failure on demand using Human Reliability Analysis is an imperfect but very necessary scientific and engineering activity. Leaving aside the numerical analysis the qualitative and analytical approach to understanding of the task often reveals important insights into psychological and ergonomic challenges to the operator. The techniques for HRA are not specifically designed for safety critical tasks with limited available time. There are, however, analytic methods that afford time estimation methods that can be used, and supplemented with empirical data, to develop robust estimates of the time required. It is argued that together these techniques can give sufficient insight into time and safety critical tasks to challenge designs and justify their improvement. Or, to demonstrate that they are within ALARP criteria. A structured analysis can help establish the support for the best designs in terms of human factors engineering because it ensures a claim-argument-evidence structure to debates and refutes a claim-argument-opinion based unqualified analysis. One might argue tempus narrabo and that is preferable to tempus edax rerum, if the operator fails to complete the safety critical task.

11. ACKNOWLEDGEMENTS

The author is grateful to BAE SYSTEMS for supporting the development of this paper and to various staff for helpful advice and comments in developing this paper. The author confirms that this paper have been cleared for public release.

12. REFERENCES

1. Adams, M. J., Tenney, Y. J., & Pew, R. W. (1995). Situation awareness and the cognitive management of complex systems. *Human factors*, *37*(1), 85-104.

2. Amalberti, R. (2013). *Navigating safety: Necessary compromises and trade-offs theory and practice* (pp. 99-105). Heidelberg: Springer.

3. Bishop, P., & Bloomfield, R. (2000). A methodology for safety case development. In *Safety and Reliability* (Vol. 20, No. 1, pp. 34-42).Taylor & Francis.

4. Boring, R. L. (2007). Dynamic human reliability analysis: Benefits and challenges of simulating human performance. *Risk, Reliability and Societal Safety, 2*, 1043-1049.

5. Boring, R. L. (2012). *Fifty years of THERP and human reliability analysis* (No. INL/CON-12-25623). Idaho National Laboratory (INL).

6. Boring, R. L., & Gertman, D. I. (2005, July). Advancing usability evaluation through human reliability analysis. In *Proceedings of HCI International 2005.*

7. Boring, R. L., & Rasmussen, M. (2016). GOMS-HRA: A method for treating subtasks in dynamic human reliability analysis. *Proceedings of ESREL2016. Glasgow, Scotland, 25–29 September 2016.*

8. Chipman, S.F. and Kieras, D.E., 2004. Operator centered design of ship systems. Office Of Naval Research Arlington, VA.

9. Despotou, G., & Kelly, T. (2008). Investigating the use of argument modularity to optimise through-life system safety assurance. In *System Safety, 2008 3rd IET International Conference on* (pp. 1-6). IET.

10. Dekker, S. W., & Woods, D. D. (2002). MABA-MABA or abracadabra? Progress on human–automation co-ordination. *Cognition, Technology & Work, 4*(4), 240-244.

11. de Winter, J. C., & Dodou, D. (2014). Why the Fitts list has persisted throughout the history of function allocation. *Cognition, Technology & Work, 16*(1), 1-11.

12. EEMUA (2013). EEMUA 191 – Alarm Systems: A Guide to Design Management and Procurement. EEMUA : London.

13. Endsley, M. R. (1995). A taxonomy of situation awareness errors. *Human factors in aviation operations, 3*(2), 287-292.

14. Fitts, P.M. (Ed.)(1951). *Human engineering for an effective air-navigation and traffic-control system.* Columbus, OH: The Ohio State University Research Foundation.

15. Frensch, P. A., & Funke, J. (2014). for Understanding Complex Problem Solving. *Complex problem solving: The European perspective.*

16. Hollnagel, E. (1998). CREAM-Cognitive reliability and error analysis method. *2006) http://www. ida. liu. se/~ eriho/CREAM_M.htm.*

17. Hutton, R. J., & Klein, G. (1999). Expert decision making. *Systems Engineering, 2*(1), 32-45.

18. John, B. E., & Kieras, D. E. (1996). The GOMS family of user interface analysis techniques: Comparison and contrast. *ACM Transactions on Computer-Human Interaction (TOCHI), 3*(4), 320-351.

19. John, B. E. (1990) Extensions of GOMS analyses to expert performance requiring perception of dynamic visual and auditory information. In proceedings of CHI, 1990 (Seattle, Washington, April 30-May 4, 1990) ACM, New York, 107-115.

20. Jorritsma, W., Haga, P. J., Cnossen, F., Dierckx, R. A., Oudkerk, M., & van Ooijen, P. M. (2015). Predicting Human Performance Differences on Multiple Interface Alternatives: KLM, GOMS and CogTool are Unreliable. *Procedia Manufacturing, 3*, 3725-3731.

21. Kahneman, D. (2012) Thinking fast and slow. Penguin: London.

22. Kelly, T. P. (1999). *Arguing safety: a systematic approach to managing safety cases.* PhD Thesis, University of York.

23. Kelly T.P. & Weaver R.A. (2004). The goal structuring notation – a safety argument notation. Proceedings of Dependable Systems and Networks Workshop on Assurance Cases.

24. Kieras, D. E. (1999). A guide to GOMS model usability evaluation using GOMSL and GLEAN3. *University of Michigan*, (313).

25. Kieras, D. E., Wood, S. D., Abotel, K., & Hornof, A. (1995, December). GLEAN: A computer-based tool for rapid GOMS model usability evaluation of user interface designs. In *Proceedings of the 8th annual ACM symposium on User interface and software technology* (pp. 91-100). ACM.

26. Kirwan, B. (1996). The validation of three human reliability quantification techniques—THERP, HEART and JHEDI: Part 1—technique descriptions and validation issues. *Applied ergonomics*, *27*(6), 359-373.

27. Kirwan, B. (1997). The validation of three human reliability quantification techniques—THERP, HEART and JHEDI: part iii—Practical aspects of the usage of the techniques. *Applied Ergonomics*, *28*(1), 27-39.

28. Kirwan, B., Kennedy, R., Taylor-Adams, S., & Lambert, B. (1997a). The validation of three Human Reliability Quantification techniques—THERP, HEART and JHEDI: Part II—Results of validation exercise. *Applied ergonomics*, *28*(1), 17-25.

29. Leiden, K., Laughery, K. R., Keller, J., French, J., Warwick, W., & Wood, S. D. (2001). A review of human performance models for the prediction of human error. *Ann Arbor*, *1001*, 48105.

30. Niessen, C., Eyferth, K., & Bierwagen, T. (1999). Modelling cognitive processes of experienced air traffic controllers. *Ergonomics*, *42*(11), 1507-1520.

31. Pew, R. W. (1969). The speed-accuracy operating characteristic. In Acta Psychologica: Attention and Performance, (E.G. Koster Ed.) 16-26.

32. Price, H. E. (1985). The allocation of functions in systems. *Human factors*, *27*(1), 33-45.

33. Rasmussen, J. (1983). Skills, rules, and knowledge; signals, signs, and symbols, and other distinctions in human performance models. *IEEE transactions on systems, man, and cybernetics*, (3), 257-266.

34. Rasmussen, M., & Laumann, K. (2016). Decomposition level of quantification in human reliability analysis. *In Proceedings of ESREL2016. Glasgow, Scotland*, 25-29.

35. Ratcliff, R., & Rouder, J. N. (1998). Modeling response times for two-choice decisions. *Psychological Science*, *9*(5), 347-356.

36. Swain, A.D. and Guttmann, H.E. (1983). Handbook of Human Reliability Analysis with Emphasis on Nuclear Power Plant Applications, Final Report, NUREG CR 1278.

37. van der Schaaf, T., & Kanse, L. (2000). Errors and error recovery. *Human error and system design and management*, 27-38.

38. Vicente, K. J., & Rasmussen, J. (1992). Ecological interface design: Theoretical foundations. *IEEE Transactions on systems, man, and cybernetics*, *22*(4), 589-606.

39. Wickelgren, W. A. (1977). Speed-accuracy tradeoff and information processing dynamics. *Acta psychologica*, *41*(1), 67-85.

40. Williams, J. (1986). 'HEART - A proposed method for Assessing and Reducing Human Error'. Proceedings of the 9th Advances in Reliability Technology Symposium, University of Bradford, 4 April 1986.

41. Woods, D. D., Johannesen, L. J., Cook, R. I., & Sarter, N. B. (1994). *Behind human error: Cognitive systems, computers and hindsight* (No.CSERIAC-SOAR-94-01. Dayton University Research Institute, Ohio, USA.

ADAPTING SMART + CONNECTED SOLUTIONS FOR SHIPBOARD USE

J Beel and **K Jones**, Cisco Systems, Inc., USA

SUMMARY

Digitization brings advances in many different industries, across government services and throughout our daily lives through smart or Smart + Connected (S+C) solutions. S+C solutions leverage advances in cloud computing, fog computing, analytics, collaboration, etc. to connect the previously unconnected and drive enhanced productivity. S+C solutions can be adopted to make shipboard operations more efficient.

Network-connected lighting is an example of S+C solutions that directly applies to ship and unmanned vessel design. Shipboard adoption of network-connected lighting will reduce cost, enhance safety, improve security and enhance shipboard working conditions. Industrial organizations leverage data virtualization, edge analytics, asset management and remote expert capabilities to enhance system performance and improve system resiliency. These solutions are directly applicable to shipboard use and can be adopted for shipboard use based on a sound network architecture approach and a holistic security strategy.

1. INTRODUCTION

Advances in Information Technology (IT) are fueling significant digital transformation of many businesses. IT is now enabling a new era of corporate identity, purpose, competition and effectiveness. This same increase in competition and effectiveness can also benefit military organizations. Digital transformation of shipboard operations can greatly enhance mission performance and improve "competitiveness" in the maritime environment. Bringing Smart + Connected capabilities to the shipboard environment can support digital transformation of warships and enhance their crews' ability to compete – or fight and win in military terms.

"Information technology is revolutionizing products. Once composed solely of mechanical and electrical parts, products have become complex systems that combine hardware, sensors, data storage, microprocessors, software, and connectivity in myriad ways. These "smart, connected products"—made possible by vast improvements in processing power and device miniaturization and by the network benefits of ubiquitous wireless connectivity—have unleashed a new era of competition.

Smart, connected products offer exponentially expanding opportunities for new functionality, far greater reliability, much higher product utilization, and capabilities that cut across and transcend traditional product boundaries. The changing nature of products is also disrupting value chains, forcing companies to rethink and retool nearly everything they do internally."[1]

This paper will focus on reducing the workload required to operate a warship and making the crew's work environment safer. Key areas to be addressed are: using data effectively, remote assistance, asset management and lighting.

2. DATA AND ANALYTICS

Digitization and the Internet of Things (IoT) will create massive amounts of data.[2] It is noted that most IoT advances to date have been seen in manufacturing, transportation, smart city and consumer applications but, it is predicted that all industries will be making use of the IoT within by 2020. Creative use of data will fuel innovation and help drive change.

The massive amount of data to be unleashed by digital transformation will cause warship operators to be faced with data from many diverse sources (some not even onboard the warship), in many formats and residing in many places. Data virtualization helps solve these data challenges by connecting and integrating data from disparate sources without the overhead cost of physically consolidating many data sources. In fact, Cisco estimates that Data Virtualization can reduce cost by 50-75% as compared to data replication and consolidation in traditional non-virtualization approaches.[3]

Data virtualization abstracts data from multiple sources and brings it together in a unified, user friendly view of the data that can be consumed by any analytical software as if it all the data were located in a single place.

A warship operates in unique, challenging environments. Operating disconnected from the shore at times is a fact of life and when the warship is connected, bandwidth may be at a premium. The ability to analyze data in this disconnected, intermittent and high latency environment is critical. Fog or Edge Analytics is a key capability that can benefit warship operations by enabling data analytics where data is highly distributed. Fog or Edge computing can run on existing network or IoT equipment, such as the engineering plant.

Analyzing data where it is created allows faster response and minimizes the need to move data around the ship, or even between the ship and shore sites, unnecessarily. Edge analytics is very useful for real-time monitoring and diagnostics which can support proactive and predictive condition-based maintenance.

3. REMOTE ASSISTANCE

Warships operate far from their home ports and traditional support infrastructure. Manning is a major cost driver in the life cycle cost of naval vessels and navies try to minimize manning to keep costs under control. Given reduced manning, ships' crews are not likely to have the technical expertise onboard that they'd like given the complexity of today's maritime systems. Remote assistance can allow connection to a subject matter expert who located in a different art of the ship or even far from the warship. Remote assistance or access to a remote expert can be a major productivity and effectiveness enhancer for warship crews who need assistance from others on the vessel, on another vessel or ashore.

A remote expert solution leverages next generation, multimedia collaboration to deliver face-to-face interaction, document sharing, white boarding, etc. A remote expert solution maximizes the value and availability of very scarce technical experts, essentially making co-location and even geography irrelevant across your enterprise when needed.

Subject matter experts can be easily accessed via a skills-based data base allowing rapid location of the right expert skill set when needed. Experts can be accessed via chat, voice, video, white board, document exchange, etc. A remote expert system can integrate with existing maintenance and logistics applications and data bases to ensure experts have immediate access to relevant maintenance history and system data when providing assistance.

4. ASSET MANAGEMENT

Bringing the right assets to bear at the precise location where they are needed at the moment that they are needed is critical to efficient operations. Optimal orchestration of asset management requires real-time visibility of asset location and state or status. The network can serve as the foundation for a real-time location system with the wireless extension of the network serving as a visibility platform for assets.[4]

Wireless access points can deliver very accurate location-based services that, when coupled with RFID tags and exciters, can provide an umbrella of real-time visibility. Wireless connectivity can enable an asset management system to serve as a real-time reporting and alerting system. RFID tags are very small and can be used for real-time tracking of critical supplies, tools and personnel.

5. NETWORK CONNECTED LIGHTING

LED lighting is clearly more energy efficient than conventional lighting in standard configurations however, many LED luminaires are also now capable of being powered by low voltage Power over Ethernet (PoE). PoE is the ability for the Local Area Network (LAN) switching infrastructure to provide power over a copper Ethernet cable to an endpoint or powered device. Space, Weight and Power (SWaP) are important in ship design. PoE reduces the weight of required cabling because only a network cable is required to be routed to each connected LED light and there is no need to route heavier high voltage cables to each light. Ethernet cables also cost less to procure than high voltage cables. Finally, installation costs are also reduced through use of PoE because installation of network data cables does not require a licensed electrician as would installation of standard high voltage cables. Low voltage PoE cables also do not pose any electrical hazard to personnel and contribute to increased safety, which is very critical in the maritime environment. Enhancements have been made to PoE over the past few years. PoE+ (IEEE 802.3at) now delivers 30 watts and UPOE (Universal PoE) now delivers 60 watts per Power Sourcing Equipment (PSE) port which enables more choices in PoE powered lighting.[5]

A converged IT/OT network platform to deliver a converged facility related control system services solution can leverage enterprise IT switches to distribute power to PoE powered devices. Switches have been improved to provide specific capabilities to optimize use of PoE devices which will enable adding IoT devices without requiring additional high voltage electric power sources. Switches now include dual power supplies to ensure high availability, key to lighting services. Switches also provide converged wired and wireless access eliminating the need for a separate wireless controller to manage a mixed wireless and wired environment which may be very useful in IoT applications.

Small form factor PoE switches enable extension of enterprise-class services well beyond the wiring closet. These compact switches even include fan-less designs that minimize noise making them far less obtrusive in occupied spaces and non-intrusive for noise sensitive spaces aboard warships. These new switches are ideally suited for space-constrained environments and for applications where multiple cable runs would present challenges. Some switches enable PoE pass-through which gives the ability for a switch to be powered via uplink from another PoE device with no direct electrical power line required at the downstream switch. Again, this capability reduces required high voltage wiring.

Network connected and controlled lighting gives the warship crew the capability to incorporate lighting into operations. Lighting can be adjusted to visually indicate current operational states, highlight dangerous situations assist crewmembers in finding safe spaces during emergency conditions.

6. CONCLUSION

The Internet of Things (IoT) is driving connections never envisioned before. These connections, coupled with small sensors, are creating massive amounts of data that present an unprecedented opportunity for analysis that can lead to enhanced effectiveness. Data virtualization delivers abstraction of data from its disparate sources and locations. Analytical tools can consume data from many different sources as if the data lived in a single data warehouse. Analysis of previous unconnected data sources will enable new perspectives not possible given traditional data stove pipes.

Computing capability at the edge enables analytics at the edge – where the data lives. Edge or fog analytics will speed the benefit of analytics and reduce the amount of data that needs to be moved around the network. Faster computing will also enable greater situational awareness and better response using automation, artificial intelligence and machine learning.

Even with enhanced data management and analytics, warship crews will, at times, need expert assistance across a vessel or even beyond the onboard expertise. Remote expert capabilities will allow technicians and operators to collaborate in real time. Remote expert assistance will reduce unnecessary delays using the network to eliminate time delays caused by physical separation. Remote expert assistance will also allow critical shore-based experts to support numerous warships around the globe from a central location that makes the most business sense.

Remote expert assistance will bring the right knowledge to technicians and operators as they conduct their daily tasks but, they will also need critical assets at the right time and place to performed their duties. The network's wireless location-based services coupled with RFID tags and exciters will allow the warship's crew real-time visibility of key parts, tools and personnel. Troubleshooting, maintenance and supply functions will become far more efficient when key assets are immediately locatable.

Network-connected lighting can significantly reduce warship construction cost, reduce energy consumption, increase safety by reducing high voltage wiring and enable the crew to integrate control of lighting into key operations.

These are just a few thoughts on what the IoT will bring to warships. When the currently unconnected are connected and the network is enabled with much broader connectivity to serve as a far more capable platform to deliver applications when and where needed, innovative Sailors will design far more ingenious ways to use IT and OT to power their missions, learning, maintenance, operations, etc.

7. REFERENCES

1. Michael E. Porter and James E. Heppelmann, "How Smart, Connected Products are Transforming Competition," Harard Business Review, Nov 2014, retrieved from https://hbr.org/2014/11/how-smart-connected-products-are-transforming-competition

2. Charles McLellan, "The internet of things and big data: Unlocking the power," Mar 2, 2015. ZDNet, Retrieved from: http://www.zdnet.com/article/the-internet-of-things-and-big-data-unlocking-the-power/

3. "Data Virtualization: Achieve Better Business Outcomes, Faster," 2014, Cisco. Retrieved from: http://www.cisco.com/c/dam/en/us/products/collateral/cloud-systems-management/data-virtualization/data-virtualization-achieve-better-business-outcomes.pdf

4. "Connected Assets Manager for IoT Intelligence Location Based Services," 2017, Cisco.

5. Kelly Jones and Joe Beel, "Network-Connected Lighting in the Maritime Environment," Royal Institution of Naval Architects International Conference on Smart Ship Technology, Jan 24-25, 2017.

8. AUTHORS' BIOGRAPHIES

Joe Beel is a strategic programs manager at Cisco Systems, Inc. He develops and implements strategies to support the U.S. Public Sector focusing on defense. He is a retired naval officer and served as a helicopter pilot and acquisition professional in both the Naval Air Systems Command and Space and Naval Warfare Systems Command.

Kelly Jones is a systems engineer for the U.S. Navy Shipboard Account team at Cisco Systems, Inc. She works with the U.S. Navy and ship integrators to identify requirements, provide technical guidance on the future direction of networking technologies and assists them in implementation, testing and validation of proposed solutions. She served six years in the U.S. Navy as a Cryptologic Maintenance Technician and is a Certified Information Systems Security Professional (CISSP).

THE EFFECT OF SPEED AND GEOMETRY ON THE CHARACTERISTICS OF THE PLUME GENERATED BY SUBMARINE MASTS

A S T Conway, **D Ranmuthugala, M R Renilson** and **J R Binns**, Australian Maritime College, Australia

SUMMARY

Due to operational requirements, submarines are required to pierce the free surface with a mast whilst underway, generating a significant plume structure which greatly increases the submarine's probability of detection. Therefore, accurately modelling and understanding the flow regime which generates the plume is critical in order to improve the mast shape and configuration to reduce the plume size and its visual signature.

This paper reports on a numerical and experimental study to investigate the plume generated by a surface piercing mast. A validated numerical model is used to investigate a number of mast shapes and double mast configurations as possible options to reduce the plume size. It is seen that the use of a smaller forward mast can significantly reduce the plume size generated by the masts.

NOMENCLATURE

A	frontal submerged area of cylinder (m^2)
C_D	coefficient of drag
D	diameter (m)
F	drag force (N)
Fr	Froude number
g	gravity (m s^{-2})
T'	ratio of cylinder submerged length to diameter
U	velocity (m s^{-1})
u_τ	friction velocity (m s^{-1})
y	wall distance (m)
y^+	non-dimensional wall distance
η_b	bow wave height (m)
η_p	plume height (m)
λ	plume length (m)
ρ	density of water (kg m^{-3})
υ	kinematic viscosity (N s m^{-2})

1. INTRODUCTION

It is well-known that for operational reasons submarines are required to pierce the free surface with a vertically mounted mast or snorkel. This appendage can generate a large plume structure, which can significantly compromise a submarine's covert operations. Being able to accurately model this is critical to understanding the size of the plume structure as a function of speed, and therefore to be able to predict the contribution of the appendage to the submarine's signature. Additionally, an accurate model can be used to test variations in mast configurations in order to reduce the plume size.

This paper reports on an experimental and numerical study to investigate methods to accurately model the plume size generated from a surface piercing mast. With a validated numerical model, further investigations are conducted on streamlined and double mast configurations as possible methods of reducing the plume size and hence the resulting visual signature.

The experimental method used in this study to replicate the plume structure from a surface piercing mast was established using previously published work on surface piercing cylinders as a guideline [1-3]. The study focused on velocities below a Froude number (Fr) of 4.0 based on the diameter of the mast as defined in Equation (1),

$$Fr = \frac{U}{\sqrt{gD}} \qquad (1)$$

where U is the forward velocity and g is the acceleration due to gravity.

The effect of the submerged free end, caused by testing a cylinder secured above the water surface in a towing tank, was investigated and minimised by ensuring adequate immersion of the free end and the use of an end body, as discussed previously by the authors [4]. The experimental work in this project expands on the current published data in both the speed range, especially at the higher Froude numbers, and in quantifying the end effects.

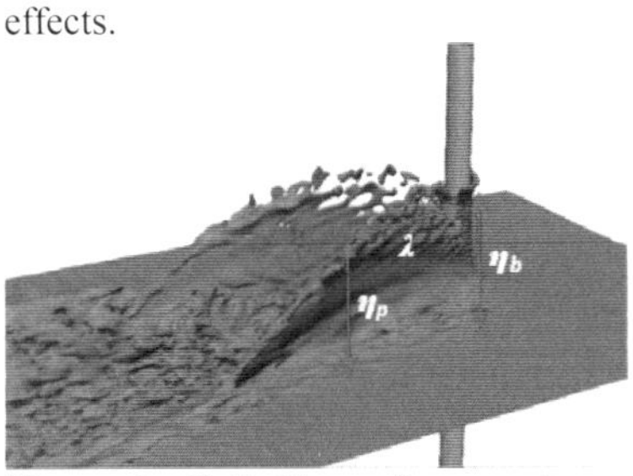

Figure 1: Definition of parameters used to measure plume structure. Bow wave height, η_b, plume height, η_p, and plume length, λ.

In order to record the experimental results, four measurements were used. The bow wave height, η_b, plume height, η_p, and plume length, λ, were utilised for measurements of the plume structure as proposed by Hay [1] and shown in Figure 1. In addition to plume measurements, the drag coefficient, C_D, was used as an additional point of validation for the numerical model
The numerical modelling was carried out using Computational Fluid Dynamics (CFD) based Large Eddy Simulations (LES). The CFD-LES model validated

against the experimental data and published results, enabled the authors to investigate the plume structure at a range of Froude numbers and mast configurations, identifying options to reduce the resulting visual signature. The combined CFD and experimental results also provide information on the development of experimental test rigs and testing programs for surface piercing masts.

2. EXPERIMENTAL METHOD

The experiments conducted for this research were completed in the towing tank located at the Australian Maritime College (AMC). The 100 m x 3.55 m x 1.5 m towing tank has a carriage capable of speeds up to 4.6 m s^{-1}, allowing tests to be conducted at typical submarine snorkelling speeds, removing any errors due to scaling effects. An overview of the experimental set-up is shown in Figure 2. Six issues were considered when designing the experimental rig. These were: plume size clearance, total drag, vortex shedding, load cells, plume size recording, and methods to quantify and minimise the effect from the submerged free end of the cylinder.

Using data previously published by the authors [5, 6] and others [1, 7] for related experimental and numerical work, the plume size was estimated to ensure there was no interference on the plume from the towing tank carriage and the test rig. In addition to published plume sizes, the theoretical maximum bow wave height was calculated using a simplified version of Bernoulli's equation given by Mentor [7] as shown in Equation (2). This also provided information to ensure no interference of the plume from adjacent structural components.

$$\frac{\eta_b}{D} = \frac{Fr^2}{2} \qquad (2)$$

where η_b is the bow wave height, D is the cylinder diameter, and Fr is the diameter Froude number as calculated using Equation (1).

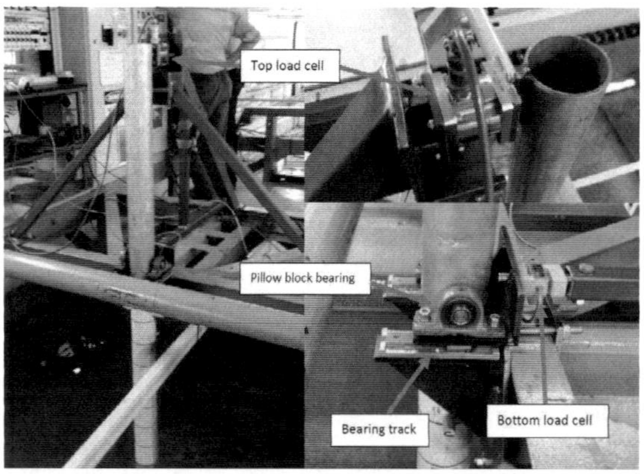

Figure 2: Overall experimental set-up (left) and detailed views of the top load cell arrangement (top right) and the bottom load cell, bearing track and pillow block bearing configuration (bottom right)

The measurement of the plume was achieved through a combination of still photography, videography, draft marks and tape measures. The overall experimental set-up detailing the location of the cameras and distance markers are shown in Figure 3. The camera used was a Digital Single-Lens Reflex (DSLR) positioned on the starboard side with the intention of capturing the bow wave height, plume height, and plume length. The video camera was a GoPro positioned on the starboard side above the DSLR camera. The GoPro camera was configured to capture 50 frames per second and was used to capture the bow wave oscillations that occur as a surface piecing cylinder is dragged parallel to the free surface. Draft markings on the cylinder were used to record the bow wave height and plume height. A horizontally mounted 3 m tape measure attached to a rigid beam was used to record the plume length.

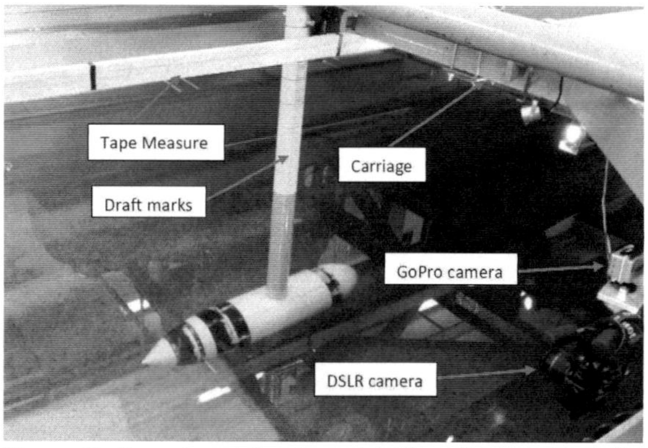

Figure 3: Experimental set-up detailing the location of the cameras, draft marks and tape measure (note the end body used to minimise the effects of the submerged free end of the cylinder).

Slaouti and Gerrard [8] concluded that the end effect on a semi submerged cylinder with the submerged end close to the free surface can decrease the size of the wake generated. Thus, in the current experimental set up the submerged length of the cylinder could be varied by vertical movement of the cylinder, enabling the free end effect on the plume size to be investigated and minimised. In addition, the end body shown in Figure 3 was trialled at different depths to further reduce the end effect on the plume structure. This resulted in investigating two cylinder lengths and two end conditions, giving a total of four cylinder configurations. All results presented in this paper are for cases where the end effects are negligible, with readers wishing to obtain further information on the submerged end effect and minimising strategies referred to the previous publication by the authors [4].

3. NUMERICAL METHOD

3.1 NUMERICAL SET-UP

The CFD-LES numerical model used for this work was developed using the open source CFD software

OpenFOAM. The authors have previously demonstrated the need for LES simulations in preference to Reynolds Averaged Navier Stokes (RANS) models to fully capture the flow structure generated by a surface piercing cylinder [9], as the inherent turbulence averaging of the latter is unable to model the plume structure. The LES model used in this study utilised the One-Equation Eddy model, which uses the Navier Stokes equations with the Sub-Grid Scale (SGS) model to allow the large eddies to be resolved whilst modelling the smaller eddies [10].

Matching the experimental conditions was critical to numerical validation, before using the model to investigate the various plume structures. Thus, the experimental domain was fully replicated, including the tow tank test section consisting of the full width and depth of the tank and an effective length to minimise boundary effects, and the geometric features below the water surface. The flow around the mast was considered to be incompressible, turbulent, and two phase.

The computational domain was discretised using the inbuilt meshing software within OpenFOAM, SnappyHexMesh, which supports hanging nodes. The latter permitted grid refinement in critical areas whilst minimising the overall size of the grid. These refined areas included: an area surrounding the plume approximately 20% larger than the largest expected plume size, the mesh adjacent to the free surface, the areas around the cylinder and end body, and an inflation layer around the cylinder and end body to capture the boundary layer. Additionally, the mesh density was greatly decreased close to the boundaries of the domain, providing two benefits. Firstly, it decreased the total mesh size, thus decreasing the computational requirements. Secondly, a course grid at the boundaries numerically dampened out the turbulent flow as it approached the domain boundaries, reducing reflection and increasing the stability of the simulations.

3.2 NUMERICAL VERIFICATION

Numerical verification was used in order to optimise the mesh size and to determine whether the numerical model selected were applicable for this study. This was completed in two stages. Firstly, a non-dimensional wall distance, y^+, investigation was performed which involved analysing the effect of the inflation layers within the boundary layer of the cylinder. The y^+ value was calculated using equation (4), where y is the distance to the wall, u_τ is the friction velocity, and ν is the kinematic viscosity.

$$y^+ = \frac{y u_\tau}{\nu} \qquad (4)$$

In order to assess the effects of different y^+ values, the bow wave height was recorded over a range of y^+ values from 150 to 20. It was found that the bow wave height began to plateau at y^+=45, and thus, all simulations were conducted with an inflation layer such that the y^+ value was approximately 35.

Secondly, a mesh sensitivity analysis was completed. This was achieved by gradually and systematically increasing the mesh density, concentrating on the plume and free surface region, until the changes to the recorded plume length and plume height were relatively insignificant. The results indicated that the percentage change in both plume dimensions became insignificant at a mesh density of 8-10 million elements, thus obtaining mesh independence [6]. Therefore, a mesh size of 10 million elements was used for the numerical model.

4. VALIDATION OF NUMERICAL METHOD

The experimental measurements obtained were compared to the predictions from the numerical model which replicated the experimental set-up. The comparison utilised the non-dimensional bow wave height, η_b', plume height, η_p', plume length, λ', and drag coefficient, C_D, defined in Equations (5), (6), (7), and (8) respectively.

$$\eta_b' = \frac{\eta_b}{DFr^2} \qquad (5)$$

$$\eta_p' = \frac{\eta_p}{DFr^2} \qquad (6)$$

$$\lambda = \frac{\lambda}{DFr^2} \qquad (7)$$

$$C_D = \frac{F}{2\rho U^2 A} \qquad (8)$$

where F is the drag force, A is total frontal submerged area of cylinder, and ρ is the density of water. In order to validate the numerical model across all conditions, the predicted results for each speed and condition were compared with the equivalent experimental measurements.

Figure 4 shows a comparison of the bow wave height between the experimental measurements and numerical predictions for an immersion ratio of T'=10 (i.e. the ratio between the cylinder submerged depth to its diameter). There appears to be a considerable spread in the data at low Froude numbers ($Fr < 1.0$). This is a result of small absolute measurement errors appearing exaggerated due to the non-dimensional representation using equation (5), with readers wishing to obtain further information on the Froude number dependent measurement errors referred to the previous publication by the authors [4]. At Froude numbers greater than 1.5, the numerical results agree well with the experimental results, with a small under prediction at a Froude number of 4.0. This region of good correlation (i.e. three to eight knots) is the region of interest for submarines operating at periscope depth. This shows that the numerical model developed in this study can be utilised in order to assess the plume structure generated by a submarine mast and potential improvements in their design.

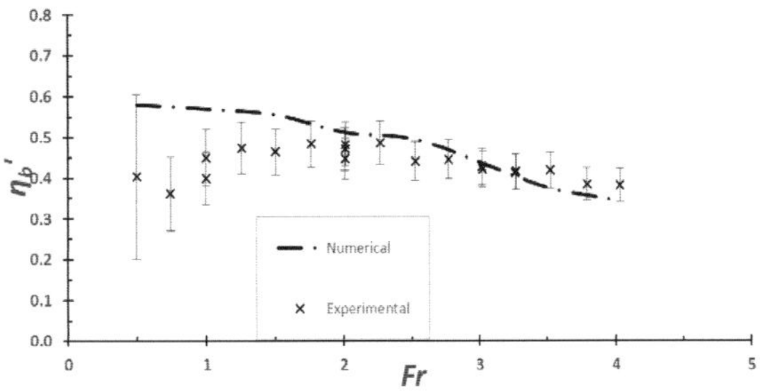

Figure 4: Comparison of non-dimensional bow wave height, η_b', as functions of Froude number, Fr, using experimental and numerical results for an immersion ratio of T'=10

Similar correlations between the numerical predictions and experimental measurements were found for the plume length, plume height, and drag coefficient, which are presented by the authors in [4]. Thus, it can be concluded that the numerical model gives sufficiently accurate predictions for the plume structure of a surface piercing cylinder over the range of Froude numbers of interest.

5. RESULTS AND DISCUSSION

5.1 PLUME SIZE AS A FUNCTION OF SPEED

Using the validated numerical model, the three non-dimensional plume dimensions predicted are presented in Figure 5 as functions of the Froude number. The plume length and plume height were only recorded for Froude numbers 3.0 or greater due to the small plume structure generated at lower speeds.

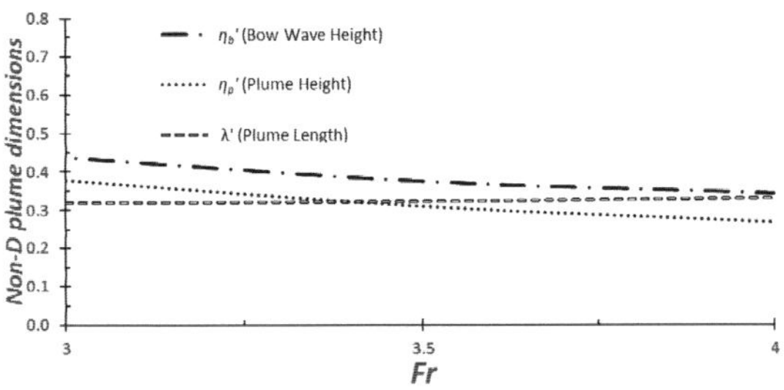

Figure 5: Predicted non-dimensional bow wave height, η_b', plume height, η_p', and plume length, λ', as a function of Froude number, Fr

It is noted for Figure 5 that the non-dimensional values remain relatively constant as functions of Froude number. This indicates that the plume size is roughly proportional to the speed squared.

5.2 EFFECT OF MAST GEOMETRY ON PLUME SIZE

Using the validated numerical model, streamlined mast sections and double mast configurations were tested in order to assess the effect of these geometry configurations on the plume size. As a basis for the comparison, the waterplane area of the single mast was retained for all configurations, as it was assumed that this would be required for operational purposes (for example to provide the same cross-sectional area for a submarine snorkel). Therefore, all the results presented are for geometries with the same waterplane area as the single cylindrical mast. It is noted that this requirement may not be valid for other situations such as for periscope masts. However, the results could easily be rescaled to maintain a constant value for any other parameter, such as the maximum width. The full scale dimensions are given throughout, as applicable to a typical submarine.

A summary of the parameters of the various masts are given in Table 1 and schematics of the mast arrangements are given in Figure 6. The double mast configurations were positioned such that the forward mast was located 0.5 m directly in front of the aft mast, measured from centreline to centreline.

The resulting plume visualisations for the single and double mast configurations are shown in Figure 7. For the tested mast configurations, it is seen that the aft mast plume is completely 'hidden' by that of the forward mast, even when the forward mast is smaller in diameter [6]. This is due to the significant wake region of the forward cylinder greatly reducing the plume produced from the aft mast. By conducting a series of simulations it is possible to determine the size and location of the forward mast in order to generate the smallest wake possible whilst still 'hiding' the plume generated by the aft mast.

Table 1: Mast geometry configurations used for numerical analysis with constant total waterplane area

	Width (m)		Waterplane Area (m^2)		
Mast Geometry	**Forward Mast**	**Aft Mast**	**Forward Mast**	**Aft Mast**	**Total**
Single Cylinder	0.100	n/a	0.00785	n/a	0.00785
Double Cylinder (equal)	0.0708	0.0708	0.00394	0.00394	0.00787
Double Cylinder (small forward)	0.060	0.080	0.00283	0.00503	0.00785
Double Cylinder (large forward)	0.080	0.060	0.00503	0.00283	0.00785
Single NACA0012	0.041	n/a	0.00785	n/a	0.00785
Double NACA0012 (equal)	0.029	0.029	0.00393	0.00393	0.00786

The results from the forward mast are shown in Figures 8 to 10 as functions of speed (in knots). Note that full scale results are given for convenience. Only the forward mast data was recorded due to the disturbed flow on the aft mast and since the plume structure from the aft mast was significantly reduced and covered by that generated by the forward mast. Since the visible plume structure is generated by the forward mast, the focus is to reduce the plume generated by that mast in order to reduce the visual signature generated by the mast configuration.

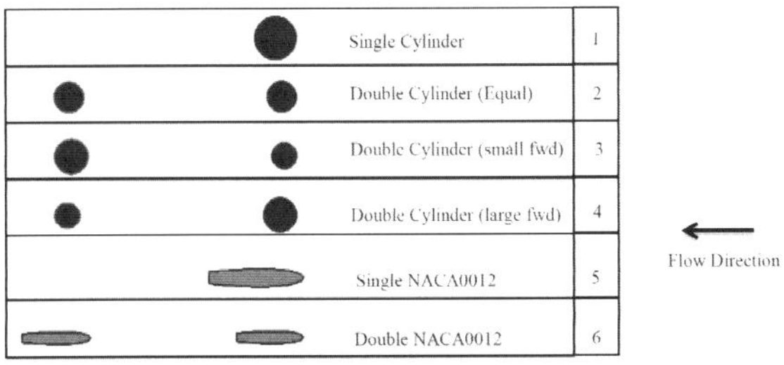

Figure 6: Visual representation of mast geometry configurations used for numerical analysis with constant total waterplane area

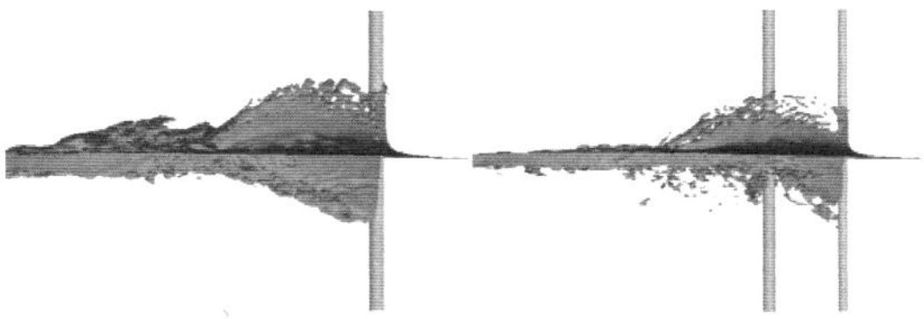

Figure 7: Comparison of the plume profile generated by a single mast (left) and double mast with small forward mast (right) at Fr=4 [6]

As seen in Figure 8, each tested configuration reduced the bow wave height by a significant amount in comparison to the single cylindrical mast. In particular, at a speed of eight knots a reduction in bow wave height of approximately 30% was recorded for the double cylinder (small forward) configuration in comparison to that for the single cylinder. Further reductions were seen in the bow wave height for the NACA0012 mast configurations, with reductions in the order of 75% and 80% for the single and double configurations respectively.

Figure 9 and Figure 10 show the plume height and plume length of configurations 1 to 4 between the speeds of six to eight knots. Again it is seen that the double mast configuration with the small forward mast gives the smallest plum size, a reduction of approximately 18% in both dimensions in comparison to the single mast.

The NACA0012 section masts generated a plume height and length too small to record accurately at all Froude numbers tested. However, they were significantly smaller than those generated by the cylindrical mast configurations.

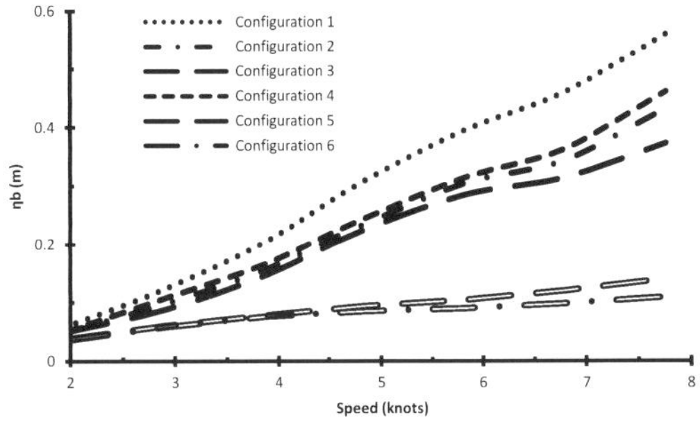

Figure 8: Bow wave height, η_b, generated by the forward mast for each tested configuration as a function of speed

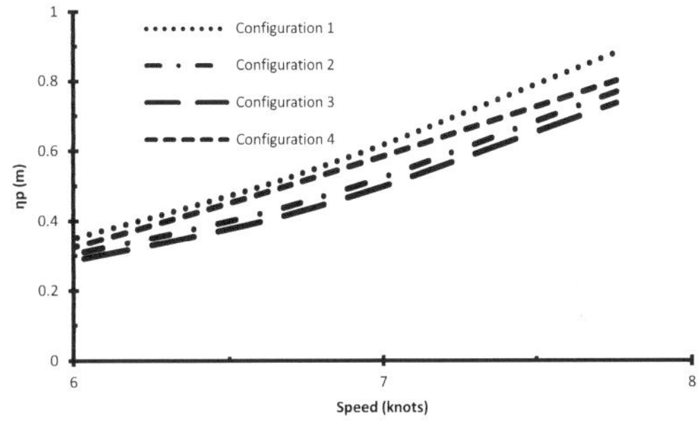

Figure 9: Plume height, η_p, generated by the forward mast for each tested configuration as a function of speed

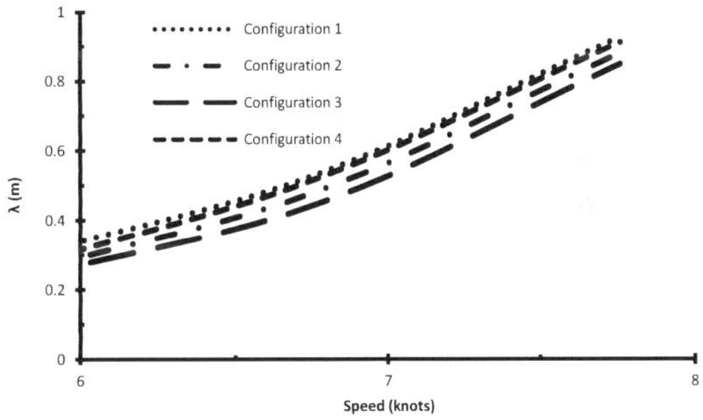

Figure 10: Plume length, λ', generated by the forward mast for each tested configuration as a function of speed

6. CONCLUDING REMARKS

This study investigates the plume structure generated by a surface piercing mast of a submarine operating close to the free surface, in order to reduce the plume size and hence the visual signature.

A CFD-LES model was developed to predict the size of the plume structure as a function of speed, which was validated using experimental measurements obtained through a test program conducted in a towing tank.

Using the validated numerical model, the size of the plumes generated by cylindrical and streamlined single and double mast configurations were investigated. It is shown that both double mast configurations and streamlined bodies can result in a significant reduction in plume size in comparison to the conventional single cylindrical mast. The use of a smaller mast, located at an appropriate distance forward of a larger mast, enables a smaller overall plume structure. This provides an option to reduce the plume signature with the use of appropriate mast shapes and configurations.

7. REFERENCES

1. Hay, A.D., *Flow about semi-submerged cylinders of finite length.* 1947, Princeton University: Princeton University. p. 178.

2. Metcalf, B., J. Longo, S. Ghosh, and F. Stern, *Unsteady free-surface wave-induced boundary-layer separation for a surface-piercing NACA 0024 foil: Towing tank experiments.* Journal of Fluids and Structures, 2006. **22**(1): p. 77-98.

3. Sumer, B.M. and J. Fredsoe, *Hydrodynamics around cylindrical structures.* Advanced Series on Ocean Engineering, ed. P.L.-F. Liu. Vol. 12. 1997, Singapore: World Scientific Publishing Co. Pte. Ltd.

4. Conway, A.S.T., D. Ranmuthugala, J.R. Binns, M.R. Renilson, and B. Anderson, *The end effect of vertical axis surface piercing cylinders with and without an end body.* (Submitted) Journal of Fluids and Thermal Science, 2017.

5. Conway, A.S.T., S. Cathcart, J.R. Binns, D. Ranmuthugala, M.R. Renilson, and B. Anderson, *Experimental analysis of surface piercing cylinders at high Froude numbers.* (Accepted) Journal of Fluids and Thermal Science, 2016.

6. Conway, A.S.T., D. Ranmuthugala, J.R. Binns, and M.R. Renilson, *The effect of geometry on the surface waves generated by vertical surface piercing cylinders with a horizontal velocity.* (Accepted) Journal of Engineering for the Maritime Environment, 2016.

7. Mentor, F.R., *Two-Equation Eddy-Viscosity Turbulence Models for Engineering Applications.* AIAA Journal, 1994. **32**(8): p. 1598-1605.

8. Slaouti, A. and J.H. Gerrard, *An experimental investigation of the end effects on the wake of a circular cylinder towed through water at low Reynolds numbers.* Journal of Fluid Mechanics, 1981. **112**: p. 297-314.

9. Conway, A.S.T., D. Ranmuthugala, and J.R. Binns, *Predicting surface wakes using LES and RANS-SST analysis*, in *2nd SIA Technology Conference.* 2013: Adelaide, Australia. p. 7.

10. Piomelli, U., A. Scotti, and E. Balaras, *Large-Eddy Simulations of Turbulent Flows, from Desktop to Supercomputer*, in *Selected Papers and Invited Talks from the 4th International Conference on Vector and Parallel Processing.* 2001, Springer-Verlag. p. 551-578.

8. AUTHORS BIOGRAPHY

Alexander Conway is a PhD candidate at the Australian Maritime College. His research is focused on the wake generated by submarine masts using both experimental and numerical methods.

Dev Ranmuthugala is a Professor in Maritime Engineering at the University of Tasmania. He is currently on secondment to the DST Group as the Head, Hydroacoustics. His previous experience includes Head, Maritime Engineering at the Australian Maritime College.

Martin Renilson is an Adjunct Professor in Hydrodynamics at the University of Tasmania. His previous experience includes as Technical Manager and Deputy General Manager for the Maritime Platforms and Equipment Group at QinetiQ.

Jonathan Binns is an Associate Professor in Hydrodynamics at the University of Tasmania. He is currently the Associate Dean of Research and the Director of the Research Training Centre for Naval Design and Manufacturing.

CYBERSECURITY AUTOMATION: SECURELY DELIVERING SMART TECHNOLOGIES AND OPERATIONS

K Jones and J Beel, Cisco Systems, Inc., USA

SUMMARY

Combat systems, weapon systems and warship control systems are rapidly evolving. This evolution is based significantly on advances in Commercial Off the Shelf (COTS) Information Technology (IT). A data center provides networked processing, storage and distribution of data – essential to warship systems. Data centers are key to delivering advanced operations, maintenance and mission performance across areas such as electronic warfare, sonar, radar, signals intelligence, etc. Networks will link systems across Operational Technology and Information Technology to extend connectivity throughout the warship and beyond to off board systems. Enhanced connectivity enables operations to move at much greater speed, requiring cybersecurity to move faster to ensure integrity, reliability and availability of critical systems. Automation will be critical to effective cybersecurity and mission accomplishment.

Warship operators will require software-defined security systems to secure advanced capabilities. Policy-based designs will replace security based on boxes positioned at a "boundary." Cybersecurity will also focus on applications and the real-time traffic that flows laterally within the data center and network to enable the ability to respond in real-time *before*, *during* and *after* cyber attacks. Warships must leverage micro-segmentation to ensure that any breach does not allow lateral movement to other systems. Software-defined segmentation will function similarly to physical ship compartmentalization which allows containment of fire or flooding and continuation of mission despite partial damage from fire or flooding.

1. INTRODUCTION

Virtualization is a primary driver of change in data centers. Software-definition of the data center is a logical extension of virtualization which continues abstraction of complexity. Reducing complexity allows operators to keep up with more and more demanding and complex operations that rapidly change the demands on the data center. Data center complexity is driving increased use of automation and orchestration. Virtual resources are now provisioned and decommissioned dynamically, optimizing the use of infrastructure and reducing operator workload. While data center management is benefitting from increased software definition and control, computing density and efficiency are also increasing to deliver very powerful small data centers. This increased efficiency and power is extremely beneficial to data intensive shipboard applications such as sonar, radar, electronic warfare, information warfare or automated HM&E systems. Data centers are becoming more flexible and better able to accept new technology insertion. Modular data center designs enable the ability to flex as demands change to meet mission requirements. Next generation data centers must be able to rapidly deploy and integrate new hardware and software.

The data center has changed dramatically, in large part due to virtualization. However, there is a lacking focus on virtual security.[1] Data center advancements have enabled improved mission system effectiveness, but the threat environment has become very complex. Threats are advancing while data center environments are becoming more complex and challenging to defend.[2] Virtualization in the data center presents new and different security challenges. Cybersecurity solutions have not kept pace with data center changes. Effective data center security must be able to address new threats that can take advantage of changes to the data center such as virtualization, cloud, etc.

It is likely that many next generation data centers will operate with a heterogeneous infrastructure. Consolidation of physical resources and virtualized resources that are assigned to applications dynamically, as needed, will replace existing application silos, resulting in significantly more agile and effective mission systems. The data center model will require open technology and Application Programming Interfaces (APIs) will enable operators to tie together various infrastructure elements.[3]

2. CHALLENGES TO SECURING THE NEXT GENERATION WARSHIP DATA CENTER

Whether on or off a ship, data centers have typically depended on laborious system controls and traditional boundary security at the network edge for protection. Communication inside the firewalled data center was assumed to be trusted, with little perceived need for protection. Perimeter security focuses on symmetric data, meaning it only secures the north-south data that enters and exits in a predictable way, typically at a network or enclave edge or boundary. While this can protect data traversing the protected boundary, today an estimated 76% of traffic actually flows laterally in an east-west fashion, never leaving the data center. Whether accidental or malicious, the majority of today's data breaches occur in this typically unprotected zone. Further, because the east-west traffic is considered trusted, a breach often leads to unwanted access across the entire data center or beyond. The current approach to securing the east-west traffic is often to route it northbound to the perimeter defense and then back into the data center. In non-network terms, this method is similar to working in a secure facility and having to walk to the main building entrance for permission every time you want to go to another room. Not only would that be inefficient, it would leave an attacker circulating undetected in trusted zones until discovered.

The data center has changed. The proliferation of mobile devices and expansion of the Internet of Things have increased the attack surface within the data center making the previously “trusted” internal network more susceptible to vulnerabilities. Attacks from within are increasing. Attackers have become smarter and have learned how to develop malware that bypasses common detection capabilities. Insider threat remains a concern even on classified networks and increased complexity of underlying network architectures has translated into an increased risk of security breaches due to simple human error. An effective next generation data center cybersecurity solution must adapt to the increased threat with an integrated threat defense.

Warship operators will require the agility of an automated, threat-centric cybersecurity model that includes comprehensive visibility and protection capable of responding in real-time *before*, *during* and *after* cyber attacks to securely deliver mission success.

- *Before* - comprehensive awareness and visibility focused on minimizing an attacker’s chances of success.
- *During* - requires a clear threat containment strategy.
- *After* - requires a visibility and forensics capability.

At the same time, effective warship cybersecurity solutions must support asymmetric traffic management, scale dynamically, secure intra-data-center traffic and allow for a mixed environment consisting of a combination of physical and virtual appliances.

How can all this be achieved while reducing complexity, increasing security posture and protecting existing infrastructure investments? The solution must follow the same trend of software-defined networking with abstraction used in the virtualization of evolving data centers themselves. The security architecture must also center on the application. Implementing an Application Centric Infrastructure (ACI) abstracts security from the physical architecture, which increases agility and flexibility. ACI scales on demand, has programmable automation that reduces complexity, and provides zero trust policy-based security for both physical and virtual environments.

3. CYBERSECURITY AUTOMATION WITH APPLICATION CENTRIC INFRASTRUCTURE (ACI)

Why make the infrastructure focus on the applications? The move toward virtualization allows deployment and re-allocation of both resources and applications to come and go much faster. Traditional hardware-based security solutions, however, remain static and can take days or even weeks to configure, certify and deploy. A physical appliance may not be able to expand or relocate resources to match changing operational requirements. Implementing box-by-box rules and controls is tedious and error prone. It is estimated that up to 95 percent of firewall-related security breaches are due to misconfigurations. When resources are no longer needed or access requirements change, manual updates are time consuming and often postponed, weakening cybersecurity defenses and increasing risk of breach

3.1. *BEFORE* AN ATTACK – CONTROL, HARDEN, ENFORCE

The *before* stage of the continuum emphasizes discovery of the devices and applications on the network, enforcement of policies among these entities and hardening the infrastructure against an attack. Some of the ways ACI addresses these requirements include a centralized policy controller, secure multi-tenancy, white listing of services and segmentation.

The traditional data center security policy model ties security to the physical topology and requires policy configuration across multiple security devices (switches, firewalls, IPSs) each time a service or application is added, deleted or moved. With ACI, a central controller manages both virtual and physical devices and has the ability to simultaneously provision network and security devices along with their associated security controls. This application centric approach decouples security policy from the underlying physical architecture, allowing for consistent security among physical, virtual and even cloud environments. It also allows for quick deployment of security devices to accommodate rapidly changing needs within the east-west traffic flow of the data center.

ACI uses endpoint groups, policy contracts and application network profiles to replace traditional network constructs like virtual local are networks (VLANs), Internet Protocol (IP) addresses and MAC addresses. Endpoints are physical or virtual entities that can still be defined by characteristics like IP address or MAC and also by Network Interface Card (NIC), virtual NIC, Virtual Machine (VM), or Domain Name System(DNS) name. Endpoint groups (EPGs) consolidate the individual endpoints into flexible, user-defined containers that could represent application, function, location in the network, organization, security zone or by any other logical parameter. EPGs can even be defined by design phases like development, test and production. Operationally, they could be defined by threat condition or mission. A warship's security posture could be changed across the entire IT/OT infrastructure by the click of a button.

Unlike traditional blacklist-based network architectures that permit all communications and linkages by default, ACI employs a zero-trust, whitelist-based model that prevents any endpoint or EPG from talking until explicitly permitted to do so. Communication between or among EPGs is defined by contracts that define rules and policies governing who (device, application or user) can talk to whom (device, application or user) and what they can talk about. Essentially, the whitelist function of ACI provides a stateless, Layer-4 firewall at line-rate. For additional security, service chaining can be used to incorporate traditional Layer 4-7 security such as firewalls and intrusion detection systems into the path. Because the service chaining is logical and not physical, ACI increases scalability. For example, if load increases, additional virtual firewalls can be activated and easily integrated into the path with no changes to physical architecture. ACI essentially offers a software version of an automated patch panel. Mission critical systems can be routed based on a "watch and alert" model while user mobile devices can be routed through a more restrictive security model. The path can be easily changed based on mission requirements; alternate mission-based cyber security profiles can even be pre-programmed.

By simplifying and abstracting policy, it becomes easier to achieve segmentation, which isolates normal traffic based on functional or application requirements. Micro-segmentation enhances security for east-west traffic by breaking the "trusted" internal zone into smaller segments to better contain threats or compartmentalize the infrastructure, both physical and virtual. Similar to the compartments aboard ship, micro segmentation can isolate damage to one space that can be quickly contained. This not only simplifies recovery, but also allows operations to continue in unaffected locations. While segmentation can be achieved with traditional constructs like Virtual Local Area Networks (VLANs), Access Lists (ACLs) and even physical devices, it could take hundreds of rules across multiple devices to segment a data center. If a user or device moves, segmentation rules would need to move as well creating a great complexity for the network operator.

The central controller of an application centric security solution creates a single point of management and visibility for the network. It controls the network and security infrastructure from one location, simplifying administration, reducing configuration time and minimizing errors. Templates allow reuse of endpoint groups and policies, which accelerates deployment. Because EPGs are not tied to the underlying hardware constructs, an endpoint can be relocated to a different physical location without changing rules.

If an application or endpoint is decommissioned, the central controller cleans the policy locally, eliminating the need for deleting at each network device. This approach greatly reduces touch labour that will help reduce configuration errors. Estimates show configuration errors are the source for up to 95% of all breaches.

3.2 *DURING* AN ATTACK – DETECT, BLOCK, DEFEND

The *before* stage of the continuum emphasizes discovery of the devices and applications on the network, enforcement of policies among these entities and hardening the infrastructure against an attack. Some of the ways ACI addresses these requirements include a centralized policy controller, secure multi-tenancy, white listing of services and segmentation.

The traditional data center security policy model ties security to the physical topology and requires policy configuration across multiple security devices (switches, firewalls, IPSs) each time a service or application is added, deleted or moved. With ACI, a central controller maintains control over both virtual and physical devices and has the ability to simultaneously provision network and security devices along with their associated security controls. This application centric approach decouples security policy from the underlying physical architecture, allowing for consistent security among physical, virtual and even cloud environments. It also allows for quick deployment of security devices to accommodate rapidly changing needs within the east-west traffic flow of the data center.

ACI uses endpoint groups, policy contracts and application network profiles replace traditional network constructs like VLANs, Internet Protocol (IP) addresses and MAC addresses. Endpoints are physical or virtual entities that can still be defined by characteristics like IP address or MAC and also by Network Interface Card (NIC), virtual NIC (vNIC), Virtual Machine (VM), or Domain Name System (DNS) name. Endpoint groups (EPGs) consolidate the individual endpoints into flexible, user-defined containers that could represent application, function, location in the network, organization, security zone or by any other logical parameter. EPGs can even be defined by design phases like development, test and production. Operationally, they could be defined by threat condition or mission.

Unlike traditional blacklist-based network architectures that permit by default, ACI employs a zero-trust, whitelist model that prevents any endpoint or EPG from talking until explicitly permitted. Communication between or among EPGs is defined by contracts that define rules and policies governing who can talk to whom and what they can talk about. Essentially, the whitelist function of ACI provides a stateless, Layer-4 firewall at line-rate. For additional security, service chaining can be used to traditional Layer 4-7 security like firewalls and intrusion detection systems into the path. Because the service chaining is logical and not physical, ACI increases scalability. For example, if load increases, additional virtual firewalls can be activated and easily integrated into the path with no changes to physical architecture. ACI essentially offers a software version of an automated patch panel. Mission critical systems can be routed based on a "watch and alert" model while user mobile devices can be routed through a more restrictive security model. The path can be easily changed based on mission requirements.

By simplifying and abstracting policy, it becomes easier to achieve segmentation, which isolates normal traffic based on functional or application requirements. Micro-segmentation enhance security for east-west traffic by breaking the "trusted" internal zone into smaller segments to better contain threats. Similar to the compartments aboard ship, micro segmentation can isolate damage to one space that can be quickly contained. This not only simplifies recovery, but also allows operations to continue in unaffected locations. While segmentation can be achieved with classical constructs like VLANs, ACLs and even physical devices, it could take hundreds of rules across multiple devices to segment a data center. If a user or device moves, segmentation rules would need to move as well creating a great complexity for the network operator.

The central controller of an application centric security solution creates a single point of management and visibility for the network. It controls the network and security infrastructure from one location, simplifying administration, reducing configuration time and minimizing errors. Templates allow reuse of endpoint groups and policies, which accelerates deployment. Because EPGs are not tied to the underlying hardware constructs, an endpoint can be relocated to a different physical location without changing rules on multiple devices.

If an application or endpoint is decommissioned, the central controller cleans the policy locally, eliminating the need for manually deleting policy at each network device. This approach greatly reduces touch labour that will help reduce configuration errors and improve cybersecurity posture as estimates show configuration errors are the source for up to 95% of all breaches.

3.3 *AFTER* AN ATTACK - SCOPE, CONTAIN, REMEDIATE

Despite all preparations, attacks can still succeed. A good security policy includes a plan to scope, contain, remediate and restore *after* an attack occurs. ACI can also leverage the micro-segmentation in conjunction with advanced malware protection to isolate and contain the threat into remediation zones where it can be analyzed and restored or replaced with known clean device.

Malware is becoming increasingly more difficult to detect. Without visibility across devices and applications tied to contextual analysis, attacks may be overlooked for months. Integrating advanced malware protection with ACI offers the ability to not only identify and block malware during an attack, but also to provide the visibility needed to scope and contain the attack. Advanced malware protection can identify where the malware originated (patient zero), what devices it infected and what that infection does, making remediation efforts far simpler to execute.

Typical malware solutions employee a point-in-time approach that scans a file once and then either block or let it pass. If permitted, the file is assumed good and not checked again. With techniques like encryption, polymorphism or hiding inside trusted protocols, a sophisticated attack may pass itself off as a good file. Once the attack launches, point-in-time protection has no visibility into what the file has done since being cleared through defenses. Advanced Malware Protection provides point-in-time protection, but then continues to check file activity and its communications on endpoints and network devices. If the cleared file exhibits malicious behavior, security administrators are alerted and can take actions to contain and remediate. Because AMP tracks the file continuously, it provides a retrospective capability that enables remediation of malware threats that were unknown at the time of penetration of the network and discovered later.

A final area of consideration in a comprehensive security design is threat intelligence that spans *before*, *during* and *after* attack. Attack vectors evolve over time and incorporating products backed by proactive detection and vulnerability research, engine development, threat intelligence, and a mechanism for communicating current trends can enhance the overall security posture of the next generation data center.

4. CONCLUSION

Next generation cybersecurity for warships must adopt new approaches and technologies if it is to evolve at the same speed as operational requirements. These new approaches must extend beyond the perimeter into the data center itself and must address security across a continuum *before, during, after* attack rather than using a point-in-time protection approach. Using a software-defined, application centric security solution that abstracts policy from infrastructure provides the agility and simplicity needed to rapidly adapt to shifting resource requirements. Security can be deployed at the same speed as virtualized applications. An application centric approach is open standards and allows integration with existing security devices, which offers increased visibility, auditing, and compliance tracking capability from a centralized management controller.

Estimates show that a large organization receives over 17,000 alerts a week, of which 19% (nearly 3400) are security related. On average, security professionals have time to investigate only 4% of the warnings received. The next-generation data center needs solutions that can correlate events and provide the context necessary to allow operators to quickly discern whether these alerts represent normal activity or an attack. It should also have the capability to detect insider threat, whether malicious or accidental.

Next Generation Data Centers will leverage increased use of automation and orchestration for both security and operations to handle increasingly dynamic adaptive environments. Successful Data Center security will be policy based and no longer depend on box-by-box management. Software-definition of the data center will enable abstraction of complexity and allow operators to keep up with more demanding operations and more complex threats.

5. REFERENCES

1. Dave Shackleford, May 2013, "Next-Generation Datacenters = Next-Generation Security, SANS Institute InfoSec Reading Room. Retrieved from https://www.sans.org/reading-room/whitepapers/analyst/next-generation-datacenters-next-generation-security-35105.

2. Calvin Hennick, Jan 27, 2016, "The New Tools Needed to Defend Next-Generation Data Centers from Cyber Attacks." BizTech Magazine, retrieved from http://www.biztechmagazine.com/article/2016/01/new-tools-needed-defend-next-generation-data-centers-cyberattacks.

3. Bill Kleyman, March 28, 2014, "Why The Next-Generation Data Center Needs to Be Agnostic." NETWORKComputing. Retrieved from http://www.networkcomputing.com/data-centers/why-next-generation-data-center-needs-be-agnostic/131950972.

GRAPHIC METHOD FOR IMPROVED INDONESIAN NAVY SUBMARINE DESIGN ACQUISITION; THE INVESTIGATION OF SMALL (MIDGET) NAVAL SUBMARINE DEVELOPMENT

M H Mukti and **R E Randall**, Indonesia Endowment Fund for Education, Texas A&M University, USA

SUMMARY

Economic growth in developing countries demands high protection from the possibility of local conflicts, yet the budget in the defense sector, particularly in submarine procurement, is very limited. Therefore, the presence of the small submarine concept seems to be a very promising solution due to its advantages over large nuclear-powered submarines. The notion of a small submarine is not new in the submarine world; the USS Holland, a small submarine, has made a significant contribution in the early submarine design stages. Several well-known submarine builders have expressed their interest in exploring the concept of small submarines to offer to prospective clients. This paper proposes an efficient graphic method to investigate the notion of developing small (midget) naval submarines for developing country such Indonesia. The method incorporates a certain stage to determine what the next generation of small submarine fleets will look like based on a survey of available knowledge. The reasons for utilizing the Joubert model hull or Defence Science and Technology Organisation (DSTO) generic conventional submarine as a base design in the investigation are explained. This graphic method approach has been drawn from the other method, e.g. the type ship and building block method, as the best approach and effective initial reverse engineering tool.

1. INTRODUCTION

1.1 BACKGROUND

Could developing countries build their own submarines? This paper proposes that an efficient graphic method would improve Indonesian Navy Engineers' capability in acquiring submarine technology in the early stage. Improvements achieved through simplified graphic approach include: better understanding of key naval architecture values of submarines, the ability to recognize local engineering capabilities for submarine equipment support, and trade-off decisions required to fulfill the character of a submarine that has both modern submarine mission capability, but also as a kick-start for research and development for local engineers, at the same time. It would not be possible to lower the cost of submarine procurement, provided that it relies on the international defense market. An effective method to ease the process for Indonesian Navy Engineers to acquire submarine technology is needed to increase in-house construction capability. Therefore, a graphic method approach has been developed to help local engineers in submarine design. The method then will be used to investigate Dradjat Budiyanto's (2006) notion of developing a small (midget) naval submarine that is affordable for Indonesian Navy.

1.2 OBJECTIVES

To consider what the design objectives of the submarine are, Roy Bucher and Louis Rydill (1994) suggest three primary factors that should be sustained throughout the whole submarine design process; the submarine should perform the functional purpose of the customer or operator (functionality), the design should be constructible within the capability of the technology and resources where the submarine will be built (constructability), and the cost should meet the owner capability (economically). Although there are many variations besides those three design objectives, in this project, the feasibility study will concentrate on satisfying design objectives using graphic method approach.

1.3 METHODOLOGY

There are three common methodologies in submarine design (Roy Bucher and Rydill (1994): Type ship method, the Building block method, and the Graphics Method. Type ship method is a very effective way to design surface vessels, where the availability of the data is enormous i.e. there are a lot of existing vessels that can be used as a comparison in the design process. Meanwhile, an iterative process and still heavily used existing practice and data in the design process is called the Building block method. This method, however, is a less constrained approach provided that a satisfactory database is available, and it is very suited to modern computational methods to design large and complex platforms. Instead of manipulating volumes to arrive at a size, a workable geometric arrangement is the main process of the graphics method or the development of arrangement drawings, which determines the ultimate size of the vessel, and provides an independent starting point of for the submarine design process. Therefore, based on the availability and capability where the submarine will be built, in this case, the developing country, Indonesia, the Graphic design method was found as the most convenient and effective tool in terms of the reverse engineering process, where the assumption of no available existing design data is made. Furthermore, the method is described as an iterative design flow as shown in Figure 1.

The iterations shown on the graphic method flow chart consist of five Decision Gates (DGs); DG0, DG1, DG2, DG3, and DG4. These DGs are expected to be the safe

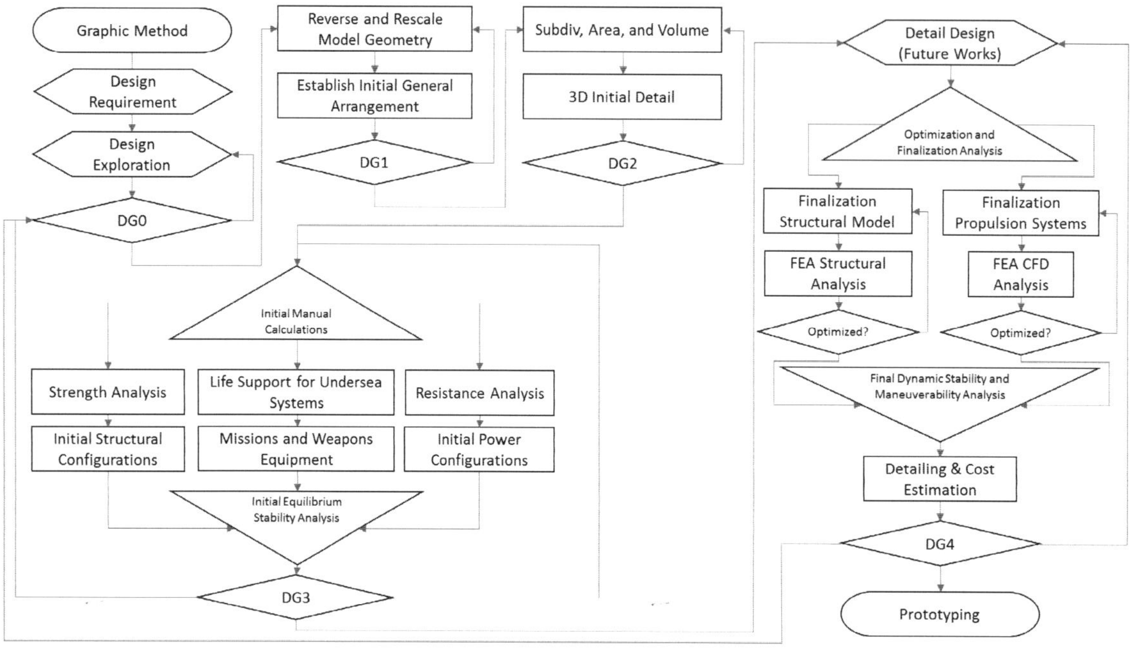

Figure 1: Graphic Method Flowchart

point for the designer to really pay attention to the key drivers associated with it prior to moving on the next DG. An optimum design that arrives at each tries DG would save time and cost in the design phase, although in some cases, iteration is not avoidable.

The first DG, DG0, is the most important phase for understanding and recognizing local capabilities. These include material availability and human resources skills. The decision in this phase might be the toughest ones since the designer has to determine what is to be expected for the submarine. It covers submarine mission requirements and definitions, which heavily influence the final principal dimensions of the submarine.

Next, DG 1 should have the initial decisions of the shape of the submarine. This is the power of the graphic method. The method allows the designer to understand the technology graphically and then apply the available knowledge to start to design a submarine in the graphic mode. As a higher DG, the design moves progressively and more detail with associated aspects.

2. CONCEPT EXPLORATION

2.1 DESIGN REQUIREMENT

The initial phase of this graphic method is to make sure that the designer gets the concept of how the submarine works, what is to be expected, and what activities need to be done. The design requirement is the important start point of the design process. Within certain circumstances, sometimes the options available are very limited, on the other hand, it gives advantages to save the cost and time to optimize the available options instead of conducting research and increasing the cost to consider other options.

- MISSION DEFINITION

It is important to conduct literature research to find out what other countries expect in designing a submarine. At some points, there are some standards that should be meeting between the navies to win the competition.

- ENVIRONMENT

Sea water density, temperature, and water depth operations are closely related to the environment where the submarine is expected to do its mission. It cannot be separated from the concept of the operations of submarines. The depth of Indonesian waters is about 200 m to 4000 m based on 1994 Arus Lintas Indonesia (Arlindo) data and the seawater temperature is about 30°C at the surface with 5 to 8°C potential temperature to a depth at about 1000 m based on Arlindo data. Furthermore, the wave height is about 2 meters to 3 meters (Kurniawan et al. 2015).

- DESIGN REQUIREMENT

At this point, the designer has to synthesize design requirements to accommodate mission definition and also consider the environmental constraints. Each of these can be related to functional capabilities required in the submarine design so that the designer is able to make a basis design as a guidance for the entire DG. Based on the basis design, the designer determines key aspects;

General Requirements, Sustained Speed and Propulsion Power, Submarine Control and Machinery Plan Automation, and Propulsion Engine and Ship Service Generator Certification. The designer has determined general requirements that include the operational capabilities as shown in Table 1.

Table 1: Example of Operational Capabilities

Description	Note
Engage airborne threats using sub-to-air armament	Future development
Conduct airborne autonomous vehicle (AAB) operations	Future development
Engage surface threats with anti-surface armaments	Future development
Engage surface ships at long range	In progress
Engage surface ships at medium range	Included
Detect and track a surface target using sonar	Included
Disengage, evade and avoid surface attack	Included
Engage submarines	Included
Engage submarines at medium range	Included
Engage submarines at close range	Included
Attack submarines with antisubmarine armament	In progress
Engage submarines with torpedo	Included
Disengage, evade, avoid and deceive submarines	In progress
Provide own unit command and control	Included
Maintain data link capability	Hardware
Provide communications for own unit	Hardware
Perform cooperative engagement	In progress
Conduct surveillance and reconnaissance	Included
Conduct mine neutralization/destruction	Included
Deploy AUVs and UUVs for mine detection and neutralization	Future development
Conduct mine avoidance	Included
Conduct magnetic silencing (degaussing, deperming)	Included
Maintain magnetic signature limits	Included
Steam to design capacity in most fuel efficient manner	Included
Prevent and control damage	Included
Perform seamanship, airmanship, and navigation tasks (navigate, anchor, mooring, scuttle, lifeboat/raft capacity, tow/be-towed)	Included
Replenish at sea	Included
Maintain health and well being of crew	Included
Operate and sustain self as a forward deployed unit for an extended period of time during peace and war without shored based support	Included
Operate in day and night environments	Included
Operate in heavyweather	Included
Operate in full compliance of existing US and international pollution control laws and regulations	Included
Operate submerged using AIP and batteries	Included
Operate and transit on snorkel	Included
Operate in littoral zones	Included
Operate covertly	Included
Support/conduct multiple cruise missile strikes	Future development

2.2 DESIGN EXPLORATION

The following sections describe Concept Exploration. Trade-off studies, design space exploration, and optimization.

2.2 (a) Concept Studies

Concept classification based on hull form is divided into 3 parts; conventional geometry, unique configuration, and radical concepts that are taken from the past (designs from WW1 and WW2), the present (existing designs that might be on active duty), and the future (conceptual designs that might be realized in the future). Author has classified three different major concepts;

- CONVENTIONAL GEOMETRY

The new construction method that was found during the cold war era by the Italian designer, Guinio Santi, allows building a perfect hydrodynamic teardrop form of a submarine. This is because the submarine hull consists of a series of doughnut-like steel pipe structures with different welded diameters. This concept was called the Gaseous Oxygen Toroidal (GST). The oxygen that is stored in the toroidal steel structure makes the submarine concept, 23 meters Maritalia LWT-23-4, to be able to have Air Independent Propulsion (AIP) capability. A conventional in-house midget submarine, also developed by Iran, was adapted from Yono class, North-Korean Submarines, namely Ghadir Class Submarines. Its advantage is the ability to penetrate shallow water over the big-subs. Likewise, Giunio Santi Engineering (GSE) is currently developing a dry Swimmer Delivery Vehicle (SDV) that is called Button 5.60 for the US Navy. Unlike wet SDV, this dry SDV allows the crew to be kept dry and due to its small size, only minor modifications are required for US Navy submarines to carry it on their deck. Lastly, Euronaut is one of the example of the home-made civilian submarine that is used for ocean exploration. It has a very simple shape, where it contains exposed single hull of the pressure Hull (PH), and deck fairing.

- UNIQUE CONFIGURATION

Piranha, a small submarine built by Russians, has a lockout chamber at the bow with a moonpool similar to the Euronaut sub. It is a unique design since the torpedo is located below the deck that will push Center of Gravity (CoG) upwards that needs to be compensated by a counter weight. On the other hand, the usage of the spaces of the PH especially at the fore end area is much more flexible and it can be optimized by adding a lockout chamber without concern for the space for the torpedo. Furthermore, the construction issues to penetrate torpedo tubes into the PH is not necessary. While, a French builder, Direction des Constructions Navales (DCNS) is currently developing a future midget submarine concept with two main thrusters, X-rudder configuration, and huge lockout chamber at the superstructure. DCNS claims it is able to remain at the seabed for a long-time period before sending a swimmer team to infiltrate the target. Although it is an interesting concept, the available information is very limited.

- RADICAL CONCEPT

DCNS is also developing large size submarines with radical shapes, SMX 24 and SMX 25. SMX 24 has a square section form, its advantages are reducing the wet surface area, and in terms of the constructability. Due to its shape, SMX 24 allows modular configurations that are able to be adjusted with different specific missions. The U-Boat is one of the most successful underwater vehicles in submarine war history. The notion of reviving an underwater vehicle that performs well on the surface like the U-Boat, is an interesting concept to be examined. Therefore, the SMX 25 radical submarine concept offers a high-performance submarine on the surface that allows SMX 25 to reach the target location as quick as possible.

However, this concept is complicated for manned operation, which requires PH and safety constraints that affect the cost and complexity of the hull design. On the other hand, it can be potentially applicable to the UUV application. Meanwhile, British defense company, BMT, is developing Ship Submersible Gas Turbine (SSGT) radical concept design that is a high mobility non-nuclear submarine. Hence, BMT claims this concept is able to reach high speed with its innovative propulsion design since it is equipped with the turbine engine, mounted at the sail. Furthermore, the high-speed submarine concept was also developed by the Florida Institute of Technology who came up with an idea to transform submarines into high-speed hydrofoil trimaran during Normal Surface Conditions (NSC).

2.2 (b) Propulsion and Electrical Machinery Alternatives

Based on the concept studies from section 2.2 (a), the utilization of the machinery plant is not homogenous. The past design was closely related to using diesel engines by surface ships for decades. The concept of using steam powered propulsion has been used prior to the invention of diesel engines. The advent of the nuclear power plants, where basically, it could not be separated from the steam engine concept, has given the new age of the submarine world. All of these alternatives are able to be in Air Independent Propulsion (AIP) with some modifications, especially for conventional engines such as diesel. The modifications include a closed cycle diesel system that enables diesel engines to do AIP mode. As a return, it demands more space that affects the submarine principal dimensions.

The designer should really pay attention to the fact that the AIP capability has an important role as the key success indicator for the submarine overall performance. Therefore, the other options have been studied to improve machinery alternatives for the submarines. One of them is the notion of utilizing fuel cell technology that can be used as the sole power plant, or as the auxiliary engine support. The fuel cell offers an unlimited power source as it only requires Hydrogen and Oxygen, which can be obtained from the seawater hydrolysis process. The only concern here is how to generate the fuel cell's fuel enormously, increase fuel cell efficiency, and maintain the fuel cell components carefully. The other important thing to consider is the battery option to manage energy storage for the submarine. This is a crucial aspect especially for non-nuclear powered submarine where the power source is very limited.

2.2 (c) Propulsors Alternative

The variant of the propulsion of the submarine may be divided into two parts, conventional and unconventional propulsion systems. The conventional ones utilize old-fashioned propellers that are common for the surface ships, but for submarine, it demands both lesser cavitation and high efficiency. The variant of the conventional propeller is by adding a shroud or nozzle. Some of the small size submarines also utilize thrusters like Remotely Operated Vehicle (ROV) for detail displacement movement assistance. The unconventional ones are being considered lately to break the domination of the conventional propeller where they may offer higher efficiency and higher stealth capability. The Wave glider, a bio-inspired engineering where it utilizes fin and wave propagation much like a fish; requires more research has to be done in the future. The past concept such as magnetohydrodynamic drive also was an interesting concept where there is no moving part in the system. This will offer a complete silence of the submarine. Again, in this case, the designer should prioritize the cost due to the limitation of the country conditions, and the important thing is, can it be built locally by the local engineers or not?

2.2 (d) Combat Systems Alternatives

Based on the limitation of the technologies of the countries, the designer must be concerned with immediately having a plan to develop complete technology in the near future, once the first experimental submarine prototype has been built. The technologies include Sonar, Combat Control, Torpedo, and Unmanned Underwater Vehicle (UUV) support. For instance, the only torpedo available in Indonesia is the Surface Underwater Torpedo (SUT) that is adapted from a German company. It has the length of about 5 meters. This means that this type of torpedo determines the next step of the graphic method described in Chapter 3.

3. CONCEPT DEVELOPMENT

3.1 REVERSE AND RESCALE MODEL GEOMETRY

In this section, the basis design has expanded to cover main hull form characteristics, for example: High-speed resistance (sustained speed): 15-22 knots, Low-speed resistance (endurance/snorkeling): 5000 nm at 12 knots, Stability and manoeuvrability: teardrop with parallel midbody, Number of decks: 1 decks, Hull size: smallest due to minimum cost, Number of hulls: 1 hull, Propulsion System: Conventional Hybrid Diesel Battery Propulsion System. At this stage, the designer should aware what the latest hull form technology that is being used, based on the past, current, and future design described in section 2.2 (a). Although some designs are not a tear drop shape, like some concepts from DCNS, the hull concept such as a U-Boat, where there are many appendages exposed on the surface hull, has been abandoned for its poor speed performance during submerged condition.

The questions then come to the designer's mind to decide how to synthesize the best hull form. In this graphic method, the designer should consider creating hull forms from the blank sketch where there are two methods available; the empirical formula from MIT, and the utilization of existing concept designs. Based on Capt. Jackson (1992), modern teardrop submarine hull geometry is divided into two types; hulls with Parallel Mid Body (PMB) and with no (PMB). Hulls with PMB has a cost advantage due to a constant diameter, such as a cylinder, compared to none- PMB configuration. However, with the no PMB type, it allows a perfect teardrop shape, where the hull resistance is minimum. The other method is utilizing existing designs described before 2.2 (a), and keep in mind that in this phase, the designer may ignore sizing calculation to determine the hull form of the submarine. This allows the designer to have unlimited options of the hull form possibilities.

The existing design in this investigation utilizes a hull with PMB, a hull form based on the latest hull form study conducted by Prof. Joubert in 2004 that is called Joubert Generic Submarine Hull form. This hull has been chosen to save time in determining the ratio and other important principal dimensions, as in the graphic method the DG0 has been reached. Furthermore, this is the strength of this graphic method, the designer has control to do reverse engineering of the desirable design. Hence, the designer shall start to reverse this hull form into a 3D model, as shown in Figure 2. Notice that the model is still raw, the designer may take personal adjustment, and some spaces need to be adjusted such as sail volume and hydroplanes in the next DG.

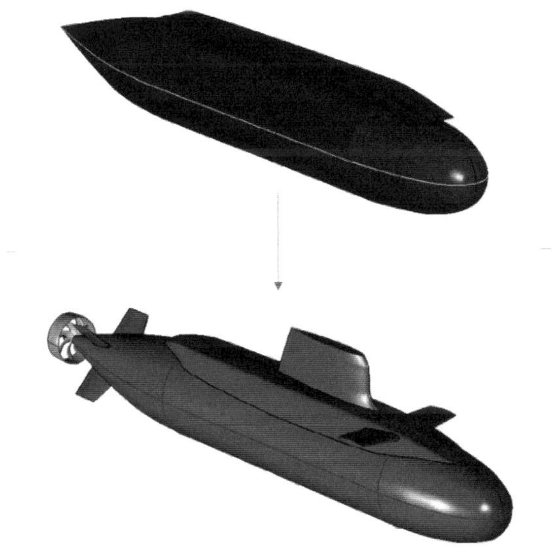

Figure 2: Example of 3D Model Reversing Process

3.2 PRELIMINARY GENERAL ARRANGEMENT

By using a cartoon flounder diagram shown in Figure 3, the designer is able to synthesize the initial arrangement of the submarine and align it with the decisions made in DG0. Therefore, the idea to put a diesel engine generator as the main power source of the submarine drives the sizing of the engine room of the submarine and also the submarine overall length. Since the diesel engine propulsion system has been chosen, it is important to have a minimum of two engines existing on the submarine to provide backup power when one is failed. Furthermore, the engine sizing itself based on DG0 must be the smallest size that is commonly used in ship industry in Indonesia. Hence, the range must be 1 to 2-meter length engine. Design requirement in DG0 is also another important factor in determining the size of the submarine. In this case, Indonesia needs a submarine that is able to do covert shallow water missions, and again, the cost is the key driver of the design process, meaning a smaller sized submarine offers a smaller cost. Therefore, based on these important considerations, 22 meters as Budiyanto has purposed can be used as the initial length of the midget submarine.

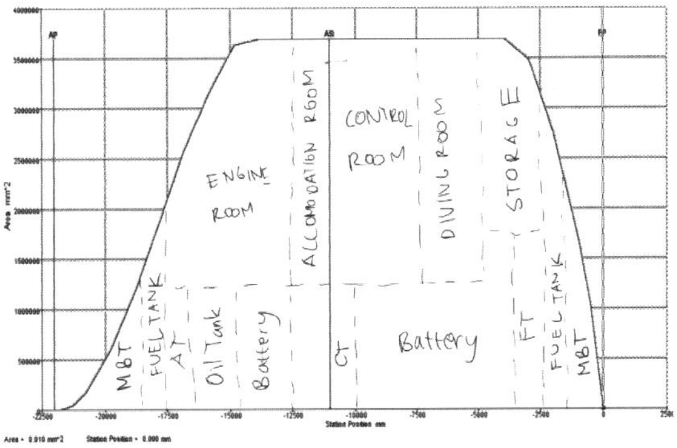

Figure 3: Example of Flounder Diagram

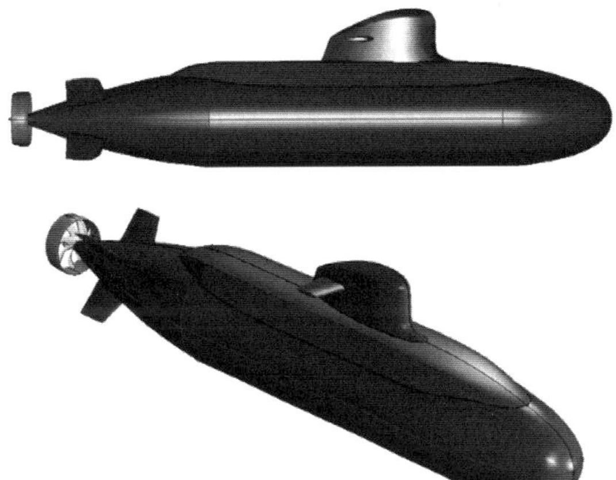

Figure 4: Example of Final Hull Form

The flounder diagram shows the preliminary PMB Hull configuration where there are other important key drivers to be considered. They are accommodation room, control room, and armament room. In chapter 2 the designer has arrived to utilize the five-meter length of the torpedo and it takes plenty spaces of torpedo penetration into the PH up to the diving room. The lockout chamber is located at the sail to save space inside the PH. Putting the lockout chamber above PH allows the possession of a higher Center of Buoyancy (CoB) when the chamber is dry. But this will be investigated more in the next phase, especially when it is filled with water to deliver to crews for special operations. It is important to model some important components in this graphic method, based on design exploration in chapter 1 and 2, in order to help sizing of each compartment inside the PH. The decision to use a single hull configuration is inevitable since the design is driven by a project cost basis for this investigation. This is an iterative process; when the optimum configuration is reached, meaning DG1 has been completed. Notice that the designer may also consider sizing for the appendage parts such as sails, and hydroplanes, based on the knowledge described in chapter 2, so that the designer will end up with the final hull form design shown in Figure 4.

4. HULL CONFIGURATIONS

4.1 SUBDIV, AREA, AND VOLUME

Based on the Flounder Diagram, DG1 has been passed, the designer may continue to detail the 3D Hull interior model into several options and estimates the volume required for each room.

For instance, in Figure 5 shows the different types of torpedo configurations, from top to bottom are three, two and single torpedo configurations. The two-torpedo configuration provides the ability to use spherical sonar as those being used in the modern submarine. Whereas, the three-torpedo one takes up space so that it eliminates access to the sail i.e. must move sail position to backward toward the stern. The torpedo configuration also depends on the budget spend to procure the torpedo for the fleet. If there is a possibility of exceeding national torpedo production, then a four-torpedo configuration may be considered for this investigation.

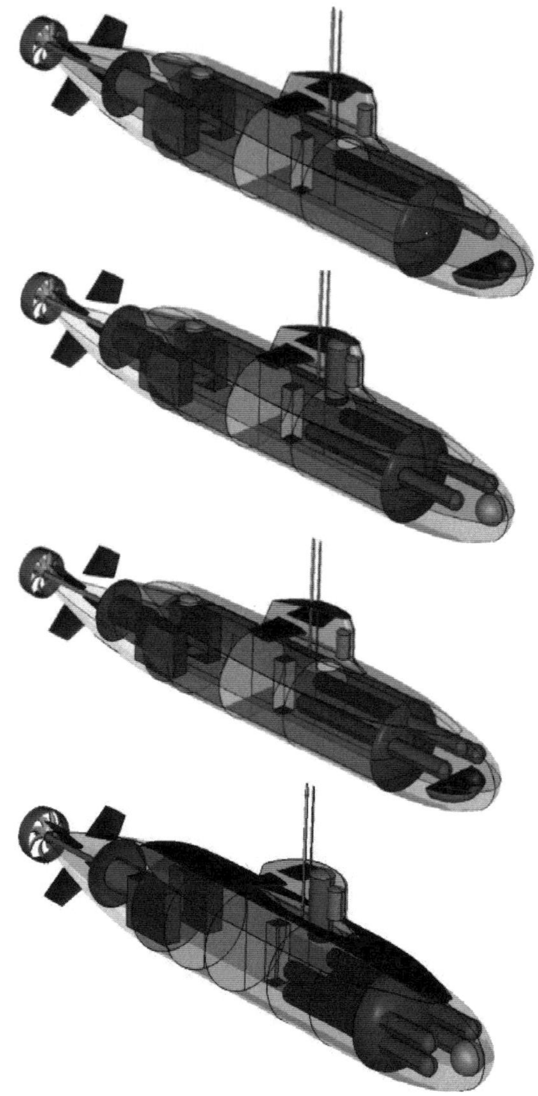

Figure 5: Example of Torpedo Options

5. PRELIMINARY DESIGN CALCULATIONS

5.1 STRUCTURAL DESIGN AND ANALYSIS

The iterative process of structural design was performed using manual calculations. The first step in the process is manually determining the shell plate thickness by using different rules and codes where the designer will choose the highest result, although the optimization in the future work is required to avoid over design and save cost. The first code is from Society Naval Architecture and Marine Engineering (SNAME) for underwater vehicle pressure. Manual calculations to verify the estimation from SNAME code are then performed using a MathCAD interpretation of Captain Jackson's structural calculation for submarines. There are additional codes from The American Society of Mechanical Engineers (ASME), American Bureau of Shipping (ABS) for PH with human operations and Det Norske Veritas Germanischer Lloyd (DNV GL) for naval submarines are included in the manual calculations as other references. The final step is to model the 3D structure of the submarine by using software.

5.1 (a) Strength Analysis

SNAME calculation determines the frame spacing, frame profile, and the thickness of the hull. Captain Jackson's method was coded into MathCAD to verify the first estimation from the SNAME code. Other hand calculations that are based on SNAME Underwater Vehicle Design Guidelines, ABS Rule, and DNV Naval Submarine Rule should be considered by the designer if time is available.

5.1 (b) Initial Structural Configuration

Based on the manual calculations that have been performed in section 5.1 (a) the 3D model of the structure can be modelled as the next level of the graphic method. Then based on DG2 the designer also has to determine how many compartments are in the PH, this is very useful in modeling the preliminary 3D structural design.

- PRESSURE HULL

Yellow frames show discontinue in the tank. White frames are for showing where the Compensation Tank (CT), Fore Trim Tank (FTT), and Aft Trim Tank (ATT) is. Blue frames describe compartments with batteries. Brown is the engine room with lube tank. Cyan frames are located in Accommodation rooms. Figure 6 shows the structure model with a stringer. This might be over designing so optimization is required in the next graphic method phase.

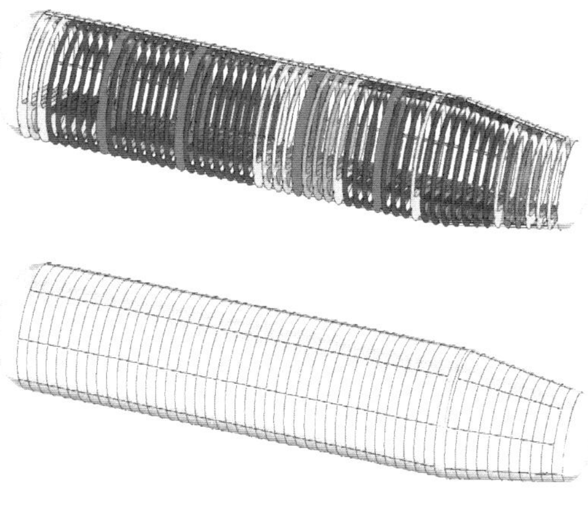

Figure 6: Example of Initial Structure Model

- EXOSTRUCTURE

The exostructure is the other important part of the underwater vehicle. In the submarine, the exostructure is used to store the compensation fuel tank and the fairings to improve the hydrodynamic performance of the submarine. As this submarine uses the Joubert Hull, it is difficult to form steel plates for such s streamlined bow within the small size of the submarine due to cost basis. Therefore, the designer may come up with the idea to use composite material for the exostructure based on the manufacture consideration and the cost that is relatively cheaper than high-grade steel. The phase in modeling exostructure is shown in Figure 7, note that this phase is also compatible to describe how the exostructure will be assembled in the future shipyard work.

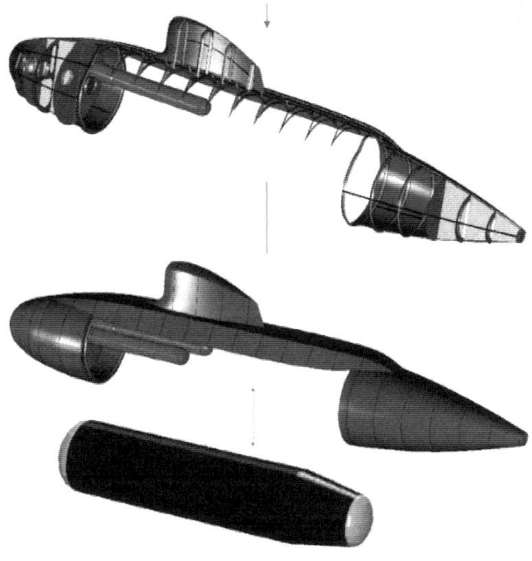

Figure 7: Example of Exotructure

5.2 LIFE SUPPORT FOR UNDERSEA SYSTEMS

The other parallel work is to make sure the atmosphere inside the PH is safe for the crews within the design operational time. The calculation is based on Randall notes (2016) that are the ventilation of submarine (large chambers) based upon carbon dioxide flushing for a submarine. Although rebreathing of expired air or breathing gas is common in diving operations, including underwater chambers, it is necessary to remove all the exhaled carbon dioxide before rebreathing the gas in order to avoid toxic effects. Therefore, the options are using carbon dioxide scrubber from Methyl Ethanol Amine (MEA), Baralyme, Lithium Hydroxide or Sodasorb.

Note that due to the limitation resources in DG0, the designer understands to choose the material available and also comply with the function needed. Sodasorb is the solution to maintain the atmosphere to be safe enough for the crews during the missions. Temperature considerations, some scrubbers sized for adequate performance at 70 F that is equal to the average temperature of Indonesian sea waters. No need for additional heater equipment. This can be done simply by hand calculation.

5.3 MISSIONS EQUIPMENT AND WEAPONS

Further future development in military technologies is inevitable for this project. However, some decisions are being prioritized as a first step of building submarine capability for developing countries such as Indonesia. The bow sonar is one of the most critical parts of the submarine's navigations since the submarine relies only on acoustic sensors in deliverable depths of mission operations. A sphere sonar should be investigated further in future works to establish an overall 3-dimensional vision to the submarine. Flank arrays and tow arrays sonars are not necessary for this feasibility study since the submarine size is very compact to save cost. A submarine is able to be equipped with several types of weapons depending on the operational capabilities. The torpedo selection has been discussed in section 2.2 (d) but not limited to the other options to develop smarter, smaller, and more powerful torpedoes in the future. As the submarine design until to this point is sufficiently small and should be cheapest among existing submarines in the world, then a strategic plan to produce numbers of this submarine should be developed in the future, as it provides small capability as an individual but could give a significant threat in acceptable quantities.

5.4 POWER AND PROPULSIONS

As determined in DG2, the propulsion system consists of two diesel engines for use during snorkel and lead-acid batteries used during submerged operations. During snorkel operation, the diesel engines power the motor through the Integrated Power Systems (IPS), while in submerged operation batteries are the only main source to power the motor through the IPS. Battery specification is also obtained by the same method as both diesel engine and DC motor selection. Based on the available decisions until this point, the designer is able to perform preliminary Electric Load Analysis (ELA).

Figure 8: Example of Software Simulations

5.4 (a) Resistance Analysis

As DG2 and DG3 are very close to each other, the process for determining the power and propulsion requirements is a back-and-forth process in this graphic method. It began with the decision of using two diesel engines at DG2, then validate and match them with a series of preliminary calculations of resistance, Shaft Horsepower (SHP), sprint speed, AIP endurance, sprint endurance, and snorkel endurance; both utilize manual calculations and the software. The calculated values must satisfy the operational requirements document, otherwise, stepping back to the previous DG must be taken.

• RESISTANCE ESTIMATIONS

Based on the design report conducted by VT, resistance calculation can be done manually by hand calculation. Based on the estimations, the AIP endurance is up to 23.131 hours where the submarine only relies on the battery power. And the longest-range capability of the submarine is up to about 4000 nautical miles at snorkel speed.

• SOFTWARE SIMULATIONS

The utilization of computer calculation provides greater confidence in the feasibility study and also one of the important aspects of this graphic method. However, this may add cost to the overall project and so the priority goes to any software that is available in the country where the sub will be built.

5.4 (b) Initial Power Configuration

This section is to specify what kind of machinery systems may be used on the submarine.

• ENGINE SELECTIONS

The selection of power generator is driven by the resistance analysis that is conducted parallel to this chapter. The engine is not a final brand or type, instead it is the challenge for the designer to push the national industry to reverse engineering such products with associated characteristics such as dimensions, performance, etc. Hence, based on the work done so far, the designer is able to select the diesel engine type and the electric motor and model them in DG2.

• GRAPHIC MODEL REVIEWS

Based on the selections, the designer may conduct a 3D model check to finalize that the size of the object fits into the hull structure shown in Figure 9.

Figure 9: Model Review

6. INITIAL STABILITY ANALYSIS

The initial stability is to estimate whether the submarine is stable during static condition. Based on the decision of the DG2, the 3D model compartment definition can be established in the software shown in Figure 10. The chamber tank at the sail is treated as the compartment as it is void. Figure 10 also describes how the compartment is connected to some intact tanks. Note that some compartments may not appear correctly due to the debug error of the program. The green color is the Main Ballast Tank (MBT) that is located at the bow and the stern. The dark red color is the diesel fuel tank that is also located at the bow and the stern next to the MBT. It also shows that it is filled with seawater so that the torpedo tube can be neutralized in weight once the torpedo has been discharged. However, if the seawater is not sufficient to replace the torpedo weight, there are three adjustable ballast tanks inside the PH shown also in green color. They are called the Compensation Tank (CT), Aft Trim Tank (ATT), and Fore Trim Tank (FTT). The purple shows non-buoyant volume that is actually a residual space to fill the empty space in the exostructure.

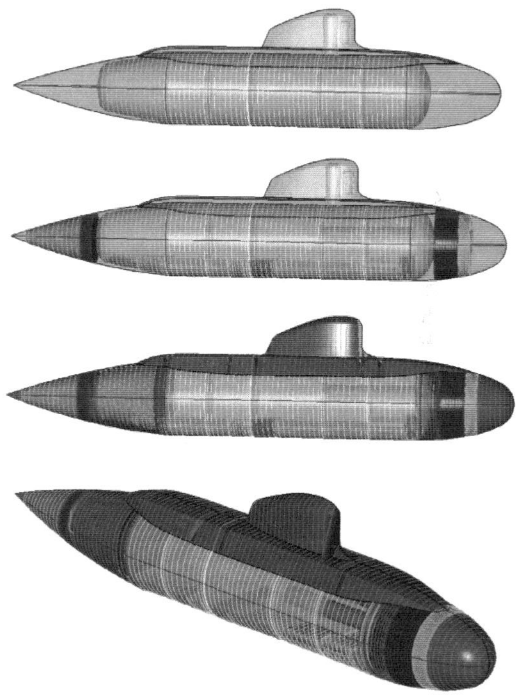

Figure 10: Compartment and Tank Definition

7. CONCLUSIONS

This paper has reviewed the graphic method which can be used to help Indonesian Navy Engineers develop a 22-meter submarine. At this point, the midget submarine notion is feasible to continue to the DG4 and prototyping stages. In terms of a reverse engineering tool, the graphic method, especially in the prototyping phase, can give more information and more confidence to local engineers with zero experience about the submarine and will be able to verify what they have done so far during the design process. Any such a commitment is essential to be made if the government of a developing country, Indonesia, wants to acquire submarine technology independently. An integrated software tool may help to make the graphic method work effectively and efficiently in the future. In a nutshell, the goal to be able to build the submarine with minimum resources is possible by using this graphic method.

8. REFERENCES

1. ARENTZEN, E. S., CAPT, USN, 'Naval Architectural Aspects of Submarine Design' *SNAME Transactions*, 1960.

2. BURCHER, ROY,LOUIS RYDILL. 'Concepts in Submarine Design'. *Cambridge University Press*, 1994.

3. CARLBERG, HENRIK, 'Concept Design of a Commercial Submarine' *MS Thesis. NTNU*, 2011

4. CHRISTLEY, JIM, 'US nuclear submarines: the fast attack', *Paperback*, 2007.

5. COMSTOCK, JOHN PAUL, 'Principles of Naval Architecture'. *New York: Society of Naval Architects and Marine Engineers*, 1967.

6. Covert Shores, *http://www.hisutton.com/*

7. Design Report, *http://www.dept.aoe.vt.edu/*

8. DODGE, RUSSEL A., MILTON J. THOMPSON. 'Fluid Mechanics'. *McGraw-Hill book company*, Inc., 1937.

9. FRIEDMAN, NORMAN. 'Submarine Design and Development'. *Naval Institute Press*, 1984.

10. FUTURE SURFACE VESSELS. 'Warship 2015', *Royal Institution of Naval Architects*, 2015.

11. GABLER, ULRICH. 'Submarine Design'. *Bonn: Bernard & Graefe Verlag*, 2000.

12. GORDON, ARNOLD L., CLAUDIA F. GIULIVI, A. GANI ILAHUDE. 'Deep topographic barriers within the Indonesian seas'. *Elsevier Science Ltd.*, 2003.

13. HUTHCINSON, ROBERT. 'Jane's Submarines: war beneath the waves: from 1776 to the present day', *HarperCollins*, 2001.

14. Indonesian MX-1, *sites.google.com/imx1midget/*

15. JACKSON, HARRY A., CAPT, USN, Ret. et. Al 'Fundamentals of Submarine Concept Design', *SNAME Transactions*, Vol. 100: pp. 419-448, 1992.

16. JOUBERT, P.N. 'Some Aspects of Submarine Design Part 1. Hydrodynamics', *Australian Government Department of Defense*, 2006.

17. JOUBERT, P.N. 'Some Aspects of Submarine Design Part 2. Shape of A Submarine 2026', *Australian Government Department of Defense*, 2004.

18. KURNIAWAN, RONI and MIA K. KHOTIMAH. 'Ocean Wave Characteristics in Indonesian Waters for Sea Transportation Safety and Planning', *The Journal for Technology and Science*, April 2015

19. NAT, CLEMENS G.J.M. VAN DER. 'A knowledge-based Concept Exploration for Submarine Design', *Delft University Press*, 1999.

20. POLMAR, NORMAN AND KENNETH J. MOORE. 'Cold War Submarines: The Design and Construction of U.S. and Soviet Submarines', *Brassey's*, 2004

21. RANDALL, ROBERT E. 'Underwater Acoustics for Ocean Engineers', *Texas A&M University*, 2016.

22. RANDALL, ROBERT E. 'Underwater and Moored System Design', *Texas A&M University*, 2016.

23. Safety Standard for Pressure Vessels for Human Occupancy. *American National Standard*, 1977.

24. SEWELL, ELI A. 'Open Architecture Framework for Improved Early Stage Submarine Design' *MS Thesis. Massachusetts Institute of Technology*. 2010

25. Submersible Vehicle Systems Design. *The Society of Naval Architects and Marine Engineers*, 1990.

26. ZIMMERMAN, STAN. 'Submarine Technology for the 21st Century', *Trafford Publishing*, 2006.

9. AUTHORS BIOGRAPHY

M H Mukti is a M.Eng student in Ocean Engineering at Texas A&M University, fully funded by the Indonesian

Presidential Scholarship by the President Susilo Bambang Yudhoyono in 2014. Although he is not associated with Indonesian Military, he is a sophisticated technology enthusiast. From his undergraduate study, he was eager to develop a platform for students interested in advanced technologies in ocean engineering applications and hence became a founder and general manager of the Autonomous Marine Vehicle (AMV) Universitas Indonesia (UI) team in 2012. During his leadership, he successfully brought AMV both in National and International recognition. One of them initiated a Military Roadmap with MABES TNI in the Sea Ghost Project to create a roadmap for Unmanned Surface Vehicles in the Indonesian Environment. Furthermore, he is actively participating in SNAME/MTS event, becoming SUT officer and will undertake the Ship Technology course at MIT this Summer.

R E Randall received a Bachelor of Mechanical Engineering degree in 1963 from Ohio State University. After graduation, he served in the US Navy as a submarine officer from 1963 to 1967. He received an MS degree in 1969 and a Ph.D. degree in 1972 in Ocean Engineering from the University of Rhode Island. From 1972 to 1975, he worked as civilian for the US Navy and for a private oceanographic institution, Harbor Branch Oceanographic Institute. In 1975, Dr. Randall began an academic career in Ocean Engineering at Texas A&M University in College Station, Texas and is now Professor of Ocean and Civil Engineering. In 1976, he became a registered professional engineer in the state of Texas. He has served as the Director for the Center for Dredging Studies since 1994. He teaches courses in Ocean Engineering Design, Underwater Systems and Mooring Design, Underwater Acoustics and Marine Dredging.

PARASITE DRAG ON SUBSTRUCTURES OF SUBMERGED BODIES

T Nietzschmann, Schiffbau-Versuchsanstalt Potsdam (SVA), Germany

SUMMARY

The interference resistance, i.e. the total resistance minus the sum of the resistances due to substructures can reach up to 10 % of the total resistance of a submerged body. That for it is usually called parasite drag. This problem relates to fin and sensor systems of submarines, ROVs and AUVs. The reduction of the interference resistance is one aim of every design process of submerged bodies.

SVA Potsdam has initiated the R&D project "Interference phenomes of substructures of submerged bodies" to investigate this parasite drag. In this project systematic numerical and experimental investigations were carried out with bodies of a submarine and an AUV. The most influencing parameters were variated: speed and depth of the submerged as well as the chord length, the span, the aspect ratio and the position of the substructure on the body

The numerical calculations were validated by model tests. Numerical calculations were performed for full and model scale to evaluate the scale effects. The model tests included the determination of the forces of the submerged bodies as well as detailed flow visualization of the flow around the substructure with PIV measurements.

As a result, the interference resistance can be quantitatively determined using systematic sets of parameters for a submerged body with substructures. The flow visualization shows details of the flow on its transition from the substructure(s) to the body. This gives an indication of the causes for increasing or decreasing interference resistance as "parasite drag".

NOMENCLATURE

c_T	total resistance coefficient
FP	*forward perpendicular*
R	Resistance (N)
λ	aspect ratio

1. INTRODUCTION

Common commercial and military submerged bodies (e.g. Submarine, ROVs or AUVs) consist of the main body and substructures for manoeuvring (like fins and rudders) and sensors. The flow around the body will be influenced be the substructures and the flow around the substructures will be influenced by the body. This results in interferences phenomes like vortex development, wake formation or inhomogeneous inflow for other substructures. These interferences produce an additional resistance, the so called interference resistance or parasite drag. So the parasite drag of substructures on submerged bodies may be defined as the difference of the summary of resistance of base and the substructure itself minus the resistance of the whole body.

$$1)\quad R_{parasite} = (R_{T\ base} + R_{T\ sub}) - R_{T\ base+sub}$$

That means the overall resistance of the combination may be increased or decreased by changing the position.

Usually the parasite drag of substructures on submerged bodies can be assumed as a function of the size and the position of the substructure, see Figure 1 [1]. Moreover, the shape of the junction has an influence.

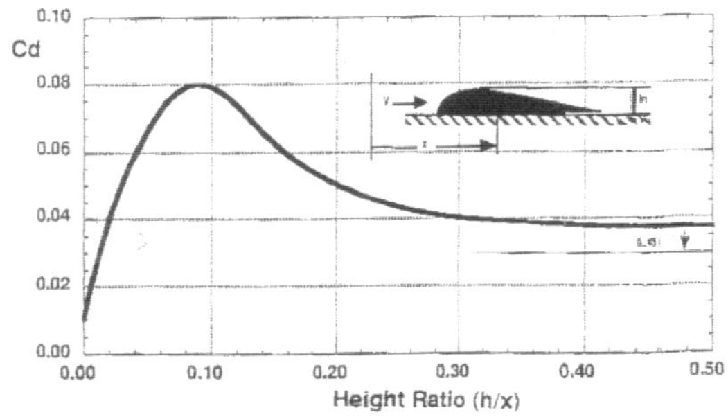

Figure 1 parasite drag indicated by Hoerner [1]

A research project of Potsdam Model Basin (SVA) was targeted at the question if the parasite drag can be determined by viscous numerical calculations (RANSE) accurately enough to compete with model tests. Additionally, the capability of PIV to reveal different flow behaviour under these circumstances was assessed. .

This project follows the line of research on submerged bodies which has been conducted at SVA for several years, leading from GeoSim resistance tests to scale effects on manoeuvring.

2. AIM OF THE PROJECT

Three aims were defined for this project:
a) To increase the accuracy of resistance prediction
b) To show the possibilities to reduce drag by optimization of the appendages' position on submerged bodies

c) To validate CFD calculations and visualisation with PIV

a) SVA usually conducts model tests for submerged bodies with a planar motion device (SUBPMM) made by Hydronautics Research Institute. The model is fixed to the device by two struts jacking the model hull. On those struts the model is fixed by measurements gauges inside the hull of the model, see Figure 2.

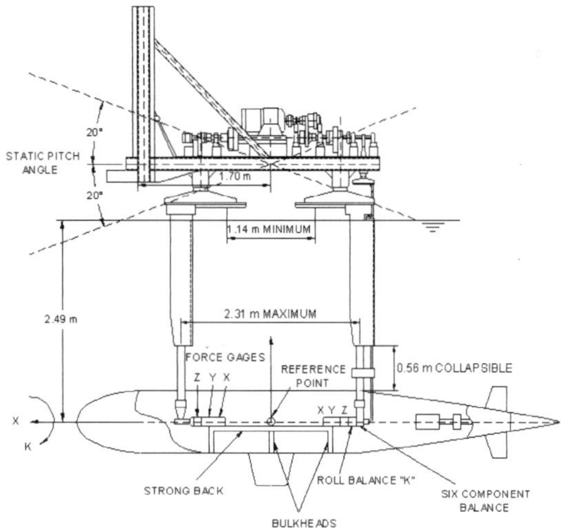

Figure 2 SUBPMM at SVA

In all cases of force measurements there is a parasite drag to be assumed due to this tests conditions. The smaller the model the higher is the relative drag, and so is the inaccuracy. In step a) of the project this was investigated using different types of struts as connection between towing carriage and model.

b) The parasite drag obviously gives the possibility to influence the resistance by changing the relation between base and substructure: size, position and fairing between both was the subject of research for submerged vessels and boats with deeply submerged appendages (substructures), for example SWATH and sailing yachts [2, 3, 4]. Mostly for SWATH and AUV the cruising range is important for low fuel margins because there is low buffer of buoyancy and capacity.

c) Due to potential flow boundary conditions it is not expected to have any results for deeply submerged conditions: The – non viscous – pressure difference between fore and aft ship is expected to be the same for joint and not joint conditions of base and substructure. It is assumed that flow separation depends on the position and size of the substructures and the behaviour of the flow around the substructure in the boundary layer of the base. That for all CFD simulations have been done using RANSE code with sst-simulation. The comparison and validation criteria of those calculations were the force measurements on one hand and PIV measurements for some test configurations on the other hand.

3. SCOPE OF WORK

The R&D project of SVA set focus on a cylindrical base, to cover for submarines and SWATH, and another base with planar bottom and deck, covering for AUV. Parameters for the test setup followed the matrix in table 1 showing the master configuration c6h4 in bold letters:

Table 1: Test matrix

constant λ		c3h2	***c6h4***	c12h8
aspect ratio	-	1.33	***1.33***	1.33
span	m	0.20	***0.40***	0.80
chord	m	0.30	***0.60***	1.20

constant span		c3h4	***c6h4***	c12h4
aspect ratio	-	2.67	***1.33***	0.667
span	m	0.40	***0.40***	0.400
chord	m	0.30	***0.60***	1.20

constant chord		c6h2	**c6h4**	*c6h8*
aspect ratio	-	0.667	***1.33***	2.67
span	m	0.20	***0.40***	0.80
chord	m	0.60	***0.60***	0.60

For all these conditions force measurements and RANSE calculations with three different speeds were executed.

The two models were built of FRP with appendages of foam material. The position of the substructures to be tested were marked with a hole pattern on the keel side. The models were 5.6 m long and the diameter of the cylindrical one was 0.6 m. The breath of the second model was 1.0 m

The model tests started with different strut lengths to evaluate the influence of the strut's interference resistance. Following this, several types of turbulence stimulators were compared to evaluate their influence in comparison to RANSE calculations. Finally the test setup for the parametric variation of substructures was conducted.

Model scale RANSE calculations with and without struts were done to evaluate their influence. Furthermore, calculations with and without tank wall effect and water surface were done to evaluate the influence of the model test setup. Finally, a comparison between model and full scale was done using RANSE simulation.

To speed up the pre-processing of this large matrix of simulations the grid was customized: The numerical mesh of the body was divided in several sections of same length covering the span of the substructure. One section contained the substructure. For every case to be investigated only the grid blocks had to be transposed. The connection of the grid blocks was accomplished using a general grid interface.

4. RESULTS

A dependency of the total resistance related to the position of the substructure could be shown for all cases within measurements and calculations.

To capture the maximum achievement in power saving all results are explained as the ratio of the total resistance of the configuration and the minimum possible total resistance for the varied parameter.

$$2) \quad \frac{R_T}{R_{T\,min}}$$

Figure 3 shows the parasite drag resulting from the strut's position during the measurements. The need for constant test setup during the sessions is obvious and the need for calibration of the setup as well.

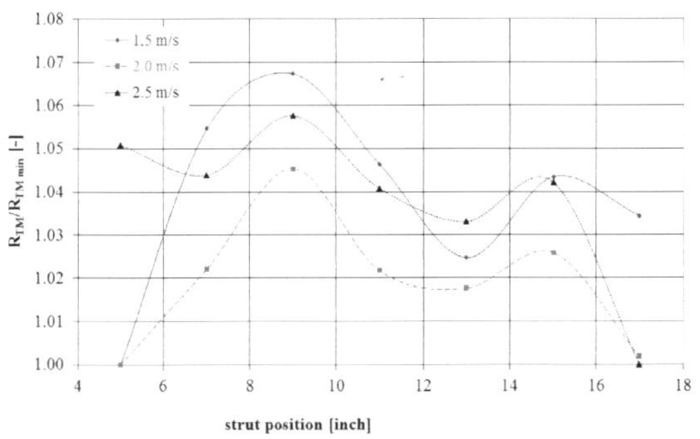

Figure 3 Influence of strut position

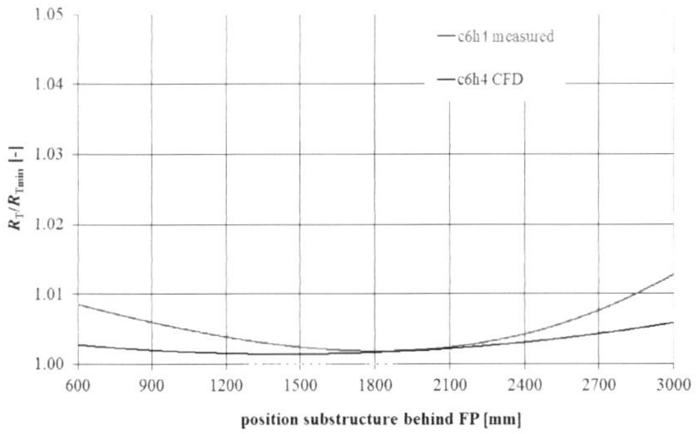

Figure 4 Comparison of CFD and model tests

The comparison between the RANSE calculations and the model tests shows a good compliance. For the master configuration a parasite drag of about 1% could be indicated for shallow water condition (0.7 m), see Figure 4.

The results of some measurements for constant span on deep water condition (2.0 m) are summarized in Figure 5:

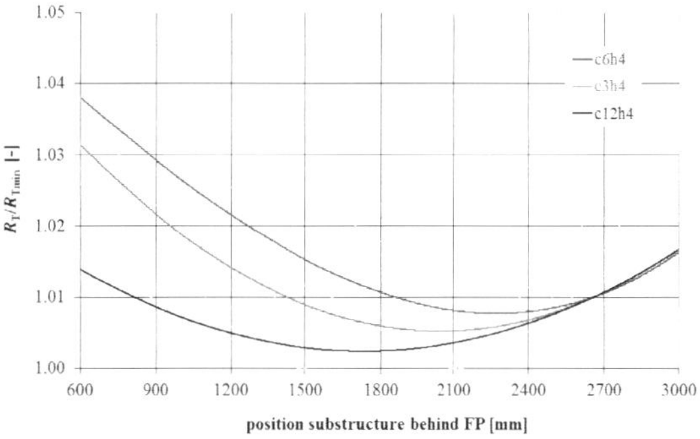

Figure 5 Measurements for constant span, deep water

The dependency of the parasite drag of size and position of the substructures becomes obvious.

The overall perception was a minimum parasite drag for high chord length and medium span for a position at about midship. This means a lower resistance when using one big substructure covering all functions needed for the ship. The deeper the body is submerged the higher is the effect of parasite drag by reducing influence of wave pattern.

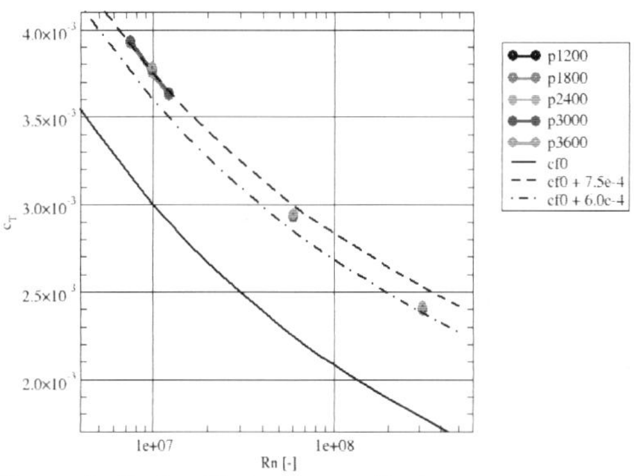

Figure 6 Scale effects on parasite drag

Figure 6 describes the results of RANSE calculations showing a $C_T = C_{F0} + 7.5E\text{-}4$ for a scale = 10 and a $C_T = C_{F0} + 6.0E\text{-}4$ for a scale = 1. The results for a scale = 3 are intermediate. That means the parasite drag depends on the scale and is some lower for full scale conditions.

With the help of the PIV measurements the different flow behavior behind the substructures near the surface of the base hull could be clarified: the more aft the position the higher is the wake. That indicates an interaction of boundary layers between base and substructure. It came as a surprise that the total wake behind the body is not influenced by the parasite drag effects.

Figure 7 shows an example for the inflow on the substructure: The more aft the position the thicker is the boundary layer of the base and due to this the lower is the Reynolds number of the substructure. This results in increased resistances for the aft positions as indicated in Figures 4 and 5.

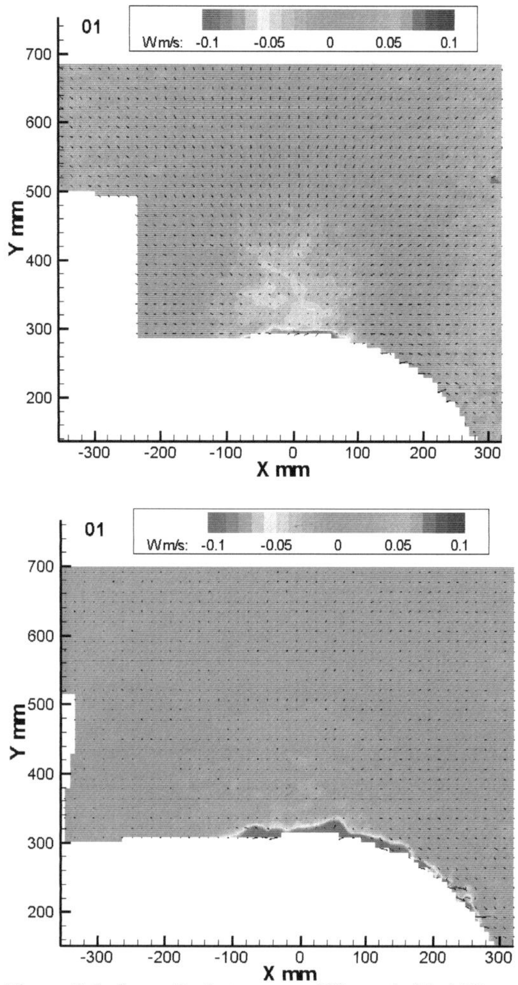

Figure 7 Inflow of substructure, 600 mm behind FP (top), 1800 mm behind FP (bottom)

5. CONCLUSIONS

Substructure interference resistance in the height of up to 5% could be proved. The interference resistance decreases by abt. 20% at full scale conditions.

RANSE simulations proofed to calculate the parasite drag in same quality and same quantity as in the towing tank tests. For the calculations a pre-processing procedure for quick setup variation was developed. The flow behaviour could be investigated successfully with PIV measurements showing the same results.

The influence of test setups for evaluation and calibration of towing tank tests on submerged bodies could be verified.

6. REFERENCES

1 S. Hörner, "Fluid-dynamic drag: practical information on aerodynamic drag and hydrodynamic resistance"; Midland Park, NJ 195

2 G. Holbach, Zum Entwurf von Flossensystemen für Swath-Schiffe, A-Dissertation, TU Berlin, 1993

3 H. Oberembt, Potentialtheoretische Strömungsuntersuchungen an einem Doppelmodell der "Meteor", Institut für Schiffbau der Universität Hamburg, Bericht 267, Hamburg, 1970

4 S.Schuster/ H.Schwanecke, Über den Einfluss der Wasserfläche auf die Auftriebsverteilung von Tragflügeln, Schiffstechnik (4)

8. AUTHORS BIOGRAPHY

Thomas Nietzschmann holds the current position of head of test facilities at Potsdam model basin. He is responsible for test preparation and measuring equipment. His previous experience includes initial design as well as form optimisation and resistance and propulsion testing for all types of ships.

THE EFFECT OF AFT-CONTROL-SURFACE DEFLECTION AT ANGLES OF INCIDENCE FOR THE BB2 GENERIC SUBMARINE

D.A. Pook, RMIT University/Defence Science and Technology Group, Australia
G.J. Seil, Advanced VTOL Technologies/Defence Science and Technology Group, Australia
M.R. Renilson, Australian Maritime College, University of Tasmania, Australia
D. Ranmuthugala, Australian Maritime College/Defence Science and Technology Group, Australia
M. Nguyen, Advanced VTOL Technologies/Defence Science and Technology Group, Australia

SUMMARY

The effectiveness of the Aft-Control-Surfaces (ACS) of a submarine is influenced by the angle of incidence of the boat and the resulting crossflow. This paper quantifies the rudder and sternplane control derivatives for a generic submarine geometry having an X-plane ACS arrangement using Computational Fluid Dynamics. For the boat running straight-ahead, the rudder and sternplane control force derivatives are nearly identical and are linear functions of ACS deflection. However, with the boat at an angle of drift or attack, the control derivatives for small control deflections are found to increase non-linearly with incidence. Further deflection reduces the control force to similar levels obtained when running straight-ahead. The paper discusses the control forces at different angles of boat incidence and ACS deflection, in order to improve the effective ACS deflection range and to limit undesired forces orthogonal to the control force.

NOMENCLATURE

a_1	Surface lift curve slope
ACS	Aft-Control-Surface
AoA	Angle-of-attack
AoD	Angle-of-drift
c	Chord length
K	Roll moment
L	Boat length
M	Pitching moment
N	Yaw moment
ρ	Density of fluid
s	ACS span, no mirror
S	Area
SA	Straight-ahead
U	Free-stream velocity magnitude
x	Body-fixed coordinate
X	Surge force
y	Body-fixed coordinate
Y	Sway force
y^+	Non-dimensional wall distance
z	Body-fixed coordinate
Z	Heave force

Subscripts

A	ACS area including outer and inner sections, and hull blank
δ	Control angle deflection
i	Inner (headbox) section of ACS
o	Outer (moveable) section of ACS
R	Rudder
S	Sternplane
0	Load with no control deflection

Superscripts

'	Non-dimensional value

1. INTRODUCTION

When a submarine is at an angle of incidence to the oncoming flow, the resulting flow over each Aft-Control-Surface (ACS) is influenced by crossflow velocity, boundary layer changes, and off-body vortical structures such as the sail wake. The local velocity seen by each ACS is altered, changing the incidence seen at each ACS, and possibly altering the lift curve slope and control force derivatives. Control characteristics of a boat at incidence are expected to differ compared to running straight-ahead.

Control surface arrangement, such as cruciform or X-plane, influences the control characteristics of the submarine. The upper rudder of a cruciform arrangement operates in the wake of the sail. Surfaces in an X-plane arrangement likely operate in the open-separation from a hull at incidence, which produces a thickened boundary layer. The cross flow velocities experienced by the cruciform and X-stern would also be expected to differ. Therefore, it is important to quantify changes in terms of the underlying hydrodynamics when designing control surfaces and their operational control strategy. This paper uses Computational Fluid Dynamics (CFD) to characterise the control derivatives of the ACS on the BB2 generic submarine geometry.

The BB2 geometry (see Figure 1) was developed by DST in conjunction with international partners, to facilitate hydrodynamic studies on an unclassified geometry that resembles a diesel-electric submarine [1]. The BB2 features ACS in an X-plane arrangement, with the outer 60% of the ACS moveable (see Figure 2).

Bettle [2] performed a detailed study of the BB2 ACS sizing and found they provide both horizontal and vertical stability. Boat stability derivatives were calculated using empirical methods and CFD, with a difference of approximately 20%. Manoeuvring simulations were performed but control force derivatives were not published.

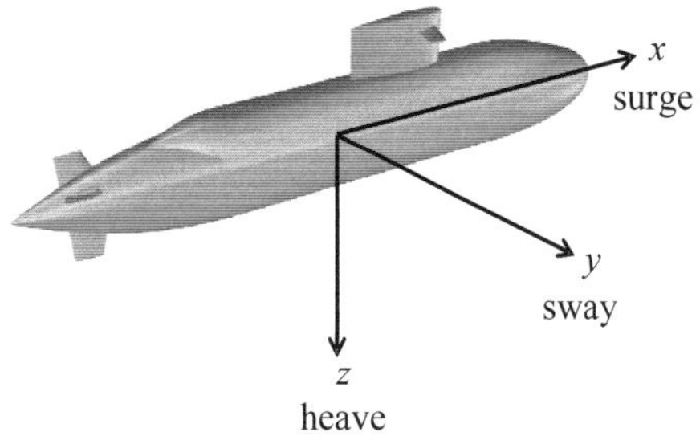

Figure 1: BB2 generic submarine geometry and coordinate system.

Traditional semi-empirical methods for estimating derivatives of cruciform control surfaces can be adapted for the X-plane arrangement [3]. However, the X-plane is known to have different control characteristics. It can offer more control force for a given control surface area [4], but the linear control deflection range is limited for control deflections orthogonal to the boat incidence, e.g. rudder when the boat is at an angle-of-attack [5]. Despite reduced linear control deflection ranges, the X-plane can be designed to handle plane control jams [6]. However, for a given desired control force, the total force that must be generated by an X-plane is greater than a cruciform arrangement, and therefore increases drag [7].

While each ACS in an X-plane may be geometrically similar, their control derivatives may differ substantially (Crossland et al. [6] states around 8% for their geometry) due to the sail, hull-casing, and hydroplanes altering the incoming flow. Although the movable sections of the BB2 ACS are a significant distance from the hull (see Figure 2), at typical Reynolds numbers achievable in wind tunnels or towing tanks, the hull boundary layer can extend to the moveable sections affecting the ACS control derivatives.

To allow comparison with experimental data at model scales, CFD simulations are conducted to calculate the rudder and sternplane control force derivatives for the BB2 ACS in a deeply submerged condition without a propeller. Simulations are conducted at a Reynolds number of 5.2×10^6 with the boat straight-ahead (SA) and at angles-of-drift (AoD) or angles-of-attack (AoA) up to $10°$.

2. ACS GEOMETRY AND DEFLECTION

A body fixed axis system with the origin placed at the amidships, as shown in Figure 1, was used in the simulations and analysis. For a positive drift angle, the freestream crossflow approaches from the port side. For a positive AoA, the freestream crossflow approaches from the lower side of the hull.

All areas are reported as non-dimensional quantities relative to L^2. Forces or moments, e.g. X or M, are made non-dimensional as follows,

$$X' = \frac{X}{0.5\rho U^2 L^2}, \quad M' = \frac{M}{0.5\rho U^2 L^3}.$$

The BB2 ACS geometry is defined in Figure 2. A small gap, approximately 1.5% of the local chord length, is left between the fixed headbox (inner) and the moveable outer section to allow for CFD meshing. The outer section pivots about an axis normal to the gap, located at approximately 26% of the chord from the leading edge at the gap. The ACS aspect ratio is four, which includes the area projection through the hull and the ACS on the other side of the hull, which together forms a wing for semi-empirical methods to calculate the control surface derivatives.

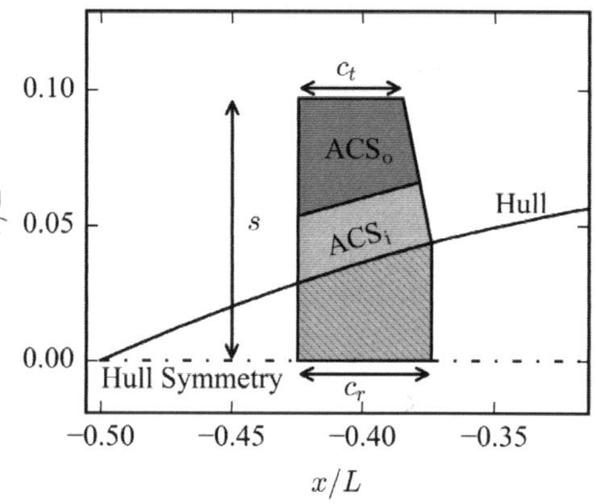

Figure 2: ACS geometry. Red is moveable outer section labelled ACS$_o$. Grey is inner headbox labelled ACS$_i$. Hatched is hull area used in derivative estimation. Mirrored ACS used in derivate calculations not shown. $c_t / L = 0.040$, $c_r / L = 0.060$, $s / L = 0.097$.

The tail cone half-angle in the vicinity of the ACS is approximately $16°$. For $\mathrm{Re} = 5.2 \times 10^6$, the boundary layer displacement thickness in the vicinity of the ACS, obtained from an axisymmetric CFD simulation of the bare hull, is approximately $3.1 \times 10^{-3} L$.

All four ACS deflect through the same angle magnitude. Positive rudder deflection of each ACS is defined to produce a positive (starboard) sway force and a moment turning to port. Thus, the ACS trailing edge of each ACS moves towards port. Positive sternplane deflection of each ACS is defined to produce a negative (upward) heave force and a bow down moment. Thus, the trailing edge of each ACS moves downwards.

3. CFD SIMULATION MODEL

ANSYS Fluent (version 16.2) is used to solve the CFD Reynolds Averaged Navier–Stokes (RANS) equations, with closure provided by the SST $k\omega$ turbulence model. Second order differencing is used for pressure terms, third order MUSCL for momentum terms, and gradients are evaluated with the Green–Gauss cell based method. Pressure-velocity coupling uses the SIMPLEC method with default relaxation values.

The boat is placed centrally in a rectangular domain of length $11.7L$, width $5.8L$, and height $5.8L$. The far field boundary conditions are specified as either a velocity-inlet or outflow, depending on the direction of the freestream required to produce the desired boat incidence. All boat surfaces are non-slip, smooth walls.

The grid is divided into three regions joined by non-conformal interfaces. The first grid region is multiblock structured and includes the entire submarine except for the ACS region (see Figure 3*a*). This grid section contains 37.8 million hexahedral cells.

The second grid region (see Figure 3*b*), consisting of 48 million cells, contains the four ACS and is embedded in the first grid region. It is a hybrid grid of hexahedral, pyramid, triangular prism, and tetrahedral cells that resolve the boundary layer with a y^+ on all surfaces of $O(1)$. The majority of the hull and ACS boundary layer are in the hexahedral grid regions. The third grid section is a multiblock structured grid that surrounds the first grid region containing the submarine. This region allows the use of structured grid out to the inlet and outflow boundaries with a reduced cell count. The total cell count for all three grid sections is 106 million.

Seil et al. [8] provides a detailed grid refinement study of the BB2 hull mesh used in this study. The grid used here produced loads within 2% of the estimated grid independent value, which is sufficiently accurate for a hull–ACS interaction study. A refinement of the second grid region (ACS) was performed in Fluent, for the straight-ahead case with $15°$ of rudder. The ACS sway loads changed less than 0.3%, indicating the grid resolution is adequate.

4. CONTROL DERIVATIVE ESTIMATE

The method documented in Lyons and Bisgood [9] for predicting the control derivative of a small aspect ratio fin is reported to have an accuracy of 10%. This method predicts the fin lift curve slope as a function of trailing edge angle, control gap chord location, after body shape, taper ratio, and aspect ratio. The aspect ratio includes a fin's streamwise extents (see Figure 2). For BB2, the predicted ACS lift curve slope (a_1) is 3.05 rad^{-1} based on the mirrored ACS area including hull projection. This value is converted to an X-plane sway force control derivative ($Y_{\delta R}$) due to rudder by,

$$Y_{\delta R} = a_1 \times \left(\frac{S_o}{S_A}\right)(2)(\cos 45°)\left(\frac{S_A}{L^2}\right),$$

where the first bracketed term accounts for only the outer ACS sections moving, the factor 2 accounts for two sets of mirrored ACS in an X-plane, the factor $\cos 45°$ accounts for the X-plane lift force inclination to the sway force direction, and the last bracketed term changes the reference area to the boat reference area

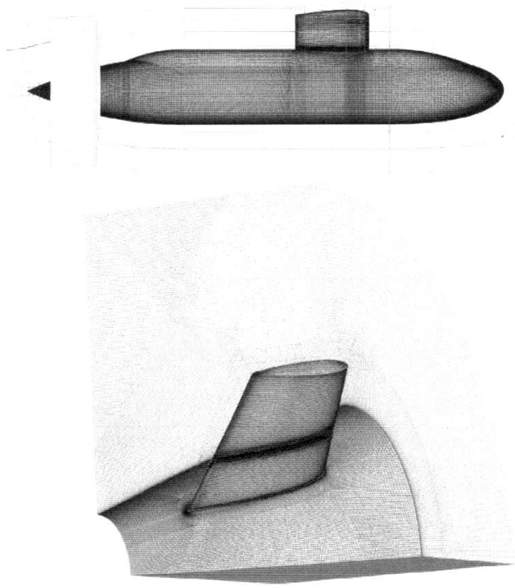
(a)
(*b*)

Figure 3: CFD Grid. (*a*) Structured grid in the first region. White region near stern indicates the second region containing the ACS. (b) Structured and unstructured grid around an ACS in the second region.

The predicted non-dimensional rudder control derivative is 1.39×10^{-2} rad^{-1}. The heave force due to sternplane $(Z_{\delta S})$ is predicted to have the same value, with a sign change to fit the boat coordinate system. Adding the boundary layer displacement thickness to the hull radius in the Lyons and Bisgood [9] method reduces the predicted rudder and sternplane control derivatives to 1.35×10^{-2} rad^{-1}.

The yaw moment due to rudder ($N_{\delta R}$) and pitching moment due to sternplane ($M_{\delta S}$) are estimated by assuming the sway or heave force are created at the quarter chord of the outer ACS sections. The quarter chord is estimated by assuming the outer section is a straight tapered wing (ignoring the non-parallel inner and outer chord). The estimated quarter chord location is - $x = -0.3817L$, giving the estimated value of $N_{\delta R}$ and $M_{\delta S}$ as -9.26×10^{-5} rad^{-1}.

5. CFD SIMULATION RESULTS

Whole boat loads with the ACS deflected (e.g. X') are reported relative to the whole boat with no deflection of the ACS (e.g. X_0'),

$$\Delta X' = X' - X_0',$$

for a given boat incidence. Loads are calculated for rudder and sternplane angles of $\pm 2°$, $\pm 5°$, $\pm 10°$, $\pm 15°$, $\pm 20°$, $\pm 25°$. Results for negative rudder angles when the boat is running straight-ahead (SA), or at an AoA, are not calculated. The value reported for these cases is the value for positive rudder, with sign changes made based on the symmetry in the $x-z$ plane.

The ACS rudder control force derivative ($Y_{\delta R}$) and sternplane control force derivative ($Z_{\delta S}$), computed using the CFD simulations are reported in Table 1. Control force derivatives are calculated using a control deflection of $\pm 2°$.

For straight-ahead, the BB2 rudder and sternplane control derivatives differ by less than 1%, indicating the upstream hull and appendages cause minimal inflow asymmetry.

For all boat angles of incidence, both the rudder and sternplane control force derivatives calculated with a control deflection of $\pm 2°$, increase in comparison to the straight-ahead condition. Although the increase can be up to 26%, it is typically closer to 10% and is not a linear function of the boat's angle of incidence.

The changes in boat loads due to rudder, over the computed deflection range, are shown in Figure 4 for the values of AoA and AoD investigated. For straight-ahead, the sway force is a linear function of rudder deflection up to $\pm 20°$. A linear regression fit between $\pm 20°$ has a coefficient of determination (r-squared value) of 0.9998.

No significant heave force is created for a rudder deflection less than $\pm 25°$, where the predicted heave force is approximately 10% of the sway force. The sway force is also smaller at this rudder angle. Examination of surface flow topology (not shown) shows significant reversed (separated) flow on the lower, moveable ACS section with its trailing edge moving downwards (starboard for positive rudder). This indicates the ACS has stalled.

Care should be taken interpreting loads when an ACS has stalled. The stall of an ACS can be inferred from the whole boat loads, by the control force (sway for rudder, heave for sternplane) significantly deviating from a linear trend with control deflection. Loads predicted at angles higher than this, are likely to be correct qualitatively, but may be in error quantitatively. This requires further investigation with physical experiments.

From figure 4, it can be seen that the maximum positive rudder angle before stall is smaller for larger drift angles; where the effective rudder stall is interpreted as a significant change in the sway force trend with rudder deflection. For an AoD of $5°$ the effective stall occurs for a rudder angle between $+15°$ and $+20°$, while for an AoD of $10°$ the effective stall occurs between $+10°$ and $+15°$, indicating the positive rudder deflection angle for ffective stall is reduced by approximately the drift angle. Rudder deflection for angles greater than the effective stall produces either smaller or approximately constant sway force.

Table 1: ACS control force derivative with boat incidence, calculated with $\pm 2°$ control deflection. % Diff is relative to the straight-ahead derivative.

Boat Incidence	$Y_{\delta R} \times 10^2$ rad^{-1}	% Diff	$Z_{\delta S} \times 10^2$ rad^{-1}	% Diff
Straight-Ahead	1.24		-1.23	
AoD 5°	1.30	5.1	-1.35	10
AoD 10°	1.50	21	-1.55	26
AoA −5°	1.42	15	-1.35	9.5
AoA +5°	1.37	11	-1.34	9.1
AoA −10°	1.44	16	-1.41	15
AoA +10°	1.34	7.7	-1.35	9.5

The predicted heave force due to rudder also becomes significant when the rudder effectively stalls, reaching 40% of the sway force with $+25°$ rudder. For angles less than stall, the heave force generated with positive rudder deflection is minimal. For negative rudder angles, no effective rudder stall is seen in the computed rudder deflection range. However, the heave force magnitude increases with negative rudder deflection. For an AoD of $10°$, the heave force due to negative rudder is 10% to 15% of the sway force.

With the boat at an AoA, Figure 4 shows the maximum rudder angle before significant change in the sway force trend (designated an effective stall) is reduced by approximately the magnitude of the AoA. Unlike for drift, the effective stall occurs for either positive or negative rudder deflections. Also different, is the ability to create greater sway force beyond the effective stall,

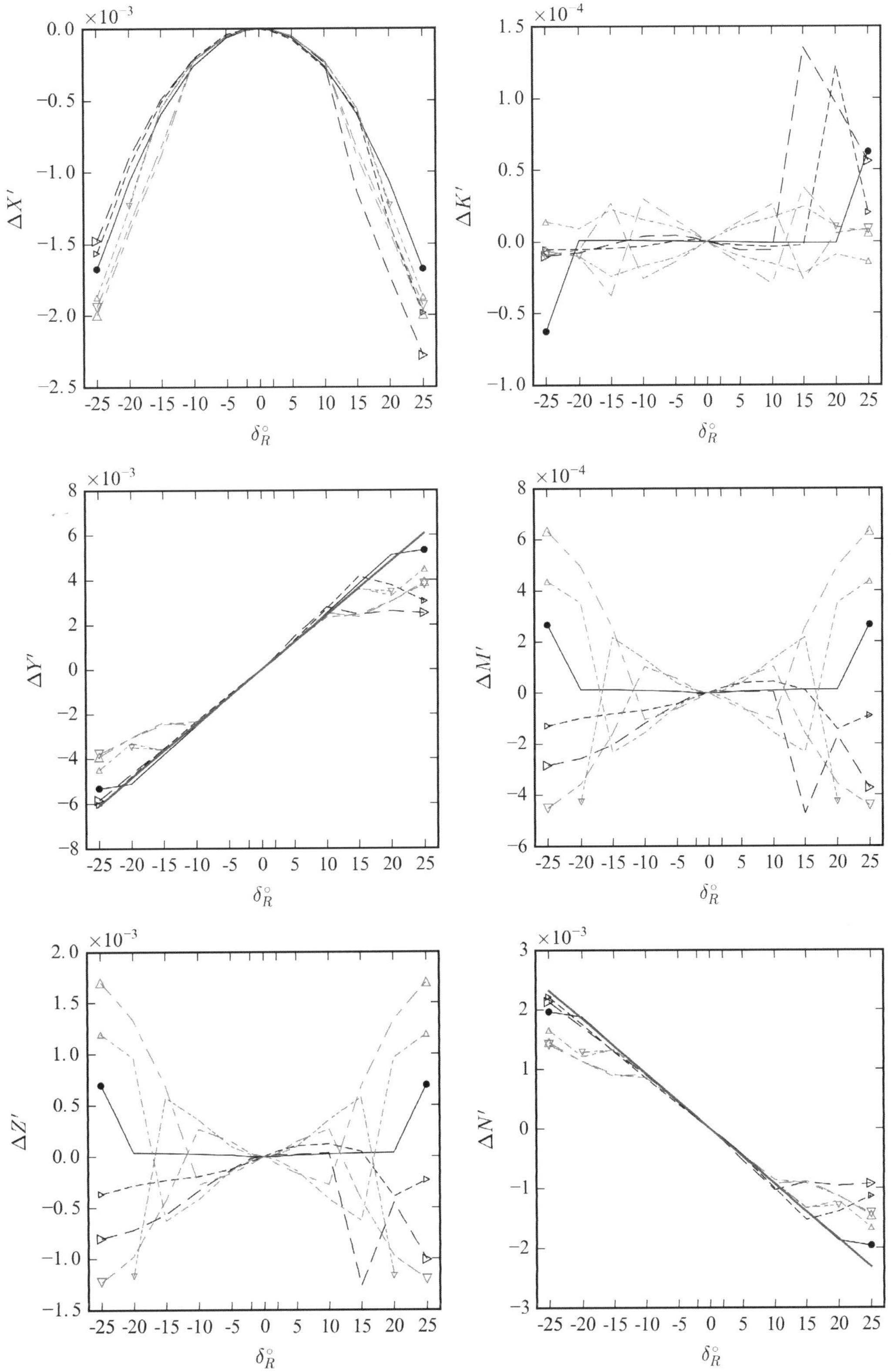

Figure 4: Change in whole boat loads with rudder deflection. Straight-ahead, •—•; AoD 5°, ▹--▹; AoD 10°, ▷— ▷; AoA −5°, ▿---▿; AoA −10°, ▽— ▽; AoA +5°, ▵---▵; AoA +10°, △— △.
For $\Delta Y'$ and $\Delta N'$; straight blue-line with no marker is from estimated derivative.

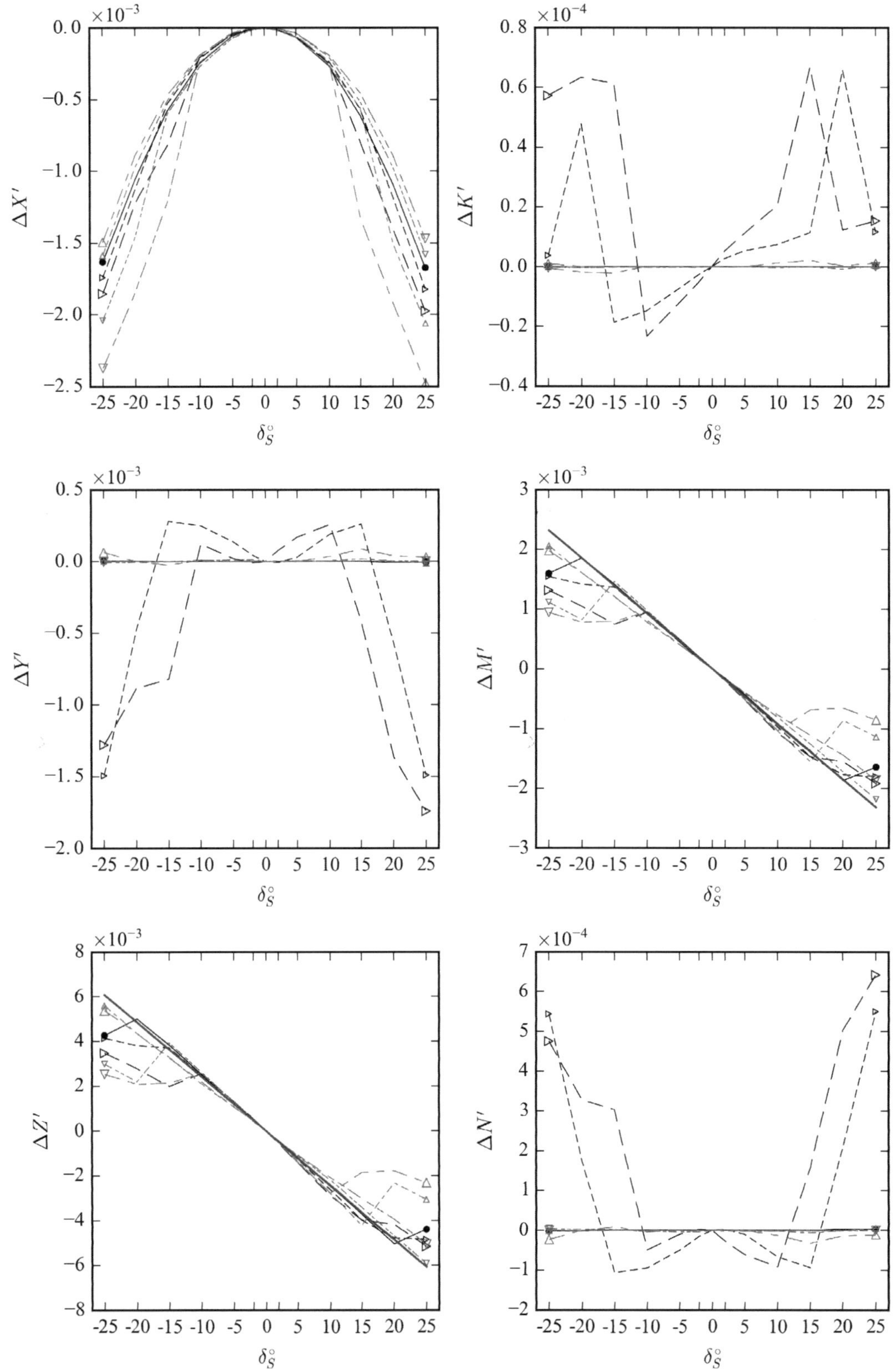

Figure 5: Change in whole boat loads with sternplane deflection. Straight-ahead, •—•; AoD 5°; AoD 10°; AoA −5°; AoA −10°; AoA +5°; AoA +10°.
For $\Delta Z'$ and $\Delta M'$; straight blue-line with no marker is from estimated derivative.

but the sway-force/rudder-deflection gradient is smaller than before the effective stall. However, the heave force created becomes significant and its direction changes compared to pre-stall. The heave force magnitude beyond the effective stall is approximately 30% to 40% of the sway force.

Figure 4 also details secondary force and moment changes due to rudder. The surge force due to rudder deflection when the boat is at incidence can vary by approximately 30%, with the largest differences occurring for rudder deflections beyond 10°. Rudder deflection with the boat at incidence creates a roll moment. For 10° rudder with ±10° AoA, the roll moment is a few percent of the yaw moment. With the boat at an AoA, the roll moment is a near linear function of rudder deflection until the rudder effectively stalls. Pitching moment created by rudder is also an approximate linear function of rudder deflection until the rudder effectively stalls.

The whole boat loads changes due to sternplane deflection are shown in Figure 5. For the straight-ahead condition, the heave force is a linear function of sternplane deflection between ±20°. A linear regression fit has a coefficient of determination (r-squared value) of 0.9997. Effective stall of the sternplane occurs between 20° and 25° sternplane deflection, but unlike for rudder, no orthogonal force (sway) is predicted. This is because an ACS on each side of the hull stalls at the same angle due to symmetry. Examination of surface topology (not shown) reveals both upper ACS stall at a lower angle with positive sternplane deflection, while both lower ACS stall at a lower angle for negative sternplane deflection.

With AoA, the sternplane deflection angle for effective stall, indicated by a significant change in the heave force trend, is lower in one direction. For positive AoA, the positive sternplane angle for effective stall is reduced compared to straight-ahead. For negative AoA, the negative sternplane angle is limited. As for drift, the reduction in the effective stall angle is approximately equal to the boat AoA. Due to symmetry, no significant sway force develops. At larger sternplane angles, a small sway force is generated when the boat is at an AoA. This is due to the separated flow on the ACS, and the sensitivity of load prediction to small variations (possibly due to small numerical errors) in separation location.

For drift, the effective stall angle of the sternplane is lower for positive and negative sternplane deflections. For ±5° AoD, the effective stall angle of the sternplane is between ±15° and ±20°, and for ±10° AoD the sternplane effective stall occurs between ±10° and ±15° sternplane deflection. Significantly greater heave force can be achieved for sternplane angles greater than the effective stall angles, however significant out of plane sway force and yaw moment is also generated.

From Figure 5, it is seen that the roll moment due to sternplane deflection is only created with the boat at an AoD. For angles less than sternplane effective stall, the roll moment is 3% of the pitching moment. A change in yaw moment is also created with drift, its magnitude is a few percent of the pitching moment before sternplane effective stall occurs.

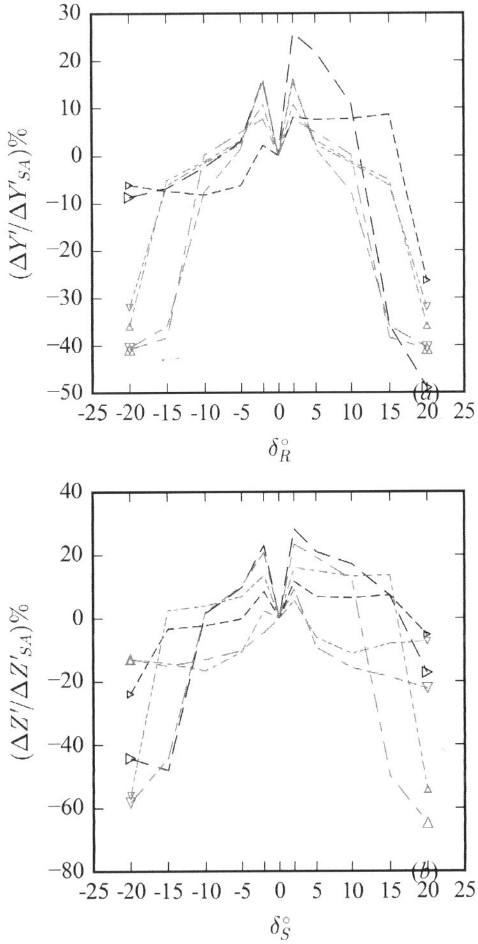

Figure 6: Relative control force for the boat at incidence relative to the control force for straight-ahead: (*a*) sway force due to rudder; (*b*) heave force due to sternplane. Lines: Straight-ahead, •—•; Drift 5°,▷--▷; Drift 10°,▷— ▷; AoA −5°,▽---▽; AoA −10°,▽— ▽; AoA +5°,△--△; AoA +10°,△— △.

Figure 6 compares the relative sway force due to rudder deflection, or heave force due to sternplane deflection, for the boat at incidence relative to the straight-ahead (SA) condition, i.e. $\Delta Y'/\Delta Y'_{SA}$ and $\Delta Z'/\Delta Z'_{SA}$. This figure demonstrates the non-linearity of the sway or heave force when the boat is at an angle of incidence. At these orientations, neither the sway due to rudder nor heave due to sternplane, is a linear function of ACS deflection. At angles lower than the effective stall of the rudder or sternplane, the relative load compared to the straight-ahead condition is generally within $\pm 20\%$. An ACS deflection of $\pm 2°$ generally produces a greater sway (rudder) or heave (sternplane) force when the boat is at incidence. Hence the larger control derivatives reported in Table 1. However, for greater deflection angles, the sway (rudder) or heave (sternplane) force relative to the straight-ahead condition will begin to decrease, and often fall below the force created with the boat moving straight-ahead.

6. DISCUSSION

The straight-ahead condition rudder and sternplane control derivatives estimated using the method of Lyons and Bisgood [9] agree favourably with the CFD-RANS computation, differing by about 9% when allowing for the hull boundary thickness. The BB2's clean hull design and forward hydroplanes placed high on the sail do not produce significant flow asymmetry at the ACS when running straight-ahead.

When the boat is at incidence, the X-plane control force derivatives increase, up to 26% for the sternplane with the boat at an AoD of $10°$, when evaluated with the load changes for $\pm 2°$ control deflection. Greater control derivatives with the boat at an angle of incidence can also be found in the experimental data of Mackay [5] for both an X-plane and cruciform arrangement. However, the sway due to rudder or heave due to sternplane, when the boat is at incidence, are non-linear functions of ACS deflection. The sway (rudder) or heave (sternplane) can reduce to levels close to straight-ahead, or even less, with rudder deflections beyond $\pm 2°$, as shown in Figure 6.

Effective stall of the X-plane differs for the case of a control deflection in the direction of boat incidence (rudder for AoD, sternplane for AoA), and a control deflection orthogonal to boat incidence (sternplane for AoD, rudder for AoA). Figure 8 demonstrates examples of how the incidence seen by each ACS differs with boat incidence and ACS deflection.

For example, Figure 7(*a*) shows that for negative AoA and positive rudder, the effect of the crossflow is to increase the incidence of the upper starboard and lower port ACS, while lowering the incidence of the remaining ACS. Stall will occur at a lower rudder deflection angle on the upper starboard ACS. With further rudder deflection, the next ACS to stall is the lower port ACS. The order can be explained by hull blockage that accelerates flow on the windward side, and slows it on the leeward side. CFD results (not shown) indicate the stall of the lower port ACS is less dramatic than the upper starboard, possibly due to circulation created around the hull when the upper starboard ACS stalls.

Figure 7(*b*) shows the boat with drift and positive rudder. The effective incidence of each ACS is increased relative to the straight-ahead condition. Considering the hull blockage, stall would be expected on the windward ACS at a lower rudder angle than the leeward ACS. However, the CFD simulations show the two lower ACS stall at the lowest rudder deflection angle. Both stall at the same deflection angle, within the resolution of the current study ($5°$), although it is likely that the windward ACS will stall at a slightly lower rudder angle than the leeward ACS. The lower ACS stall at a lower rudder angle due to the anti-clockwise circulation around the hull, created by the sail trailing vortex [10]. For an AoD of $10°$, the CFD simulations show that a further increase in rudder angle will stall the upper windward ACS, and even further rudder deflection will stall the upper leeward ACS.

Figure 7(*c*), the boat at a positive AoA and with a positive sternplane angle, shows that all ACS experience a relative increase in the AoA due to the crossflow. As the sail creates no circulation around the hull, both the windward (lower) ACS are found to stall at the same sternplane angle. They stall at a smaller sternplane angle than the two leeward ACS due to the hull blockage accelerating the crossflow on the windward (lower) side.

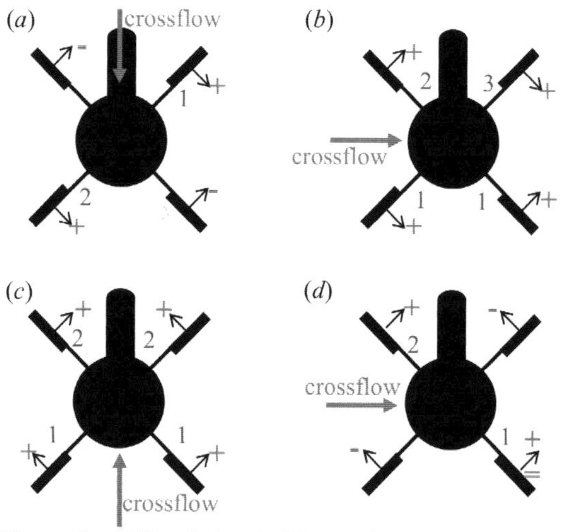

Figure 7: ACS relative incidence decrease or increase due to crossflow velocity and ACS deflection. View is from the stern. Small black arrows indicate change in ACS control force with deflection. + or - symbol indicates increased or decreased ACS incidence due to crossflow. Numbers indicate order of ACS stall with control deflection, to the accuracy of computed deflection range and resolution. (*a*) negative AoA with positive rudder; (*b*) AoD with positive rudder; (*c*) positive AoA with positive sternplane; (*d*) AoD with positive sternplane.

For positive sternplane and drift, Figure 7(*d*), the leeward (starboard) lower ACS stalls before the other ACS due to the increased local incidence from the sail induced circulation. The windward (port) upper ACS then stalls with a further increase in sternplane deflection. This indicates the incidence change induced by the sail circulation is greater than the incidence induced by the hull blockage, as they accelerate the flow in opposite directions relative to the upper windward (port) ACS.

The reduced effective control range for control deflection orthogonal to the boat incidence is because two ACS are always increasing their relative incidence with control deflection. Increased control force (sway for rudder, heave for sternplane) beyond this range is possible with the two remaining, unstalled ACS. However, as the examples in Figures 7(*a*) and 7(*d*) demonstrate, the force in the plane of incidence will always increase. The two unstalled ACS produce a force in the direction of the boat incidence (or opposite to) that is no longer cancelled by the stalled ACS. Hence, the large undesired force created orthogonal to the control force.

The effective control range could be increased to some degree, by limiting the deflection of the two ACS that stall first. This would maximise the control force that could be created. This would require, the designer to have detailed knowledge of each ACS stall angle for a given boat state (e.g. in a turn). However, the undesired orthogonal force would still be present at control angles greater than the effective stall, but its magnitude would be minimised.

7. CONCLUSION

The ACS control forces for the BB2 generic submarine geometry have been computed using CFD-RANS simulations. Due to the clean hull and sail mounted forward hydroplanes, when the boat is running in the straight-ahead condition, the rudder and sternplane control derivatives are nearly identical. The agreement between CFD and an estimate based on the method of Lyons and Bisgood [9] was approximately 9%.

With the boat at incidence, all control force derivatives for the X-plane are increased for small control deflections. However, the control forces generated are non-linear functions of ACS deflection. For larger ACS deflections, the control forces reduce to levels similar to that generated when the boat is running in a straight-ahead condition.

The effective control deflection range for a control deflection orthogonal to the boat incidence is limited, but could be maximised by preventing individual ACS from deflecting past their stall. This requires the designers to know the respective angles of stall at various angles of incidence.

8. REFERENCES

1. OVERPELT, B., NIENHUIS, B., ANDERSON, B., 'Free Running Manoeuvring Model Tests On a Modern Generic SSK Class Submarine (BB2)', *Pacific 2015,* Sydney, Australia, 6–8 October 2015.

2. BETTLE, M.C., 'Validating Design Methods for Sizing Submarine Tailfins', *Warship 2014*, Bath, UK, 18–19 June 2014.

3. SEIL, G., 'The Effect of the Bridge Fin Tip Vortex on the Aft Control Surface Sway Derivative for a Submarine at an Angle of Drift', *DST Technical Report*, Under Review

4. BURCHER, R., RYDILL, L.J., 'Concepts in Submarine Design (Cambridge Ocean Technology Series)', *Cambridge University Press*, 1994.

5. MACKAY, M., 'Wind Tunnel Experiments with a Submarine Afterbody Model', *DRDC Atlantic TM 2002-194,* March 2003.

6. CROSSLAND, P., KIMBER, N., THOMPSON, N., 'Understanding the Manoeuvring Performance of an X- Plane Submarine in Deep Water and Near the Free Surface', *Pacific 2012*, Sydney, Australia, 31 Jan–2 Feb 2012.

7. RENILSON, M., 'Submarine Hydrodynamics', *Springer Briefs in Applied Sciences and Technology*, 2015

8. SEIL, G., NGUYEN, M-C., POOK, D.A., LEONG, Z.Q., 'CFD Study of the Hydrodynamics of the Evolved DST-Group Generic Submarine (BB2)', *DST Technical Report,* Under Review.

9. LYONS, D.J., BISGOOD, P.L., 'An Analysis of the Lift Slope of Aerofoils of Small Aspect Ratio, Including Fins, with Design Charts for Aerofoils and Control Surfaces', *ARC R&M No 2308*, 1950.

10. LEONG, Z., PICCOLIN, S., DESJUZEUR, M., RANMUTHUGALA, D., RENILSON, M., 'Evaluation of the Out-of-Plane loads on a submarine undergoing a steady turn', 20th *Australasian Fluid Mechanics Conference*, Perth, Australia, 5–8 Dec, 2016.

9. AUTHORS BIOGRAPHY

David Pook is a Research Fellow at RMIT University, currently working within the DST Group in the area of naval hydrodynamics. His previous work has focused on boundary layer transition and wind tunnel flow quality.

Gregory Seil is a Contractor to DST Group working in the area of naval hydrodynamics. He has previously worked at Rolls-Royce (Sweden & UK) on marine propulsion.

Martin Renilson is an Adjunct Professor in Hydrodynamics at the University of Tasmania. His previous experience includes as Technical Manager and Deputy General Manager for the Maritime Platforms and Equipment Group at QinetiQ

Dev Ranmuthugala is a Professor in Maritime Engineering at the University of Tasmania. He is currently on secondment to the DST Group as the Head, Hydroacoustics. His previous experience includes Head, Maritime Engineering at the Australian Maritime College.

Mai-Chi Nguyen is a Contractor to DST Group working in the area of naval hydrodynamics. Previously, she developed fibre optic sensors at the Institute of Photonics and Advanced Sensing, University of Adelaide.

ON MATHEMATICAL MODELLING OF MAIN BALLAST TANK SYSTEM OF A SUB-SURFACE PLATFORM

Z Riaz, M A Waqas, W Ahmed, H Khalid, A R Hashmi, Centres of Excellence in Science & Applied Technologies (CESAT), Pakistan
P G Wrobel, University College London, UK

SUMMARY

Purpose of Main Ballast Tank (MBT) system of a subsurface platform is twofold; to bring the platform to surface from any operating depth (Blowing System) and secondly to submerge the platform below water surface (Venting System). Designing of MBT system of a subsurface platform involves complex engineering and rigorous mathematical modelling. Special care is required to precisely and accurately model the requisite sub-systems as it is unambiguously understood that MBTs system is one of the most important systems of any underwater platform. MBT system can be divided into the modelling of two subsystems, namely, blowing and venting as discussed above. In blowing, High Pressure (HP) air through an air nozzle is directly blown into the ballast tanks in order to push water out of the tanks via flood ports. Whereas in case of venting system, water through flood ports enter inside the ballast tanks to occupy the space created by air that is allowed to escape via vent valves. Paper comprises of a detailed mathematical modelling of both these sub-systems in terms of their various components in time domain encompassing the transient and steady states. Main design criteria for blowing system considered in this paper was the resultant differential pressure across the outer hull of MBT. This differential pressure thus calculated through designing is then utilized as design pressure for the structure of the outer hull of MBT. Whereas Design factor for venting system is time taken by platform to submerge below surface. In the end, a case study is carried out using the developed mathematical model for both blowing and venting system.

NOMENCLATURE

ρ_{sw}	Density of sea water (kg m^{-3})
g	Acceleration due to gravity (m s^{-2})
ρ_{air}	Density of air (kg m^{-3})
P_{hp}	Pressure in HP air bottle (bar)
A_{fb}	Plan area of platform (m^2)
D_{fh}	Diameter of flood hole (m)
t_0	Valve opening time (s)
$\dot{z}$	Rise rate of platform (m s^{-1})
DDD	Deep Diving Depth (m)

1. INTRODUCTION

There are two types of sub-surface platforms; manned military submarines and unmanned commercial submersibles. Unlike submersibles, where surfacing and diving is mostly carried out either by combination of forward and aft hydroplanes or by using gimbaled thrusters (i.e. thrust vectoring control system), military submarines are equipped with several Main Ballast Tanks distributed along the length. MBT system is used to change the overall buoyancy of a sub-surface platform from surface condition to submerged condition [1]. It serves two purposes; *Venting*; while on surface, MBT system kills reserve buoyancy of platform by taking in additional weight (Ballast) to just submerge it under sea surface where platform settles to its new equilibrium state. Second purpose is *Blowing*; by pushing out the same additional water ballast, it helps in unbalancing the state of equilibrium between weight and buoyancy (Figure 1). Thus, increasing the upward thrust to bring the platform to surface from any depth. In any condition whether surfaced and submerged, an equilibrium state known as neutrally buoyant state of a sub-surface platform is when the upward buoyant force acting on the centre of mass of the body is equal to the weight of the fluid displaced by the platform.

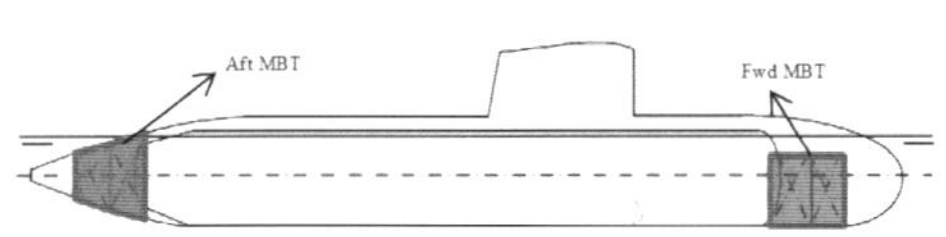

Figure 1(a) Schematic of Aft and Fwd MBTs while empty (Submarine at surface)

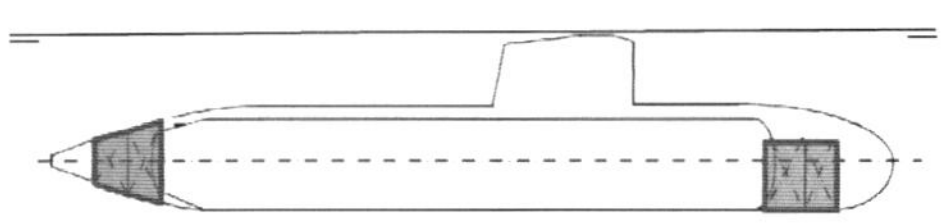

Figure 1(b) Schematic of Aft and Fwd MBTs while filled with water ballast (Submarine submerged below sea surface)

Understanding the significance of this system in transiting the platform from submerged to surfaced state, precise control over the platform's displaced volume is achieved by critically and carefully designed MBT system. Following are the major components of the system: (as shown in Figure 2).

a. Ballast tanks
b. Air vents/venting valves
c. Air nozzles/blowing valves
d. Flood ports/opening

The principle operation of this system is as follows

a. Ballasting/Air Venting System: For submergence, vents at the top of MBT are opened; air inside the tank will be pushed to escape under the sea water pressure, entering through the flood ports located in the bottom of each tank.

b. De-ballasting/Air Blowing System: For surfacing, same vents are closed. High pressure air stored in HP air bottles onboard is injected through the air inlet valves located on the top of MBT. This pressurized air blows the water out of tanks through the flood ports. It is obvious that pressure of air entering the ballast tank must be kept greater than the pressure of water surrounds the flood port.

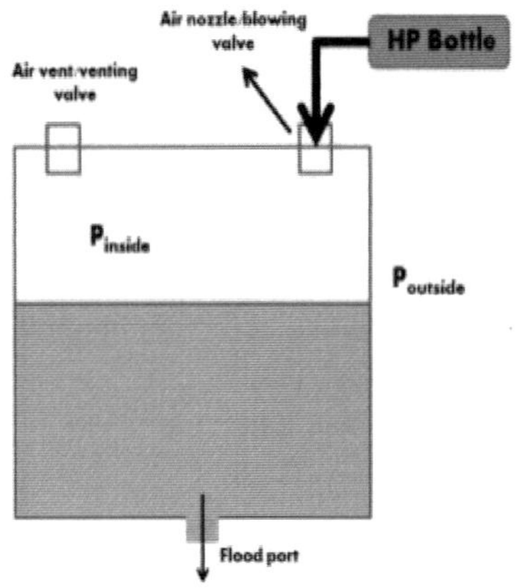

Figure 2- Component wise representation of MBT system

With this introduction to what MBT system is, motive behind this study is summarized in this paragraph. Authors of this paper are well aware of the fact that MBT system designing is not entirely a new subject and considerable amount of work has already been undertaken by many others. Science of MBT systems has been amply matured and there is no reason of why this fundamental work has to be re-done and presented in this paper. However, with a literature review, it was revealed that due confidentially issues, open literature on MBT system is very limited rather unavailable. This is due to very significance and strategic importance of the final application where this system can potentially be used. Information available is either very preliminary in nature or is a presentation of final results with very little description of important steps involved. Therefore, a need was felt to study and thus present a step wise approach in formulating a mathematical model of complete MBT system. Study may not be regarded as completely novel albeit a review study with a focus to consolidate the jigsaws of information on the title. Nonetheless, at times, equations have to be developed from abinitio where there was no previous reference available. Work carried out by [2] and [3] has been a major reference throughout this work.

2. DESIGN CONSIDERATIONS/CRITERIA

In case of combat submarines, time of ascending and diving is one of driving factors for designing MBT system as these platforms may need to rapidly dive and rise as per the tactical requirements. However, considering the operation of commercial submersibles, it is obvious that these platforms do not require to surface or dive quickly. Notwithstanding above, even for submersibles equipped with MBTs, time to surface must be of a sensible value. Therefore, time taken by platform to surface from deep diving depth (DDD) is one of the design criteria to be kept in mind while designing MBT system.

The most fundamental design driver is to limit the pressure exerted by HP air on soft structure of ballast tanks during the blowing process. The stresses developed due to pressurized air entering in the tank are to be kept lower than the structural strength of exterior wall of tanks. This implies that air blowing system/de-ballasting is to be designed prior to design of air venting/ballasting system.

3. AIR BLOWING SYSTEM

De-ballasting or air blowing system involves replacement of water from ballast tank by air so as to create additional buoyancy. This can be done in several ways, but the most common is the use of pressurized air.

Important criteria for the design of the system are:

a. The flow rate of HP air entering in the tank through air inlet valve especially when it has to counter the accidental flooding. This is called as emergency blowing. As discussed above due to structural limits of exterior walls of tanks as well as the time to surface, air flow rate is also required to be critically chosen even for normal blowing system. In this study, accidental flooding is not catered, thus air flow rate is to be chosen so as to keep in consideration the time required for surfacing only from deep operating depth.

b. The air pressure in the main ballast tank during blowing because of the limitations of structural strength.

There are two types of pressure loading inside ballast tanks during the process of blowing which are to be considered for blowing system designing; the initial transient pressure that is resulted due to trapping of HP air for a very small instant of time. This occurs with the sudden discharge of HP air trying to push sluggish water out of the tank [3]. Second phenomenon is steady state pressure loading that is when air flow rate becomes equal to rate of water going out of the tank through flood ports. For the sake of mathematical modelling, following assumptions were made to simplify the calculations:

a. Air is a perfect gas.
b. Area of flood ports is considerably smaller than the tank cross sectional area. Thus, the velocity and acceleration of the water in the tank can be neglected.
c. Pressure (P_1) from the reducer will remain constant throughout the blowing operation.
d. Frictional losses in cases of piping are not catered.
e. Density of sea water is assumed to be constant for all sea conditions.
f. In reality, platform will rise with a trim angle due to the difference in capacities of tanks as well as the relative distances of tanks' centre of mass from the centre of floatation of platform. However, to keep the mathematics simple, it is assumed that platform is equipped with single ballast tank located on the centre of mass of the platform. Therefore, the platform will rise with level trim.
g. Tank has single flood port located in its centre (directly below the centre of mass of platform) and a single air nozzle for HP air.

3.1 INPUT DATA

Following input data is used for the development of equation for air blowing system:

Table 1 - Input data for blowing system

Description	*Symbol*
Pressure in HP air bottle	P_{hp}
Pressure after reducer	P_1
Area of air inlet valve	A_{inlet}
Capacity of all tanks	V_T
Density of Sea water	ρ_{sw}
Specific heat ratio of air	ξ
Length of platform	L
Diameter of platform	D
Plan area platform	A_{fb}
Submerged mass of platform	m_{sb}
Acceleration due to gravity	g
Surface mass of submarine	m_{su}
Density of air at atm. pressure	ρ_{air}
Universal gas constant	R
Height of Main Ballast Tank	h_a
Discharge coefficient of flood port	C_D
Area of flood port opening	A_{fl}
Diameter of flood port	D_{fh}

As already discussed, in case of commercial submersibles, depth control is primarily achieved by dynamic forces i.e. with the help of thrusters. However, for submarines, both static and dynamic mechanisms are employed for depth variation. In normal operations, hydroplanes (bow planes and stern planes) control the depth variation of platform from maximum operating depth (as high as 300 m) to shallow depths (typically 10 m) while the MBTs are required to be blown to surface the platform from this shallow depth. However, in case of an emergency, main ballast tanks can be blown at the deep operating depth to quickly bring the submarines on surface. Therefore, in designing, main ballast tanks are designed to surface the platform from the deep operating depth. Therefore, the highest pressure required to push the water out of the tank must corresponds to the hydrostatic pressure at the deep operating depth.

3.2 TRANSIENT PRESSURE

With the activation of emergency blow valve, high pressure air enters directly into the ballast tank. Air flow rate starts increasing gradually with time of air valve opening, thus reaching to maximum flow rate in a very small instant of time. However, until a steady state condition is arrived, an overpressure exist in the tank due to the sluggish response of water in the flood port, not accelerating at the same rate as that of air. This overpressure is dependent mainly on the rate of increase of the air flow and the relative size of the mass to be accelerated through the flood port. Transient pressure is relatively higher than steady state pressure, therefore, in designing, this pressure has to be made lower than the steady state pressure and outer hull of platform is designed against the value of steady state pressure.

3.3 DERIVATION OF TRANSIENT PRESSURE LOADING

Considering again the simplest main ballast tank shown in Figure 2. Let's denote the different terms related to transient pressure calculations: t_0 is valve opening time, Q_w is flow rate of water, Q_a is a flow rate of air, Q_{a_0} is the maximum air flow rate, τ is time delay i.e. is the lag time of water flow rate to reach to the maximum air flow rate value.

Assuming exponential function for the increase in the flow rate:

$$Q_a = Q_{a_0} e^{(t-t_0)} \qquad (1)$$

and

$$Q_w = Q_{a_0} e^{(t-\tau t_0)} \qquad (2)$$

Exact value of τ can only be determined by carrying out experimental simulation. The flow rate of air and water can be represented graphically as given in Figure 3.

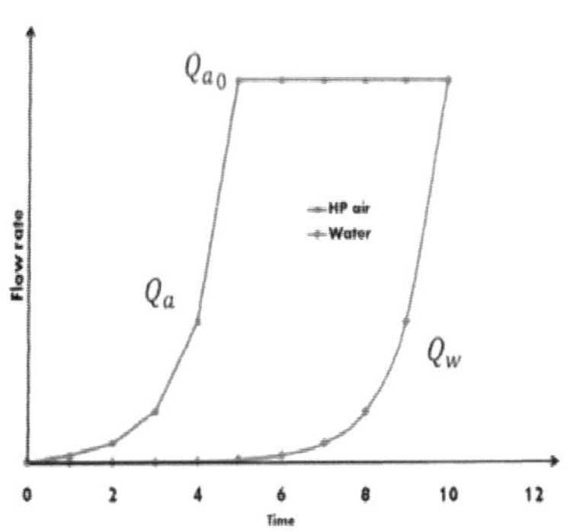

Figure 3 - Mass flow rate of air and water

Area under this curve shows the energy stored in the system. Now, if the valve opening time t_0 reduces by an amount ζ i.e.($t_0 - \zeta$), the red curve (air flow rate curve) will shift to left hand side, thus increasing the amount of energy stored inside main ballast tank. This is shown in Figure 4. This behavior shows that valve opening time of HP air plays an important role in building up the transient pressure inside main ballast tank. Therefore, during designing an appropriate value of time for valve opening should be chosen in order to minimize the exceeding pressure crossing the structural limits of the outer hull of the platform.

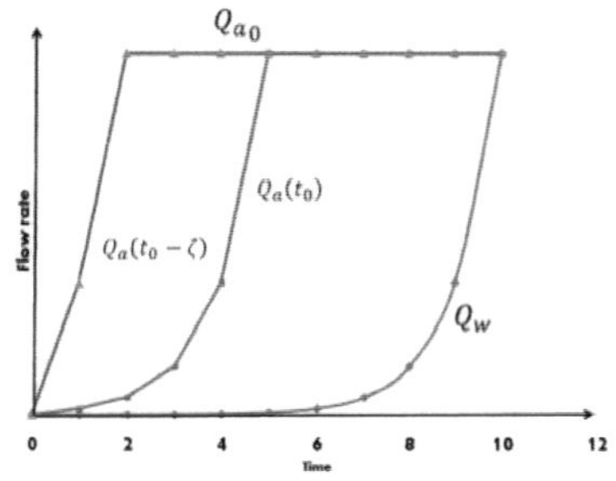

Figure 4 - Mass flow rate of air and water with different valve opening times

The pressure inside MBT can be simplified as;

$$P_B = P_{atm} + \rho_{sw} Z_T + \rho_{sw} h_a + \delta p \tag{3}$$

where

P_B = Pressure inside main ballast tank,
Z_T = depth of water to MBT top,
h_a = vertical distance from tank top to water level and
δp = the pressure head to generate the flow of water

If flow velocity of water is represented by v_w, then the dynamic pressure head, δp is given by;

$$\delta p = \frac{\rho_{sw} (v_w)^2}{2} \tag{4}$$

where

$$v_w = \frac{Q_w}{C_D A_{fl}} \tag{5}$$

ρ_{sw} = density of sea water,
C_D = discharge coefficient of flood port and
A_{fl} = flood hole cross-sectional area

By inserting (5) into (4), we get the dynamic head due to the given flow rate of water

$$\delta p = \frac{\rho_{sw} \left(\frac{Q_w}{C_D A_{fl}} \right)^2}{2} \tag{6}$$

The rate with which water was supposed to flood out of the main ballast tank is the rate of air entering inside tank, Q_a and the corresponding dynamic head due to this flow rate would be

$$\delta p_a = \frac{\rho_{sw} \left(\frac{Q_a}{C_D A_{fl}} \right)^2}{2} \tag{7}$$

Subscript a in (7) is used for air. This can be interpreted as the energy which should have been taken out of the system (MBT). However, the actual amount of energy coming out of the system is lesser than this due to the sluggish response of water to get accelerated. This actual energy is given by (6). Therefore, the instantaneous total energy stored in the system (MBT) can be given by:

$$\delta p^*(t) = \delta p_a(t) - \delta p(t) \tag{8}$$

and the energy stored for the duration of τt_0 will be;

$$\delta p^*(t) = \sum_{t=0}^{t=\tau t_0} \left\{ \frac{1}{2} \rho_{sw} \left(\frac{Q_a(t)}{C_D A_{fl}} \right)^2 - \frac{1}{2} \rho_{sw} \left(\frac{Q_w(t)}{C_D A_{fl}} \right)^2 \right\} \tag{9}$$

Inserting the above in (3), the transient pressure generated inside main ballast tank can be worked as;

$$P_B = P_{atm} + \rho_{sw} Z_T + \rho_{sw} h_a + \sum_{t=0}^{t=\tau t_0} \left\{ \frac{1}{2} \rho_{sw} \left(\frac{Q_a(t)}{C_D A_{fl}} \right)^2 - \frac{1}{2} \rho_{sw} \left(\frac{Q_w(t)}{C_D A_{fl}} \right)^2 \right\} \tag{10}$$

Hence, the differential pressure across the tank structure is merely the difference between internal and external pressure, or

$$\Delta P = P_B - P_{ambient} \tag{11}$$

3.4 STEADY STATE PRESSURE

With the admission of HP air in ballast tank, water is forced out of the tank, through the flood ports. The internal pressure in tank would be same as that of outside sea pressure if water leaves the tank freely with no obstruction. However, in reality, substantial restriction is provided by the flood ports, primarily due to its size to the flow of water passing through it. Thus, a pressure differential develops across the flood port that is augmented by the continuous influx of HP air through air inlet valve. This pressure is known as steady state pressure and thus to be calculated as a design value for structural analysis of outer hull of main ballast tank.

3.5 DERIVATION OF STEADY STATE PRESSURE LOADING

For derivation, following steps were followed:

3.5(a) Calculation of Instantaneous Volume of Air in Ballast Tank, $V_A(T)$

Compressed air from HP air bottle P_{hp} flows via a reducer at pressure, P_1 through blowing valve into the tank. Water inside main ballast tanks are at ambient pressure, P_2, which is given by:

$$P_2 = P_{atm} + \rho_{sw} Z_T + \rho_{sw} h_a \qquad (12)$$

where flow of compressible fluid is governed by equation;

$$d\frac{P_1}{P_2} = \left[1 + \left\{\frac{\xi - 1}{2}\right\} M^2\right]^{\frac{\xi}{\xi - 1}} \qquad (13)$$

in which M is the Mach number and by definition is

$$M = \frac{v_a}{C}$$

and C is the speed of sound which is 330 m/s. v_a is the velocity of air entering in the tank, which can be calculated by using above relations. This is generally supersonic speed. Then the volume flow rate of air can be calculated using;

$$Q_a = A_{inlet} v_a \qquad (14)$$

where A_{inlet} is the area of inlet air inlet valve. This depends on the diameter of nozzle which is kept small for tanks. Then, by applying Boyle's law, volume of air in each tank at ambient pressure P_2 can be calculated as;

$$V_{P_2} = \frac{P_1}{P_2} V_{P_1}$$

Under steady state conditions, volume of water displaced out of tanks should be same as that of the incoming air. Thus, for volume of water going out, $V_{w_{out}}$

$$V_{w_{out}} = V_{P_2} \qquad (15)$$

3.5(b) Calculation of instantaneous rise rate of Platform, $\dot{z}(t)$

Assuming no external force being applied by either hydroplanes or thrusters (in case of submersibles), vertical rise of a fully submerged platform will then depend on the gravitational force, buoyant and other forces that appear because of motion. Let $F_B(t)$ is instantaneous value of buoyant force, Δ is the gravitational force, $F_D(t)$ is instantaneous drag force, f_a is force that appear because of acceleration, m_a is the added mass, m_{su} is surface mass and m_{sb} is the submerged mass of platform, then the force equation acting on a platform is given as;

$$m_{sb} g - V_{w_{out}}(t) \rho_{sw} g = F_B(t) - F_D(t) \qquad (16)$$

where

$$F_D(t) = \frac{1}{2} C_d A_{fb} \rho_{sw} \dot{z}^2(t)$$

where $\dot{z}$ is the vertical velocity or the rise velocity and C_d is the drag coefficient. A conservative approach is to use the value of drag coefficient of a submerged flat plate of equivalent platform (L/D), perpendicular to the flow. The plot for this is shown in Figure 5.

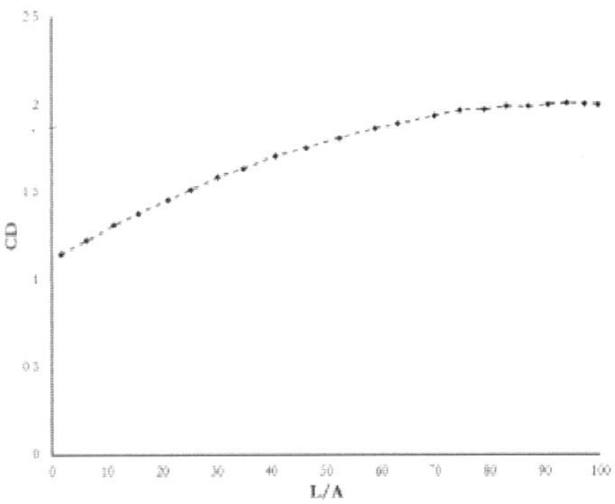

Figure 5 - Drag coefficient versus L/D for submerged flat plate

By re-arranging, an expression for the instantaneous velocity of the platform is:

$$\dot{z}(t) = \sqrt{\frac{2\{B_f - (m_{sb} - V_{w_{out}}(t)\rho_{sw})g\}}{C_d A_{fb} \rho_{sw}}} \qquad (17)$$

where B_f is buoyancy force.

This value of depth thus obtained will be subtracted from the present value of depth to get the new depth and iteration goes on. The new value of depth will be used to calculate the new ambient pressure, P_2 corresponding air velocity v_a, and flow rate Q_a. Consequently, the volume of air, V_a will increase to give rise the new vertical velocity, $\dot{z}(t)$.

3.5(c) Calculation of instantaneous steady state pressure of air inside the ballast tank

Starting with Bernoulli equation given above (3),

$$P_B = P_{atm} + \rho_{sw} Z_T + \rho_{sw} h_a + \delta p$$

Since,

$$\delta p_w = \frac{1}{2}\rho_{sw}\left(\frac{Q_w}{C_D A_{fl}}\right)^2$$

and for steady state condition,

$$Q_w = Q_a$$

Thus,

$$\delta p_a = \delta p_w = \delta p$$

and

$$Q_a = {dV_a}/{dt}$$

Re-arranging above relations to get

$$P_B = P_{atm} + \rho_{sw} Z_T + \rho_{sw} h_a + \frac{1}{2}\rho_{sw}\left(\frac{{dV_a}/{dt}}{C_D A_{fl}}\right)^2 \quad (18)$$

The last term in (18) represents the pressure drop across the flood hole. It can be derived from ideal gas equation, the rate of change of volume of air in the tank is given by:

$$\frac{dV_a}{dt} = \frac{V_a \rho_{sw}(-\dot{z})}{\xi P_B + \rho_{sw} h_a}$$

where h_a is the vertical distance from tank top to water level.

Substituting above into (18) and re-arranging gives a cubic equation in pressure [3];

$$\begin{aligned}
&\xi^2 P_B^3(t) \\
&+ [(2-\xi)\xi\rho_{sw} h_a(t) \\
&- \xi^2 P_{amb}(t)] P_B^2(t) \\
&+ [(1-2\xi)\rho_{sw} h_a(t) \\
&- 2\xi P_{amb}(t)]\rho_{sw} h_a(t) P_B(t) \qquad (19)\\
&- [\rho_{sw} h_a(t) + P_{amb}(t)]\rho_{sw}^2 h_a^2(t) \\
&= \frac{\rho_{sw}^3}{2g}\left(\frac{V_a(t)}{C_D A_{fl}}\right)^2 \dot{z}^2(t)
\end{aligned}$$

The pressure inside tank, P_B can then be calculated by solving the above cubic equation. Hence, the differential pressure across the tank structure is merely the difference between internal and external pressure, or

$$\Delta P(t) = P_B(t) - P_{amb}(t)$$

4. AIR VENTING SYSTEM

In contrast to blowing system, venting system is relatively less critical from design point of view. This is fundamentally due to the absence of high pressure air in this case, which causes the high stresses on the outer hull structure, thereby becoming the major design criteria in blowing system. Therefore, the only design criteria in this case will be the time taken by platform to submerge. In venting system, an air vent of suitable size on the top of the MBT is to be opened to release the air inside MBT and thus allowing the water to enter into the MBT through flood ports in the bottom of tank. Air flow is generated because of the difference in pressure internal and external of the tank. The pressure inside MBT will be the pressure of the sea water entering flood holes, while the pressure outside MBT will be the ambient pressure head at the level of vent valve. Thus, the differential pressure in this case would roughly be the difference in the pressure heads on the top and bottom of the platform.

4.1 ASSUMPTIONS

a. Air is a perfect gas.
b. Area of flood ports is considerably smaller than the tank cross sectional area. Thus, the velocity and acceleration of the water in the tank can be neglected.
c. Frictional losses in cases of piping are not catered.
d. Density of sea water is assumed to be constant for all sea conditions.
e. Starboard and port halves of both forward and aft MBTs have one vent. This means that there are two vent valves in total to blow out all MBTs.
f. To keep the calculations simple, it is assumed that there is a single vent located at the top of the tank.
g. Flow rate of incoming water is equal to that of air going out of MBT.
h. Vent valves open into the air exactly at the top deck and the flood ports and vent valves are located exactly on platform fore and aft centerline. This means that the distance between the vent valves and flood ports will be equal to the diameter of platform at that point.
j. Again, as mentioned in the case of blowing system that platform will dive with a trim angle due to the difference in capacities of tanks as well as the relative distances of tank's center of mass from the center of floatation of platform. However, to keep the mathematics simple, it is assumed that submergence will occur with level trim.

4.2 CALCULATION OF TIME OF SUBMERGENCE

Similar to air blowing system, a step wise iterative approach is taken to calculate the time to dive i.e. in this time, the platform will get submerged just below the water surface. Following inputs will be required:

Table 2 - Input data for venting system

Description	***Symbol***
Area of vent valve	A_v
Tonnes per centimeter immersion of the platform	TPC

Following derivations were carried out:

4.3 INSTANTANEOUS MASS AND VOLUME FLOW RATE OF AIR THROUGH VENT

Again, referring to (13), where P_1 and P_2 in this case will be pressure inside MBT and the pressure in the vent valve respectively. Re-arranging this governing equation to calculate the Mach number M and air flow velocity, $v_{a_{out}}$,

$$M = \sqrt{\frac{2}{\xi - 1}\left(\frac{P_1}{P_2}\right)^{\frac{\xi-1}{\xi}} - 1} \qquad (20)$$

The air velocity in this case is subsonic. Volume flow rate through the vent valve will involve the diameter of vent valve. The value for Mach number and air velocity are dependent on the location of vent valve and flood port.

$$Q_{a_{out}}(t) = v_{a_{out}} A_v \qquad (21)$$

This is the volume of air coming out of the MBT in one second. This volume of air going out, $V_{a_{out}}$, would be equal to the volume of water coming in, $V_{w_{in}}$. Thus,

$$V_{w_{in}}(t) = V_{a_{out}}(t)$$

For mass flow rate, density of air at given condition will be multiplied by this flow rate.

$$\dot{m}_{a_{out}}(t) = Q_{a_{out}}(t)\frac{P_2}{RT} \qquad (22)$$

4.4 INSTANTANEOUS SUBMERGENCE RATE OF PLATFORM

The instantaneous submergence rate, $\zeta(t)$ in this case is calculated using below equation:

$$\dot{\zeta}(t) = \frac{V_{w_{in}}(t)\rho_{sw}}{TPC} \qquad (23)$$

where TPC is Tonnes per centimeter Immersion of the platform.

5. CASE STUDY – DARPA SUBOFF

In order to check the behavior of above mathematics, a case study on DARPA suboff is carried out. A scaled-up model of DARPA suboff is developed as shown below:

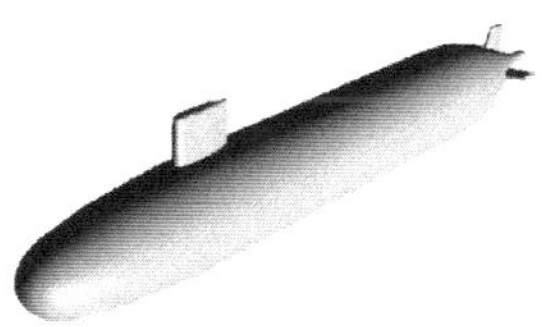

Figure 6 – Scaled up model of DARPA suboff

Following input data as mentioned above were taken:

Table 3 - Input taken for the case study - DARPA suboff

Parameters	***Values***
P_{hp}	250 bars
P_1	50 bars
D_{air}	4 mm
V_T	7.23 m^3
ρ_{sw}	1026 Kg/m^3
ξ	1.4
L	18 m
D	2.3 m
A_{fb}	36.81 m^2
m_{sb}	60000 Kg
g	9.81 m/s^2
m_{su}	58000 Kg
ρ_{air}	1.293 Kg/m^3
R	287 JKg/K
h_a	2.3 m
C_D	0.8
A_{fl}	0.0324 m^2
D_{fh}	203 mm
DDD	150 m

Two main ballast tanks are considered in the DARPA suboff model, each divided into two sub-tanks starboard and port. Detail of each MBT is given below:

Table 4 - Details of Main Ballast Tanks

Description	***Capacities***
Forward Tank stbd	2.021 m^3
Forward Tank port	2.021 m^3
Aft Tank stbd	1.592 m^3
Aft Tank port	1.592 m^3
MBT gross volume	7.23 m^3

Arrangement of MBTs is shown in Figure 7 below:

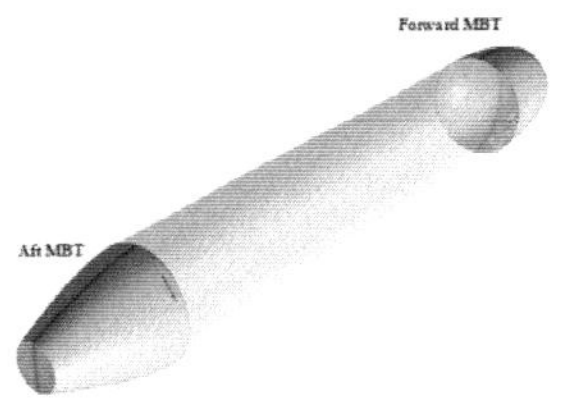

Figure 7 – Arrangement of forward and aft MBTs

5.1 RESULTS & DISCUSSION

5.1 (a) Transient Pressure – Blowing system

Different values of transient time are chosen to estimate the expected transient pressures inside MBT. Calculations were carried out for three different transient time values, 10, 8 and 6 sec. Values of differential pressure, ΔP for different combinations of flood hole diameters, D_{fh} and discharge coefficients, C_D, are shown in following tables.

Table 5 - ΔP for transient time 10 s, C_D=0.6, D_{air}= 0.4mm

D_{fh}(inches)	τ (sec)	t_0 (sec)	ΔP(bar)
	6.67	1.5	8.28
3	5	2	7.81
	1.12	9	0.98
	6.67	1.5	2.62
4	5	2	2.47
	1.12	9	0.31
	6.67	1.5	1.07
5	5	2	1.01
	1.12	9	0.13
	6.67	1.5	0.092
8	5	2	0.087
	1.12	9	0.011

Table 6 - ΔP for transient time 8 s, C_D=0.8, D_{air}= 0.4mm

D_{fh}(inches)	τ (sec)	t_0 (sec)	ΔP(bar)
	5.33	1.5	6.32
3	4	2	5.85
	1.33	6	1.96
	5.33	1.5	2
4	4	2	1.85
	1.33	6	0.62
	5.33	1.5	0.82
5	4	2	0.76
	1.33	6	0.25
	5.33	1.5	0.07
8	4	2	0.065
	1.33	6	0.022

Table 7 - ΔP for transient time 6 s, C_D=0.8, D_{air}= 0.4mm

D_{fh}(inches)	τ (sec)	t_0 (sec)	ΔP(bar)
	4	1.5	4.36
3	3	2	3.9
	1.2	6	0.98
	4	1.5	1.38
4	3	2	1.23
	1.2	6	0.31
	4	1.5	0.57
5	3	2	0.51
	1.2	6	0.13
	4	1.5	0.049
8	3	2	0.043
	1.2	6	0.011

Above results show that both valve opening time and flood hole area plays a major role in influencing the value of differential pressure, ΔP . It is observed that, for flood port size of 127 mm and above, no matter how long the transient time is, as long as the valve opening time, t_0, approaches the value of transient time, τt_0, the differential pressure, ΔP , will not exceed the value of 0.15 bar. Secondly, increase in the size of flood port will decrease the excessive pressure inside MBT. Although, as it was mentioned earlier that it is analytically difficult to estimate the actual transient time, however, this could be said with certain amount of confidence that transient state will last for fraction of a second. Assuming a worst transient state value of 10 sec, which seems to be reasonably higher value in this case, the valve opening time may be kept to value of 9s to keep the differential pressure as low as 0.13. Alternatively, if the discharge coefficient, C_D, is increased from 0.6 to 0.8, the value of differential pressure decreases drastically and the limitation on valve opening time can be reduced.

5.1 (b) Steady state Pressure – Blowing System

Using the methodology mentioned in 3.5(a), the volume flow rate of HP air (50 bar) entering through air nozzle in each tank is 0.00573 m^3/s. Thus, by Boyles law, the volume occupied by air in each tank at ambient pressure in a second is 0.01781 m^3. Under steady state conditions, same volume of water will be flowing out of the flood port of each tank per second.

Using (17), for the first iteration, the instantaneous velocity of rise $\dot{z}(t)$ is 0.174 m/s. This value of depth will be subtracted from the present value of depth to get the new depth. This new value of depth will be used to calculate the new ambient pressure, P_2, new air velocity v_a, new flow rate Q_a, and consequently the volume of air, V_a will increase to give rise the new vertical velocity. The time domain plot of this volume of air and depth change is shown in Figure 8 and Figure 9 respectively.

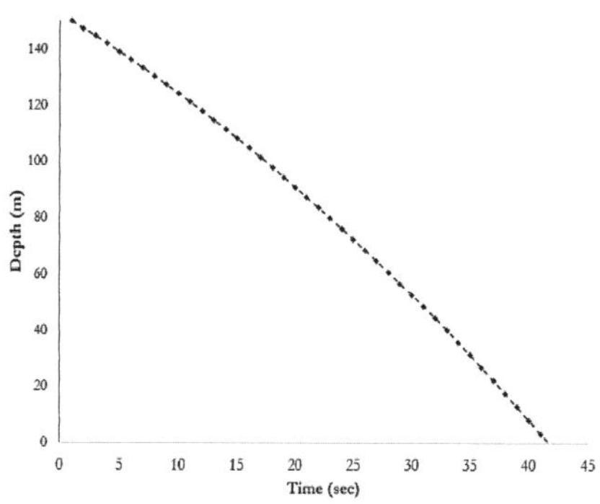

Figure 8 - Instantaneous depth change of platform during blowing operation

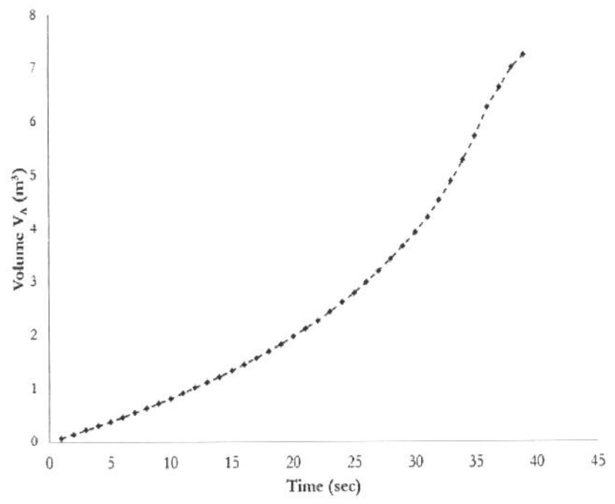

Figure 9 - Instantaneous volume of air change

Time taken by the platform to surface is 47 *sec*. While the time taken to empty different tanks are given in Table 8.

Table 8 - Time taken to blow tanks of platform

Tank	*Time(s)*
Forward tank stbd	42
Forward tank port	42
Aft tank stbd	39
Aft tank port	39

Solving cubic equation given in (19), instantaneous values of steady state pressure generated inside tanks were calculated. Maximum values of differential pressure across the outer hull for different tanks are given in Table 9.

Table 9 – Maximum differential pressure across the outer hull of different tanks

Tank	*ΔP(bar)*
Forward tank stbd	0.41
Forward tank port	0.41
Aft tank stbd	0.345
Aft tank port	0.345

Above results shows that the differential pressure across the soft structure of all the tanks is less than 0.5 bars. With the safety factor of 2, structure of tank structure can be designed for 1 bar. It may be noted that values of transient pressure for all tanks has been kept intentionally lower than the values of steady state pressure, the designed pressure. As predicted, it was also observed that the highest differential pressure occurs as the platform approaches surface i.e. when the difference between ambient pressure P_2 and P_1 is maximum.

5.1 (c) Instantaneous dive rate of platform – Venting System

The instantaneous dive rate as given in Eq. (23):

$$\dot{\zeta}(t) = \frac{V_{w_{in}}(t)\rho_{sw}}{TPC}$$

The value of TPC for different volumes of water in MBTs are shown in Figure 10. Thus, the time taken by the platform to submerge below water level is 12 second, if all tanks are vented at the same time.

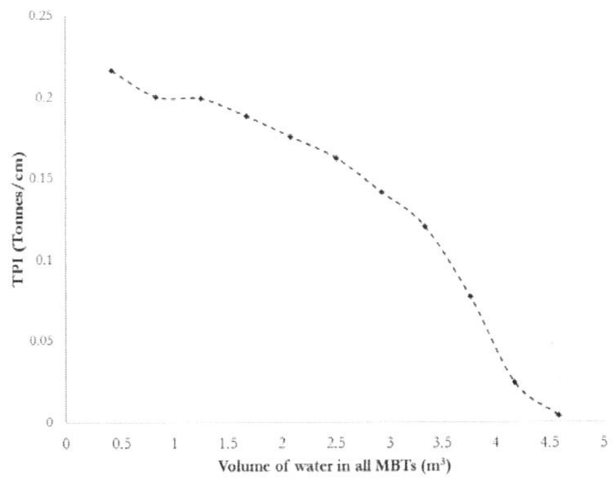

Figure 10: Variation of TPC with volume of water in tanks

6. CONCLUSIONS

Main ballast system is the most critical system with respect to transition from deep diving depth to surface of an underwater platform and vice versa. Designing of MBT involves complex engineering and rigorous mathematical modeling. Special care was required to precisely and accurately model the requisite sub-systems as it was unambiguously understood that MBTs system is directly related to diving safety of a sub-surface platform. Due to confidentiality issues involved in sub-surface projects, open literature on this system is therefore, either very limited or unavailable.

MBT system comprises of two sub-systems i.e. blowing system and venting system. However, the fundamental design criteria is the maximum pressure differential developed across the outer hull structure of tanks during blowing system. Therefore, the design of blowing system has to be considered prior designing venting system. This study comprises of a detailed mathematical modelling of both these sub-systems in terms of their various components in time domain encompassing the transient and steady states. The differential pressure thus calculated through designing of blowing system will be utilized as design pressure for the structure of the outer hull of MBT. The above mathematical model is applied to calculate the differential pressure across the soft structure of MBTs of DARPA suboff model to assess its performance/compliance with theory.

7. FUTURE WORK

Mathematical modelling for both blowing and venting systems achieved above has been derived through abinitio albeit based on many assumptions. In order to validate the mathematical model, it is recommended that data regarding MBT system of a known submarine is to be obtained. Values for various components are to be inserted in equations to compare the results i.e. time to surface from deep diving depth and the internal tank pressure both in transient and steady states. In case of absence of such data, a CFD simulation using multi-phase modelling may be carried out to compare the above results.

8. REFERENCES

1. BURCHER, R., & RYDILL, L. J., 'Concepts in submarine design'. *Cambridge University Press*, 1995.

2. FONT, R., PELAEZ, J. G., MURILLO, J. A., & PERIAGO, F., Modelling, Analysis and Control of Blowing, Venting Operations in Manned Submarines, *Journal of Control Engineering and Technology,* 2014.

3. POHLER, C. H., BEMENT, A. A., WILSON, D. S., & SKINNER, W. A., submarine main ballast tanks - theory and methods for refined structural design, *Association of Senior Engineers, Bureau of Ships - Third Annual Technical Symposium,* 1969.

4. RENILSON, M., Submarine Hydrodynamics, *Springer International Publishing*, 2015.

5. WILGENHOF, J. D., GIMENEZ, J. C., & PELAEZ, J. G. , Performance of the Main Ballast Tank Blowing System. *Undersea Defence Technology Conference*, 2015.

9. AUTHORS BIOGRAPHY

Zeeshan Riaz completed his doctorate from University College London, UK in 2015. Currently, he is working as a full time researcher in a design house. His areas of interest are computational fluid dynamics, UUV designing and submarine hydrodynamics.

Muhammad Abi Waqas earned his Bachelor degree in Mechanical Engineering with specialization in Mechatronics and controls from University of Engineering and Technology, Lahore, Pakistan. He is currently working as research engineer. His skill sets include structure design using ANSYS and is currently involved in design and development of various marine platform.

Waqas Ahmed graduated from UET Taxila in Mechanical Engineering and masters from SMME, NUST in industrial & Manufacturing Engineering. His area of interest are advanced materials in ship production and Systems Engineering.

Hassan Khalid has received his PhD from Strathclyde University UK. As research fellow at Strathclyde, he has worked on IMPROVE, EUMOP, SILENV and SEAHORSE EU projects. His areas of interest are ship design, seakeeping and ship structures.

Atta ur Rehman Hashmi has received his Masters degree in Naval Architecture from Harbin Engineering University China. Presently, he is working as a full time researcher in a design house. His areas of interest are marine hydrodynamics and submersible designing.

Paul Geoffrey Wrobel has been the Professor of Naval Architecture at University College London since 2008. He is a graduate from Cambridge University and University of London. During his career, he has worked on wide range of Ship and Submarine projects during his service at Sea System of MoD (PE), VSEL, Thales and QinetiQ.

INTEGRATED PERFORMANCE MODELLING OF MODERN CONVENTIONAL SUBMARINES TO SUPPORT STRATEGIC DECISION MAKING

K Slater, H Diab and S Harrison, Defence Science and Technology Group (Department of Defence), Australia

SUMMARY

Integrated performance modelling is concerned with the understanding of the performance of a platform from a whole of system perspective. Subsequently, numerical and systems modelling techniques are applied to understand the variation of platform performance with various concepts of operations, technology, and mission parameters. Of key importance are the interactions between constraints placed on major subsystems of the platform, which must be captured if the platform is to provide a level of performance that is operationally acceptable throughout its life-cycle.

Integrated performance modelling has been used by the Maritime Division of Australia's Defence Science and Technology (DST) Group to provide informed and evidence-based analysis in support of strategic decisions related to Australia's procurement of a future submarine fleet. This paper provides an overview of the approach, capabilities and tools developed by DST in integrated performance modelling.

1. INTRODUCTION

The intricate design, complex roles and criticality of many Defence assets make it essential to consider platform performance in an integrated and systematic manner. Otherwise, it would not be possible to assess whether complex platforms, such as Australia's future submarines, meet extensive technical and operational requirements.

Prior to 2009, the Maritime Division of the Defence Science and Technology (DST) Group – Department of Defence – conducted analysis of platform performance using a variety of methods, one of which was using empirical based equations in an Excel spread sheet applied in support of submarine applications. Whilst this spread sheet was able to undertake appropriate tasks at the time of development, it was not a robust method for providing decision makers with objective quality evidence, nor was it repeatable or simple to integrate other tools; and as with all large spread sheets was prone to user error.

Considering the complex nature of submarine operations; coupled with the extensive requirements stipulated by Australia's Future Submarine Program (FSP) it became apparent to DST that a capability was required that could:

- understand top-level requirements and their impacts on submarine platform designs
- assess the overall performance of various platform concepts/designs whilst simulating the influences of different mission scenarios
- integrate the capability of sub-systems/technologies to assess the impacts on platform performance parameters, enabling trade-off studies to be completed
- highlight key design and performance drivers
- validate manufacturer's claims
- support the maturation of performance and operational requirements for the FSP.
- ensure any assessment was auditable and repeatable; and was conducted using a 'level playing field'.

The above challenges required a suite of system modelling tools; however there were no suitable tools available. Additionally, this capability would need to be deployed in early years of the procurement phase of the FSP, where any outcomes would provide the most benefit, as illustrated in Figure 1.

As such DST, in support of the Commonwealth of Australia, embarked on a task to develop an in-house software tool known as the 'Integrated Platform System Model' (IPSM). The IPSM is a platform systems analysis tool that has been and continues to be developed; and has become a critical modelling tool to support whole of platform performance studies during the FSP.

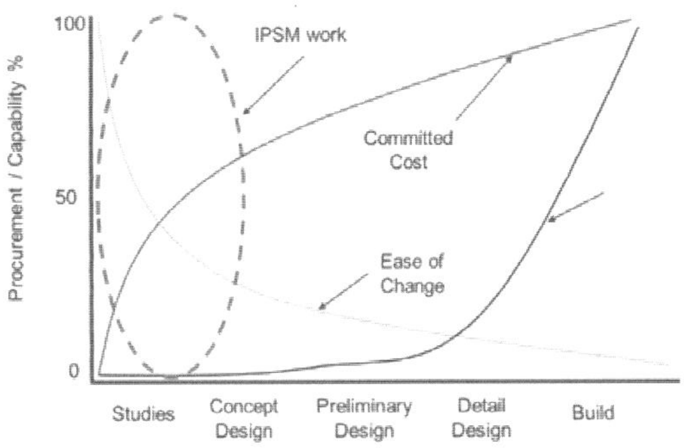

Figure 1: IPSM's Value in the Life-cycle

This paper provides an overview of DST's development of the IPSM tool. Due to the sensitive nature of the FSP, specific outcomes of DST's work in the development and use of the IPSM cannot be provided; however this paper will demonstrate how the tool works. For open literature details on the operation and capability of the IPSM refer to [1], [2] and [3].

2. SOFTWARE AND MODEL OVERVIEW

2.1 THE CORE OF THE IPSM

The Platform Systems Analysis team of DST's Maritime Division have applied the following principles in the development of the IPSM over the past decade:

- continual software development with continual feedback from stakeholders to ascertain current and future needs
- continual improvement of processes for software development, numerical modelling and simulation
- adoption of 'SysML' (systems modelling language) principles [2] without creating constraints/techniques imposed by propriety software
- platform system models are to be validated - validation efforts should be considered upfront
- quality control and pre/post configuration tools are essential.

A conceptual framework showing functional inputs, system models; and various infrastructure and support tools of the IPSM [4], is shown in Figure 2. The blocks inside the box labelled 'Modelica Environment' represents the core of the IPSM, consisting of various 'system models', systems requirements and design rules, all of which are written in an object oriented modelling code Modelica (discussed in more detail in Section 2.2).

The following physics based models (i.e. 'platform systems') are currently included as part of the 'Modelica Environment':

- power and energy systems (e.g. battery technologies; diesel systems and air independent propulsion)
- propulsion system
- atmosphere management system
- potable water production, consumption and storage capacity
- victual consumption and storage capacity (including cool, cold and dry stores)
- waste production rates, storage capacity and management (grey water, black water and garbage)
- habitability spaces.

The 'Modelica Environment' is capable of modelling the integrated effects of these platform systems on mission performance, enabling an assessment on whether particular measures of success or requirements are being met. The algorithms and equations underpinning several of these platform system models are derived from research, development and analyses conducted by DST subject matter experts over several decades, in addition to external sources such as those considered by Burcher and Rydill [5].

The platform representation (i.e. 'design model'), is derived from the desired inputs and as a minimum, requires descriptions of the size of the platform (displacement) and basic power and energy parameters. If, for example, habitability related analysis is desired, then parameters representing those aspects of the design are also included.

The 'mission model' enables platform models to be exercised through a deterministic simulation of realistic mission. A mission scenario is described as a state machine where each state represents a discrete segment of the scenario describing the submarine's 'behaviour' for the present state. This may include patrol times, segment lengths, submarine speed, battery use profile, and relative priorities and availabilities of the installed power systems.

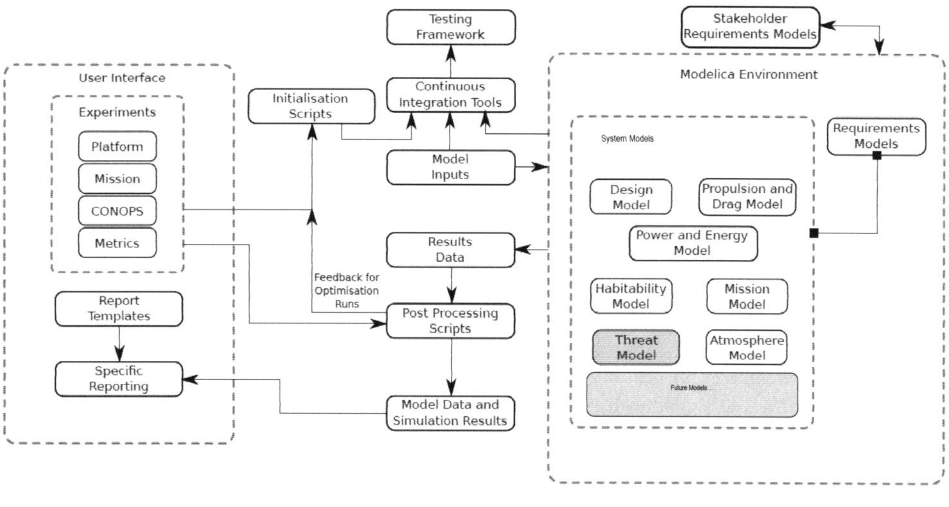

Figure 2: The 'IPSM Framework' (shaded blocks are those components planned for future software builds)

An idealised representation of an example submarine mission modelled for analysis in the IPSM is shown in Figure 3 and an expanded view of the Operational Area (OA) is shown in Figure 4. The OA encompasses the movement of a submarine through a patrol area, transit to a safe area for battery charging using diesel engines (snort box), and returns to the patrol area. The ability to model this level of mission detail capability has become a feature of IPSM Phase 3.0, as will be highlighted in Section 3.

The remaining blocks in the 'IPSM Framework' shown in Figure 2 represents components and tools of the software infrastructure, as well links to the stakeholder requirements and data products. These products (visual, tabular and textual results) are derived from performing model simulations. If IPSM simulations are conducted via single experiment, parametric studies (i.e. multiple simulations are performed across a range of values); and in the most recent version - Phase 3.1 - optimisation techniques have been used to determine the overall mission time and transit speed combination that maximises the time on task. Optimisation methods are currently implemented via the Dakota software package [6], which allows the flexibility of exercising different optimisation options.

Additional details of the software infrastructure will be briefly discussed further in this paper.

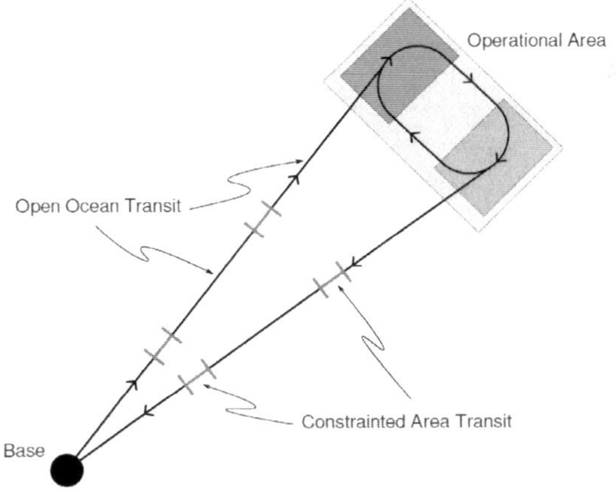

Figure 3: Idealised Submarine Mission Schematic

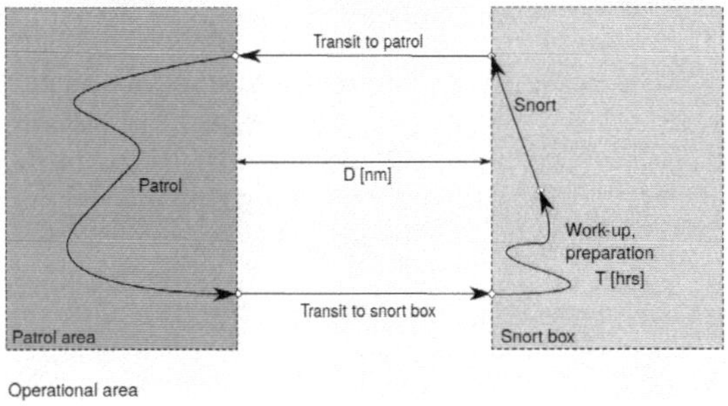

Figure 4: Expanded view of the Idealised Submarine Operational Area (OA)

2.2 MODELICA SOFTWARE LANGUAGE

The IPSM tool has been developed using the Modelica language, a systems-based modelling language with no upfront or ongoing licencing costs. The key features of Modelica include:

-
- object-oriented language with replaceable classes
- equation-based modelling (also referred to as a 'system dynamic simulation')
- capable of modelling complex physical systems such as those contained in a submarine
- widely used in the modelling community.
- The main advantage with an equation based modelling approach is that it gives more flexibility to re-configure and optimise the system (Figure 5), in comparison to a traditional time-step approach. Modelica implements system models internally as a nonlinear, constrained system of equations which are solved during simulation execution, given a set of initial parameters. This enables parametric analysis to be conducted, by altering these input parameters over several simulation runs, allowing the IPSM to determine the sensitivity of performance to the parameters of interest.

The flexibility of Modelica also enables the efficient application of external optimisation algorithms, which can automatically iterate through parameters without the need to recompile the model.

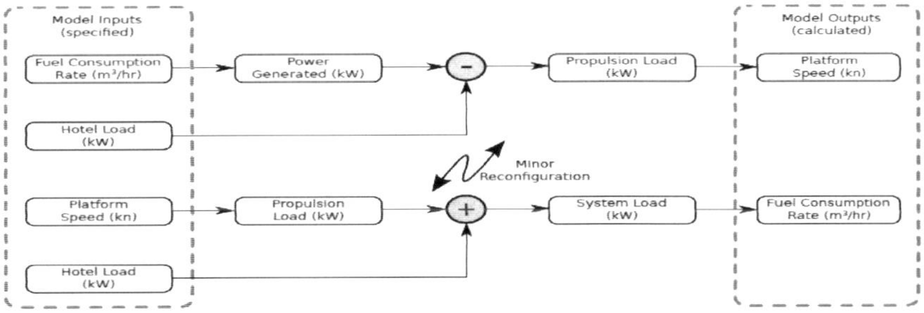

Figure 5: Modelica Offers Ease of Re configuration

3. PHASED DEVELOPMENT APPROACH

The IPSM has been developed in a phased approach over several years as shown in Figure 6. The IPSM development team has generally comprised of 3 - 5 staff, with details of the various phases are provided in the following sections.

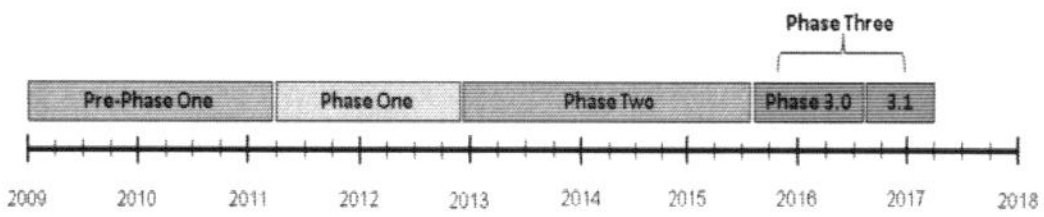

Figure 6: IPSM Development Timeline (future phases will be developed from 2017 onwards)

3.1 PRE-PHASE ONE

Prior to the IPSM, modelling was undertaken in the group using custom software written by DST staff, mainly investigating the influence of scaling submarine displacement with performance. This approach was very limited in its capabilities and wasn't suited for overall systems analysis of different submarine options. Later, a spreadsheet using empirical based equations, as previously mentioned, was developed for analysing the alternate submarine options. This approach resulted in a large cumbersome spreadsheet that was difficult to use, manage and would not allow robust or repeatable results.

The experiences learned from these early attempts demonstrated that a different approach and method for analysing platforms systems and concepts were needed. Investigations were undertaken to find a more suitable method that was robust but flexible; repeatable but always auditable; and capable of system-based modelling, hence Modelica was adopted.

3.2 PHASE ONE

Development of the IPSM Phase One started in 2011 with a view of providing a platform systems analysis capability that could evaluate whole of platform performance of future submarine concepts. The initial focus was on power and energy aspects of the submarine design (diesel, batteries and air independent propulsion), followed by the victualling parameters (cold, cool and dry stores) and atmosphere control consumables (CO_2 canisters and oxygen candles). The power and energy modules were developed first as the impact of variations in the power and energy mediums were considered to have a significant influence on the relative performance of the concept submarine designs.

A visual-based version of Modelica – *ModelicaML* – was used to develop IPSM Phase One. However, upon release of the IPSM Phase One, it was deemed that any future enhancements would be done using the language-based version of Modelica. The visual-based package wasn't powerful or flexible enough and as such modellers devoted significant effort editing the code generated by the in-built code generator.

Phase One also included a basic Graphical User Interface (GUI) to demonstrate its functionality, and was complete in mid-2012.

As part of the exploratory process involved in developing the IPSM Phase One, a collaborative program was established between DST and QinetiQ GRC (UK based) to link the IPSM with Paramarine, with a view to assess the trade-space between platform performance and naval architecture attributes [4]. The intent was not to create a submarine concept design tool, but to use the advantages of each tool to complement each other to assess potential design configurations. This exercise was deemed successful as not only did it demonstrate the success of IPSM to interrogate further aspects of the design space, it also showed the flexibility of IPSM to interact with other modelling packages. As this activity was conducted as a demonstrator, no additional efforts were expended in linking Paramarine with IPSM.

3.3 PHASE TWO

Phase Two of the IPSM added a parametric modelling capability (instead of only single experiment simulations), enhanced submarine atmosphere models; and waste and victualling models. It also included a significant redesign of the GUI to make it more intuitive and user friendly (Figure 7). Phase Two was delivered in late-2013.

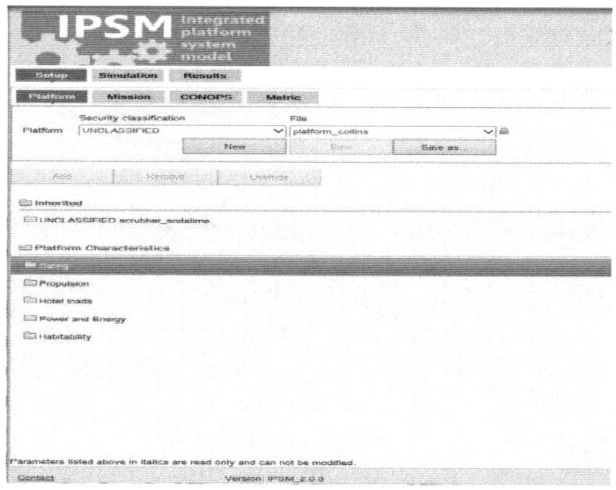

Figure 7: IPSM GUI Screenshot

As part of this phase, the development team was modelling the impact of habitability aspects of submarine design. New IPSM modules were developed to investigate aspects such as waste, grey and black water; and refinements were made to the atmosphere management and food storage elements.

Phase Two was used to support analyses of data provided to DST in relation to initial military off-the-shelf designs considered by the FSP in 2012. The use of the IPSM and its outcomes were received very well by the FSP stakeholders and hence further investment into the IPSM's development was made.

3.4 PHASE THREE

Phase Three (includes 3.0 and 3.1 – see Figure 6) was developed and used to support the FSP Competitive Evaluation Process (CEP) in 2015/16 and most recently, to provide support and evidence for developing and maturing effective requirements in the Preliminary Function and Performance Specification (PFPS). As part of the development of this phase, extensive validation activities were undertaken to validate the IPSM against 'real-world' Collins Class data. Upon successful validation, Phase 3.1 was deployed to staff embedded in the FSP, to allow them to run their own simulations.

Phase Three added additional functionality that included an enhanced battery model and an increased mission management module. This functionality allowed the user to model missions of any format so long as they can be decomposed for representation by discrete phases. This enhancement has been especially useful as it allows for significantly more flexibility in the way in which missions are modelled, for example the ability to add any number of choke points and mission leg lengths.

During the development of IPSM Phase Three the functionality of the IPSM grew significantly and as such maintaining the GUI to remain 'in-step' with the IPSM became a resource drain. Consequently, due to the prioritisation of work, the GUI is obsolete for Phase Three. Rather, the IPSM development team have developed fit-for-purpose training guides to assist any new users to access the IPSM via command lines in Unix, providing greater functionality and flexibility to build platform models and run simulations. Additionally, using command lines allows users to interrogate the tool, providing an in-sight on how the model has been developed.

4 VALIDATION AND VERIFICATION APPROACH

4.1 IPSM VERIFICATION

The purpose of model verification is to verify that the model is consistent with its specification and fulfils its intention. This differs from model 'validation', which is primarily concerned with determining the degree to which the model is an accurate representation of the real world, from the perspective of the intended use of the model.

Figure 8 represents a software development process that demonstrates the relationships between each phase of the development life-cycle and its associated phase of testing. The IPSM used this model as a guide to support verification and validation activities, including the design of automated unit tests, system tests and acceptance tests.

Software testing is conducted in the IPSM using Continuous Integration practices, which allows isolated IPSM code changes to be immediately tested and reported on when added to the main code base. Continuous Integration provides rapid feedback so that if a defect is introduced into the model, it can be identified and corrected as soon as possible. The IPSM development team have created a testing framework that has been integrated into the Jenkins software, which is used for implementing the Continuous Integration process.

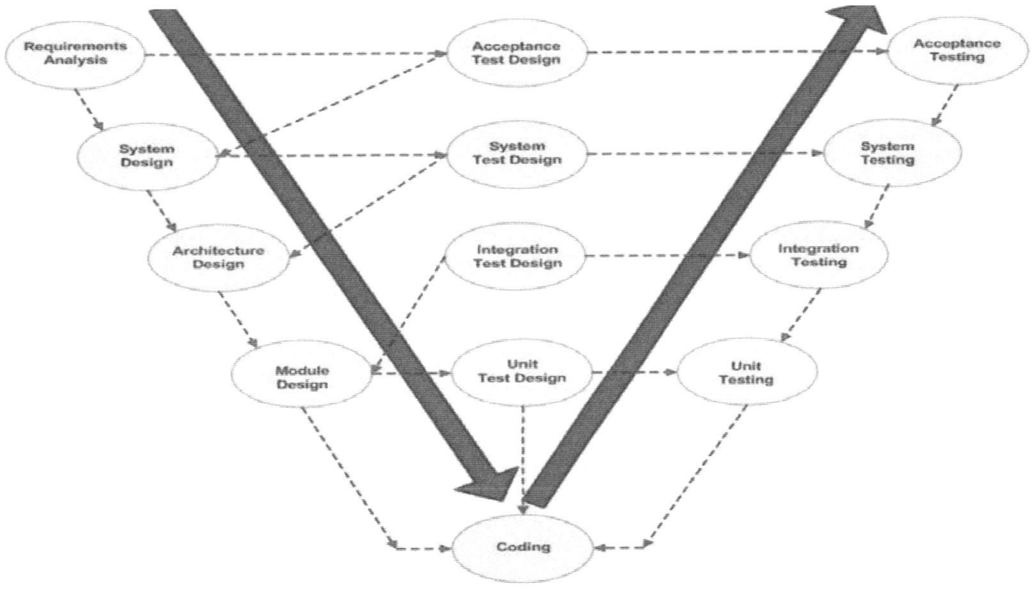

Figure 8: The 'V-Model' for IPSM Verification and Validation

4.2 IPSM VALIDATION

Validation of various phases of the IPSM has utilised several sources of data.

For IPSM Phase One and Phase Two validation, unclassified open literature data were used. 'Brochure' data from prospective manufacturers were also used to support the validation of IPSM Phase Two.

For Phase Three, the validation activities were much more rigorous due to the availability of data. Firstly, the ISPM was cross-validated against various outputs from various power and energy models internal to DST and the FSP. Additional validation activities were then conducted using 'real-world' Collins Class submarine fleet data, which was a pre-requisite for the deployment of the IPSM to staff embedded in the FSP.

5. OVERVIEW OF IPSM SUPPORT AND INFRASTRUCTURE

Supporting the IPSM requires hardware, software and supporting processes. Beyond model development and execution and technical support, it is essential to ensure the outputs of the model are accurate, robust, repeatable, traceable and in a form that serves the needs of the intended audience.

5.1 HARDWARE REQUIREMENTS

Early in the development of the tool it was decided that it would be advantageous to have the tool available on a dedicated Defence server rather than on a stand-alone platform. This would allow multiple users to be able to use the tool from different premises across Defence and would allow the developers to keep tight control of the versions being used.

Due to the different security classifications of data used in IPSM simulation runs, coupled with strict protocols, the IPSM is hosted on multiple blade servers within Defence premises. As the IPSM is hosted on multiple servers, an integration manager has been established that ensures the IPSM version that exist on each server are compatible, where possible – this is achieved via rigorous software testing, configuration management and version control.

5.2 SOFWTARE REQUIREMENTS

The IPSM servers are configured with a Ubuntu Linux based operating system, however putty or Xwin-32 is used for remote display of Unix from Windows workstations. The operating system uses several software and support tools to enable the model to be developed, compiled, tested, executed and managed, such as:

- Modelica: IPSM is implemented in the Modelica modelling language and is compiled using the Open Modelica Compiler (OMC).
- Git Distributed Version Control System: In the IPSM Git is used for version control of model code, data, build scripts, test files and some documentation.
- Jenkins: an open sourced Continuous Integration Server that provides continuous integration services for software development. This includes building and testing (i.e. verification) software projects continuously, and monitoring the execution of externally-run jobs. In the IPSM, various Jenkins jobs are created to monitor the IPSM test suites. When the developer makes a commit to a code repository it can trigger a build of the IPSM code base, a build and execution of the IPSM unit test suite and a build and execution of the IPSM system test suite.
- TracWiki: an open sourced, web-based project management and bug tracking system hosted on a Wiki page. In the IPSM, Trac is used to manage and track all IPSM issues, from future enhancements to existing 'bugs' and, provides wiki facility to store supporting material for IPSM user and developers, including background information.
- Make: a build automation tool that automatically builds executable programs and libraries from source code by reading files called Makefiles. Make is used to build executables for the IPSM.
- Perl: used for building scripts in IPSM for various purposes.
- Gnuplot: used in IPSM to generate output plots.

5.3 QUALITY MANAGEMENT

Without quality management processes, it would be difficult for the IPSM development team to claim that the model is producing valid, consistent, repeatable and traceable outputs. As such ISO 9000/01 standards for Quality Management Systems were used as a guide to develop fit-for-purpose quality management processes for the IPSM, which includes:

- the establishment of team roles, responsibilities and competencies
- information sharing
- collaborations external to the team
- rigorous software and configuration management
- fit-for-purpose software support requirements
- effective management of stakeholder requirements
- verification and validation activities
- change control procedures
- issue tracking and management
- software release approval and management
- data and software security measures
- configuration management.

The IPSM's quality management is agile in nature and has been updated, modified and improved significantly between the release of Phase One and Phase Three.

5.4 CONFIGURATION MANAGEMENT

The Configuration Management (CM) approach for the IPSM is as light weight as possible, whilst still retaining the ability to adequately control and manage any changes to the model code, the server software and hardware configuration, and any associated documentation. The goal is to give the IPSM developers the ability to identify (and return to) a known state at any point in time.

The IPSM consists of the model code itself, the model infrastructure, formal documentation, and numerous other project artefacts that are produced as part of the development process. Each of these items is placed under configuration control and baselined, as this information is important to reproducing any previously known state of the IPSM. The following is a list of configuration items identified for control in the IPSM:

- model source code
- model input and data, including reports
- requirements specifications
- test plans
- review evidence (e.g. validation data)
- design documents
- test procedures and procedures (unit tests, system tests and acceptance tests)
- verification and validation reports
- user guides
- security documents
- software release notes
- version description documents.

The configuration management tools used in the IPSM consist of the Git Distributed Version Control System, TracWiki and internal defence archiving systems.

6. EXAMPLE ANALYSIS

The IPSM is capable of supporting a broad range of analysis tasks, from evaluating design alternatives (to determine what capability is to be built), to developing concept of operations for the capability being introduced (to determine how the capability is to be deployed).

An example analysis using the IPSM is presented below to provide a high level overview of the type of analyses that can be conducted with the IPSM. This example also demonstrates how the model can be used to generate insights into the domain. The aim of this example analysis is to investigate the sensitivity of operational area performance to variation of operational parameters.

As discussed previously, a representation of a platform and of the scenario of interest is required. The platform representation at a minimum requires descriptions of the size of the platform (displacement) and parameters describing the power systems included in the design.

The subsystem used for representing scenarios or missions in the IPSM is flexible with the only requirement being that the desired scenario can be decomposed into discrete segments. Within the IPSM, the scenario is described as a state machine where each state represents a segment of the scenario. Each state has a configuration applied describing the submarine's behaviour for the present state, this includes: submarine speed, battery use profile, and relative priorities and availabilities of the installed power systems.

The example scenario used in this analysis is an idealised representation of a submarine area of operations and is shown in Figure 4. This scenario models the movement of a submarine through a patrol area, transit to a safe area for battery charging using diesel engines (snorting), and return the patrol area.

In this analysis the patrol box in Figure 4 represents the area of interest to the submarine; as such a measure of performance is defined as the operational area effectiveness: the proportion of time spent in the patrol box to the total time, Equation 1.

$$OA_{effectiveness} = \frac{Time_{patrol}}{Time_{total}}$$

Equation 1: Operational area effectiveness.

The IPSM offers both parametric and point study methods of running simulations. Parametric simulations may be run as either a sequential sweep, or binary search optimisation. The ability to conduct parametric analysis allows the IPSM to determine the sensitivity of performance to the parameters of interest.

This analysis will investigate the performance variation with regard to the distance between patrol and snorting boxes (in the Operational Area), investigated by means of a parametric sequential search through a specified range. The resulting operational area effectiveness will be recorded for each run of the scenario.

Figure 9 shows the results of the analysis with respect to the Operational Area (OA) effectiveness. The key points of the results are that the OA effectiveness reduces linearly with the increasing distance between the patrol box and the snort box (see Figure 4). Where the distance between the two boxes is 0nm the OA effectiveness is large ~95%, however even at a modest 25nm standoff distance the OA effectiveness has been significantly reduced to ~40%.

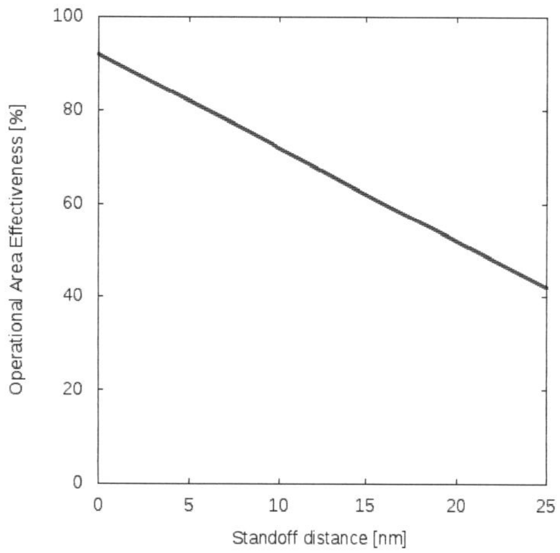

Figure 9: Operational Area Effectiveness variation with standoff to snort distance.

In the context of a study conducted to facilitate the maturation and further refinement of requirements for a future platform, the outcomes of the example analysis could include:

- A requirement should specify the minimum operational area effectiveness for the submarine.
- A minimum distance between the patrol and snort boxes should be specified based on expected platform signature levels and expected threat environment.

7. IPSM SUMMARY

7.1 KEY FEATURES

In summary the IPSM provides the following features and capabilities:

- Platform level performance parameters are calculated based on integrating multiple system models, as opposed to a coarse single-equipment approach.
- Post processing of simulations to produce graphical, tabular and textual results to analyse and compare results of multiple platform concepts/designs against the same performance and capability criteria.
- The execution of single experiment simulations, parametric simulations and optimised simulations to generate performance and operational attributes.
- Effective management and segmentation of data (input and output), coupled with strict access control, to ensure varying levels of security classification restrictions and caveats are handled appropriately within each server.
- The ability to undertake platform related sensitivity analysis of performance for existing, alternative and new technologies.
- Ease of re-configuration between inputs and outputs; and re-use of various platform and mission parameters.
- Ability to accept user defined metrics to be applied in order to test a series of conditions on simulated results (i.e. for verification and validation activities).
- An automated and flexible reporting function that is highly effective and tailored for the FSP.
- Continuous and automated software testing (at unit-level and system-level); and effective version control, quality management and configuration management.
- Extensive logs of simulations and data products for tracking and auditing purposes.
- A series of secure wiki pages used for information sharing, which provides background details of the IPSM, up to date user guides; and other supporting material and requirements for users and developers.

7.2 FUTURE IPSM EXPANSION PLANS

The success in generating auditable, repeatable, clear, meaningful and reliable results using the IPSM has shaped and matured various platform requirements for the FSP. Subsequently, the IPSM has become an exemplar modelling tool.

In the short term IPSM Phase Three, and its infrastructure, will continue to be deployed during various phases of the FSP, as more platform details emerge. As such the IPSM will continue to be maintained, verified and validated as more modelling and in-service Collins Class data becomes available. Additionally, outputs from the IPSM will be used to support the validation of deep-dive models.

The flexibility, adaptability and effectiveness of the IPSM, enables the tool to be used for in-service applications. As such, there is interest from the Collins Class submarine community in using the IPSM to support pre/post-mission planning analysis and activities.

DST, with support from key FSP stakeholders, is also interested in expanding the functionality of the IPSM to directly support assessments of measures of effectiveness. The stealthy nature of submarine operations means that platform performance parameters strongly interface with the threat domain. As such the IPSM development team are in the inception stages of developing frameworks to understand the trade-off space between platform performance, mission effectiveness and susceptibility to exposure; and are aiming to provide a demonstration of this capability over the next year.

The case study provided in [7] showed that the IPSM can be adapted to assess the benefits of particular technology concepts. As such, the IPSM could be used to assess the capability impacts of various technology demonstrators considered in the FSP. Additionally, this assessment could also consider any integration impacts that this technology has on overall performance.

7.3 OBSERVATION AND LESSONS LEARNT

Throughout the development of the IPSM tool a number of lessons have been learnt and observations made by the development team. A summary of these are listed below.

- Platform system modelling tools are highly effective and useful when representative mission scenarios and operational parameters are being considered.
  - Subsequently it is paramount that a continuous feedback loop is established between modellers and the submarine user community.
  - Using this feedback is an effective method in identifying areas of improvement and enhancements for the future.
- The time required to undertake verification and validation tasks are often underestimated and not planned for sufficiently in the inception stages.
  - However, once complete these add a significant amount of credence and trust in the use and future development of the tool.
- Developing a GUI capable of undertaking 'generic all purpose' analysis would not be cost effective. GUIs designed for conducting analysis can be of limited utility as they tend to constrain the type of analysis that can be performed.
  - The IPSM development team spent a lot of resources initially in developing the GUI for use during the development of Phase Two, but ultimately it became underutilised as the functionality of the IPSM grew.
- Effective visualisation of analysis results is critical for communicating findings to a broad audience.
- Enabling access of the IPSM using a remote login ensures that all users are using the most updated software build of the tool.
- Invest effort in reducing the run-time of the model - a single experiment simulation of the IPSM is in the order of a few seconds, making it highly attractive from a usability perspective.
- The end goal of the model cannot be forgotten i.e. it is not necessary to have a perfect or fully developed model to generate effective results.
- Analytical tools do not follow a linear environment where 'development' is followed by 'use'. For example:
  - tools can highlight limitations in the analysis being conducted, requiring further development to address these limitations. This may be the case when an analysis reveals outcomes that are extremely sensitive to a particular model, requiring that model to be replaced with more detailed logic.
  - introduction of new technologies need to be considered, requiring new system models to be developed and integrated into the overarching model.
  - different mission scenarios or platform tactics need to be evaluated, requiring a different approach or new models to be developed for operational analysis.
- A software development team must be diverse in their skills and not all focussed on the technical modelling aspects.
- Having multiple models with overlapping capability, that are independently developed, promotes trust in the development, use and validation of models.
- The capability provided by the IPSM provides a useful test-bed for novel concepts of operation as demonstrated in [7].
- Reconstructing performance estimates provided by prospective tenders has enabled the impact of uncertainty in sub-system performance to be quantified; and all tenders evaluated fairly with a common set of assumptions.

8. CONCLUSIONS

The development and use of the IPSM software tool described in this paper demonstrates DST's capabilities in integrated platform modelling.

The IPSM has been developed incrementally over the past ten years using DST's knowledge on submarine systems, coupled with their diverse experience in systems modelling and simulation. Subsequently, the IPSM has evolved from an Excel-based exploratory tool to a highly powerful, integrated, sophisticated and validated platform systems modelling tool developed in Modelica.

The IPSM is a highly effective and respected tool for providing analysis and advice on platform performance parameters and trade-off studies for Australia's FSP; one of Defence's flag-ship programs over the next few decades.

9. ACKNOWLEDGEMENTS

The authors would like to acknowledge the following personnel who have contributed to the development of the Integrated Platform System Model (IPSM): Mr Alex Bulach, Mr Simon Demediuk, Mr Jon Hornstein, Mr Rob Jarvis, Mr Josh Rodgers, Mr Mario Selvestrel, Mr Andrew Tynan; and in particular, Dr. John Wharington.

10. REFERENCES

1. HARRISON, S. et al., 'Analysis of Platform Configurations Using the Integrated Platform Systems Model', *International Maritime Conference – Pacific 2012*, Australia - Sydney, 2012.

2. HEPWORTH, G. et al., 'High Level Capability Analysis of Submarine Platforms', *RINA Submarine Conference,* UK - Bath, 2014

3. HARRISON S., Wharington, J. and Slater, K., 'An Investigation into the Penalties Incurred due to Operational Constraints Imposed on an SSK-type Submarine in an Operational Area', *International Maritime Conference – Pacific 2015*, Australia - Sydney, 2015.

4. RODGERS, J. et al., 'A Framework for Integrated Platform System Modelling', *Maritime / Air Systems and Technologies (MAST) Conference*, Europe - Malmö, 2012.

5. BURCHER, R. and RYDILL, L., 'Concepts in Submarine Design', *Cambridge University Press, 1995*

6. ADAMS, B.M., BAUMAN, L.E., BOHNHOFF, W.J., DALBEY, K.R., EBEIDA, M.S., EDDY, J.P., ELDRED, M.S., HOUGH, P.D., HU, K.T., JAKEMAN, J.D., STEPHENS, J.A., SWILER, L.P., VIGIL, D.M., and WILDEY, T.M., "Dakota, A Multilevel Parallel Object-Oriented Framework for Design Optimization, Parameter Estimation, Uncertainty Quantification, and Sensitivity Analysis: Version 6.5 User's Manual," *Sandia Technical Report SAND2014-4633*, Updated November 2016 (Version 6.5)

7. HARRISON, S., et al., 'The Operational Benefits of Covertly Refuelling a Conventional Submarine whilst at Sea', *RINA Warship Conference: Naval Submarines & UUV*, UK - Bath, 2017.

11. AUTHORS BIOGRAPHY

Karl Slater currently works in the Maritime Division at DST on several tasks, including the development of methods and tools for undertaking integrated performance analysis of maritime platforms through modelling and simulation. Before joining DST in 2011, Karl spent 7 years with BAE Systems, Williamstown working primarily with surface warships, and 8 years with the UK Ministry of Defence working in submarine naval architecture. Karl obtained a Bachelor of Engineering (Honours) in Marine Technology (Naval Architecture) from the University of Newcastle upon Tyne in 1996 and a Master of Science in Naval Architecture from University College London in 1998.

Simon Harrison is a Defence scientist and analyst at DST developing and implementing platform system models for naval platform assessments, through simulation activities. Simon joined DST in 2010 and has a Bachelor of Science in Physics and a Bachelor of Engineering in Electronics from RMIT University Melbourne.

Hamid Diab is in the Maritime Division at DST leading a modelling and simulation team in whole of platform performance analysis. Since joining DST in 2003, Hamid has worked on the F-111 Sole Operator Program as a structural analyst and fatigue test engineer; provided analysis and modelling support on the sustainment management for Defence's AIR 9000 Helicopter Program and air lift platforms; and became the lead researcher and advisor for the Air Force on strategic asset management.

AUTHOR'S CONTACT DETAILS

Alexander Conway
Australian Maritime College
a.conway@utas.edu.au

Caren Soper
QinetiQ
CSSOPER@qinetiq.com

Chris Carter
Frazer-Nash Consultancy Ltd
c.carter@fnc.co.uk

David Andrews
UCL
d.andrews@ucl.ac.uk

David Pook
Australian Defence Science & Technology (DST) Group
David.Pook@dsto.defence.gov.au

David Ritchie
BAE Systems Maritime – Submarines
david.ritchie2@baesystems.com

Fabian Pecot
DCNS Research / SIREHNA
fabian.pecot@sirehna.com

Henrique Vieira da Silva
CINAV, Portuguese Naval Research Center
henrique.vieira.silva@marinha.pt

Jan D. Wilgenhof
MecDes
jdwilg@mecdes.nl

Joe Beel
Cisco System, Inc.
jbeel@cisco.com

Karl Slater
Defence Science and Technology Group, Australia
simon.harrison@dsto.defence.gov.au

Kelly Jomes
Cisco System, Inc.
kellyjon@cisco.com

Kevin Heaney
BMT Defence Services Ltd
kheaney@bmtdsl.co.uk

Malcolm Cook
BAE Systems Maritime
malcolm.cook2@baesystems.com

Matthew Palmer
Lloyd's Register
Matthew.Palmer@lr.org

Morgan Parker
NAVSEA
morgan.parker@navy.mil

Muhammad Hary Mukti
Texas A&M University
hary.mukti@tamu.edu

Neil Horn
MoD
Neil.Horn364@mod.uk

Paul Crossland
QinetiQ
pcrossland@qinetiq.com

Robert Ball
QinetiQ
RABALL@qinetiq.com

Robert Dvorak
NAVSEA
robert.e.dvorak1@navy.mil

Simon Harrison
Defence Science and Technology Group, Australia
simon.harrison@dsto.defence.gov.au

Stefan Daum
thyssenkrupp Marine Systems
stefan.daum@thyssenkrupp.com

Thomas Nietzschmann
Potsdam model basin SVA
nietzschmann@sva-potsdam.de

Zeeshan Riaz
(CESAT): Islamabad
midshipman2k1@gmail.com

The Royal Institution of Naval Architects
8-9 Northumberland Street
London, WC2N 5DA
United Kingdom

ISBN 978-1-5108-8302-4

CURRAN ASSOCIATES INC.
proceedings.com